T0199350

Environmental Risk Assessment

A Toxicological Approach

Environmental Risk Assessment

A Toxicological Approach

Ted Simon, Ph.D., DABT

CRC Press
Taylor & Francis Group
Boca Raton London New York

CRC Press is an imprint of the
Taylor & Francis Group, an **informa** business

CRC Press
Taylor & Francis Group
6000 Broken Sound Parkway NW, Suite 300
Boca Raton, FL 33487-2742

First issued in paperback 2016

© 2014 by Taylor & Francis Group, LLC
CRC Press is an imprint of Taylor & Francis Group, an Informa business

No claim to original U.S. Government works

Version Date: 20140402

ISBN 13: 978-1-138-03383-2 (pbk)
ISBN 13: 978-1-4665-9829-4 (hbk)

This book contains information obtained from authentic and highly regarded sources. While all reasonable efforts have been made to publish reliable data and information, neither the author[s] nor the publisher can accept any legal responsibility or liability for any errors or omissions that may be made. The publishers wish to make clear that any views or opinions expressed in this book by individual editors, authors or contributors are personal to them and do not necessarily reflect the views/opinions of the publishers. The information or guidance contained in this book is intended for use by medical, scientific or health-care professionals and is provided strictly as a supplement to the medical or other professional's own judgement, their knowledge of the patient's medical history, relevant manufacturer's instructions and the appropriate best practice guidelines. Because of the rapid advances in medical science, any information or advice on dosages, procedures or diagnoses should be independently verified. The reader is strongly urge to consult the relevant national drug formulary and the drug companies' printed instructions, and their websites, before administering any of the drugs recommended in this book. This book does not indicate whether a particular treatment is appropriate or suitable for a particular individual. Ultimately it is the sole responsibility of the medical professional to make his or her own professional judgements, so as to advise and treat patients appropriately. The authors and publishers have also attempted to trace the copyright holders of all material reproduced in this publication and apologize to copyright holders if permission to publish in this form has not been obtained. If any copyright material has not been acknowledged please write and let us know so we may rectify in any future reprint.

Except as permitted under U.S. Copyright Law, no part of this book may be reprinted, reproduced, transmitted, or utilized in any form by any electronic, mechanical, or other means, now known or hereafter invented, including photocopying, microfilming, and recording, or in any information storage or retrieval system, without written permission from the publishers.

For permission to photocopy or use material electronically from this work, please access www.copyright.com (http://www.copyright.com/) or contact the Copyright Clearance Center, Inc. (CCC), 222 Rosewood Drive, Danvers, MA 01923, 978-750-8400. CCC is a not-for-profit organization that provides licenses and registration for a variety of users. For organizations that have been granted a photocopy license by the CCC, a separate system of payment has been arranged.

Trademark Notice: Product or corporate names may be trademarks or registered trademarks, and are used only for identification and explanation without intent to infringe.

Library of Congress Cataloging-in-Publication Data

Simon, Ted.
  Environmental risk assessment : a toxicological approach / Ted Simon.
    pages cm
  Summary: "Risk assessment is of increasing importance as health and safety regulations grow and become more complicated. Focusing on environmental risk assessment, this book looks at various factors relating to exposure and toxicity, human health, and risk. The book is aimed at practitioners and students who need to know more about understanding, developing, conducting, and interpreting risk assessments. It provides sufficient background to enable readers to probe for themselves the science underlying the key issues in environmental risk"-- Provided by publisher.
  Includes bibliographical references and index.
  ISBN 978-1-4665-9829-4 (hardback)
  1. Environmental risk assessment. 2. Environmental toxicology. I. Title.

GE145.S56 2014
363.73'2--dc23   2013047342

Visit the Taylor & Francis Web site at
http://www.taylorandfrancis.com

and the CRC Press Web site at
http://www.crcpress.com

This book is dedicated to Dr. Randall Oliver Manning, mentor, scholar, true southern gentleman, and constant friend. Dr. Manning served as the state toxicologist for Georgia until his passing in 2012.

Contents

Preface

As the developed world moves further into the twenty-first century, what is becoming increasingly clear is that societal decisions must be based on an honest and forthright appraisal of the state of relevant knowledge. More often than not, these decisions relate to so-called "wicked" problems for which no easy answers exist. Unfortunately, these decisions almost always become highly politicized.

Much of the methodology and knowledge generally used in environmental risk assessment today dates from the 1980s and 1990s. Environmental risk assessment has become highly dependent on the use of default values for a range of factors in both exposure and toxicity—defaults are appropriate when existing data are inadequate, and it is important for risk practitioners to understand the basis for and appropriate use of default values. For example, the linear no-threshold hypothesis for chemical carcinogens was derived from early twentieth-century work on radiation mutagenesis and has been used for the past 25 years. However, recent investigations of radiation mutagenesis and the increasing knowledge about fundamental biology and nature of cancer suggest that this hypothesis is based on flawed assumptions and is inconsistent with the biology.

This situation is changing: Recent publications from the US National Academy of Sciences, including *Toxicity Testing in the 21st Century: A Vision and a Strategy*, *Science and Decisions: Advancing Risk Assessment*, and others, have the potential to engender significant progress. My purpose in writing this book is to give the next generation of risk assessors a view of both the past and the future. In order for future practitioners to understand the effect of this coming change, they must know what was done in the past and also what the future may hold in store.

Key issues in risk assessment are those that either contribute appreciably to quantitative estimates of risks (a risk driver) or add to the uncertainty of those estimates (an unknown). Hence, this book serves to provide students with sufficient knowledge and confidence to enable them to probe the science underlying key issues in environmental risk for themselves rather than accepting so-called conventional wisdom or the opinion de jour.

Realizing the need to use relevant scientific information as the basis of decisions, having confidence in one's own knowledge and one's ability to learn, and possessing the humility to accept the limitations of humankind's knowledge are the hallmarks of a great scientist—the kind who can provide an unbiased and science-based appraisal of risk and uncertainty that best serves the decision makers in today's complex society.

The aim of this book is to provide a text and reference that will enable students of risk assessment to approach the future with confidence about the state of their knowledge. No aspect of environmental risk assessment should be a "black box" for any practitioner. Environmental risk assessment courses are taught in either public health or civil/environmental engineering programs. Students from these two diverse academic backgrounds can both become proficient in risk assessment—engineering students will likely need to put greater effort into learning about toxicology and epidemiology whereas public health students will likely need to work harder on learning about environmental sampling and analysis, hydrogeology and soil science. This textbook cannot cover all aspects of risk assessment, and one of the hallmarks of a good practitioner is sufficient engagement with the subject to seek out relevant information. For the sake of honesty, it is vitally important to acknowledge one's areas of ignorance and seek help from others as needed—doing so most often requires mustering one's humility.

Democratic societies in the twenty-first century have become invested in risk assessment as a tool for informing important societal decisions. The ability to conduct environmental risk assessments is a marketable skill, and with hard work, this skill can provide the practitioner a rewarding career. To provide practical experience in performing risk assessments as a start to developing one's skills, fully worked examples of specific human health and ecological risk assessments including the environmental sampling data are provided. In addition, an electronic workbook with more exercises/examples is provided on the publisher's website.

Regarding these examples, there are no "correct" risk assessments—there are only those that comport with existing guidance or the present state of knowledge to greater or lesser degrees. One of the central tensions in risk assessment today is the gap between regulatory guidance and the state of the science. The datasets and examples provide the opportunity for students of risk assessment to explore this science/policy gap for themselves.

CHAPTER DESCRIPTIONS

Chapter 1 presents an introduction to the field and a history of environmental risk assessment in the United States and elsewhere.

Chapter 2 discusses problem formulation and hazard identification as the two initial steps in risk assessment. Problem formulation is necessary to ensure the scope of a risk assessment matches the size of the problem addressed. Hazard identification, although inexact, provides a means of deciding when to investigate a perceived problem further.

Chapter 3 provides a narrative and examples of exposure assessment—how do receptors come into contact with contaminated environmental media? Both qualitative and quantitative aspects are discussed.

Chapter 4 deals at length with the dose–response assessment. Because this book concentrates on the toxicological aspects of risk assessment, this chapter is central to the book. It provides both students and instructors a look back and a look forward. Methods for dose–response assessment are changing rapidly, and, as noted, it is vital to know the past in order to understand the future. Many examples are provided.

Chapter 5 on risk characterization presents two detailed case studies that demonstrate complete environmental risk assessments as might be written by a regulatory agency, a regulated entity, or a contractor employed by either of the former. These two examples allow students to hone their skills in using the various tools of environmental risk assessment.

Chapter 6 on ecological risk assessment discusses the development of the guidance on ecological risk assessment and provides an example.

Chapter 7 on the future of risk assessment presents a number of issues in the realm of risk policy, societal decision-making, and the role of science in society. This final chapter should be viewed as a thought starter. The chapter discusses some thorny issues such as bias, conflict of interest, and the appropriate use of the precautionary principle. You likely have strong opinions yourself about these issues. I will consider this chapter (and the book) successful if it engenders discussions about the appropriate role for science in society.

GOALS FOR THIS BOOK

My hope is that this book will enable students of risk assessment to approach the future with confidence in their skills and knowledge. As a scientist acting in the role of risk assessor, you will be called on to answer difficult and at times impossible questions. In many situations, the most honest answer is "I don't know." It takes humility and courage to answer in this way when sitting in a meeting and everyone else there thinks you are the smartest guy or gal in the room. Indeed, many of the decision makers (your clients as a risk assessor) will be looking to you for answers. The best way to serve them is absolute honesty.

The goals of this book are fourfold: (1) to provide a summary of the history, the current methodologies and practices, and likely future of environmental risk assessment; (2) to provide the tools and opportunities for practice and thus enable students to develop and conduct their own environmental risk assessments; (3) to provide students the ability to understand and potentially address the gaps in the relevant knowledge base supporting environmental risk assessment; and (4) to imbue these students with confidence in their abilities to understand and perform complex risk assessments, humility regarding the extent of their knowledge, and a healthy skepticism that is the hallmark of any good scientist.

HOW TO LEARN RISK ASSESSMENT

In 1992, I was teaching biology at a small chiropractic college in Atlanta, Georgia. Not to mince words, it was a terrible job, and I was looking for other employment from day one. Six months later, I got a call from a government contractor that supplied onsite personnel to the Atlanta regional office of the U.S. Environmental Protection Agency, and thus began my risk-assessment career.

The first time I ever heard the term *risk assessment* was when I arrived at EPA's Atlanta regional office for work the first day. I felt as if I were drinking from a fire hose much of the time during that first year, trying to learn about environmental regulation, EPA guidance, and the underlying science. In the early 1990s, I believe this was the only way to learn risk assessment.

This textbook represents something I would have found useful—to put the things I was learning into some sort of perspective. I hope it works that way for the instructors and students who use it.

TERMINOLOGY AND UNITS

A sincere attempt was made to define all of the terms and acronyms. SI units are generally used.

EXERCISES AT THE ENDS OF THE CHAPTERS

These exercises are meant to stimulate thought and discussion. The field is changing and today's student will be tomorrow's leader. The opportunity to consider some of the issues in risk assessment in a high-trust, low-concern situation such as a university classroom will prove a valuable experience. I have taught a risk assessment class at the University of Georgia on an occasional basis since 2004. One of the exercises the students found most informative was a mock risk communication/public meeting exercise in which many divergent points of view were expressed.

ELECTRONIC WORKBOOK

After discussions with university colleagues during the writing of this book, I became aware of the need for and utility of specific examples. Hence, environmental datasets and descriptions of situations needing a risk assessment are provided in this workbook. The datasets are provided as Excel spreadsheet files with accompanying narrative. From the worked examples in Chapters 5 and 6, students will have sufficient background to work through the workbook exercises. Again, there are no right answers—only risk assessments that comport with regulatory guidance and/or scientific information to a greater or lesser extent.

Additional material is available from the CRC website: http://www.crcpress.com/product/isbn/9781466598294.

ACKNOWLEDGMENTS

For getting me started in risk assessment, I want to thank Dr. Elmer Akin, my supervisor for almost ten years at EPA. I have never met anyone wiser in navigating the shoals of the interface of science and policy. I also want to thank Julie Fitzpatrick of EPA's Office of the Science Advisor who was my coworker at EPA during the 1990s and patiently answered all my questions at that time and remains a friend and colleague today.

For their helpful comments and discussions as I was deciding to write this book, I want to express my gratitude to Dr. J. Craig Rowlands and Dr. Mary Alice Smith. I want to thank Dr. James Klaunig for making clear the need for examples. I also want to thank Ann Mason, Rebecca Simon, and Victoria Krawchek for helping me with the tone and message of Chapter 7. I also want to thank Dr. James Bus and Dr. Edward Calabrese for many helpful discussions.

Finally, I cannot find the words to express my thanks to my wife Betsy for her continuing encouragement in everything I do.

Author

Dr. Ted Simon, PhD, DABT, has 12 years experience with the Environmental Protection Agency (EPA), Region 4, in Atlanta, as the senior toxicologist in the Waste Management Division. At EPA, he provided scientific support at many public and private sector hazardous waste sites in the areas of risk assessment, soil cleanup, statistics, and toxicology. At a national level, he served as a resource for toxicological, chemical, health, and ecological data regarding exposure to hazardous substances. He provided guidance to EPA, the Department of Defense, and state and private sector managers and personnel about risk communication. He also developed national and regional guidance in the areas of probabilistic risk assessment, soil cleanup, statistics, and toxicology.

Dr. Simon also has 20 years of teaching experience and has lectured in 7 states in the United States and has given lectures in Belgium and Japan as an invited speaker. He teaches environmental risk assessment in the Department of Environmental Health Sciences at the University of Georgia.

Dr. Simon is the principal and owner of Ted Simon, LLC, and a member of the Round Table of Toxicology Consultants (RTC*). He provides scientific support in the areas of toxicology, environmental risk assessment, statistics, drug and alcohol abuse, and other issues. He is a board-certified toxicologist and has been the sole author and on various committees drafting EPA guidelines/papers on ecological risk assessment.

Dr. Simon is the author of 30 peer-reviewed articles and book chapters and has received several medals from the EPA for his science and policy work; this work includes his expert testimony in the Norman Mayes case in Knoxville, Tennessee, which led to additional UST regulations being passed by the State of Tennessee, and development of guidance for risk-based monitoring of land use controls at military facilities. In 2002, he was awarded the Science Achievement Award by EPA for his work on guidance for probabilistic risk assessment. In 2009, he received Honorable Mention from Risk Assessment Speciality Section of the SOT for the best paper in 2009.

Dr. Simon also writes mystery/suspense novels under the pen name of Wix Simon.

---

* Other RTC authors/editors we have include Ron Hood (*Reproductive and Developmental Toxicology, 3rd edn.*), Shayne Gad (*Safety Evaluation in the Development of Medical Devices and Combination Products, 3rd edn.; Animal Models in Toxicology, 2nd edn.; Toxicology of the Gastrointestinal Tract; Regulatory Toxicology, 2nd edn.*), and Robert Kapp (*Reproductive Toxicology, 3rd edn.*).

1 Introduction to Risk Assessment with a Nod to History

What a piece of work is a man, How noble in Reason, how infinite in faculties, in form and moving how express and admirable, In action how like an Angel!

William Shakespeare
The Tragedy of Hamlet Prince of Denmark, Act II, Scene 2, 1963

At the outset of the twenty-first century, technology and industrialization have provided advantages for much of the world's population—but technology is a double-edged sword. Twenty-first-century technology affords us many benefits, including advances in medical care and pharmaceutical products, cell phones, microwave ovens, and mass transit—but industrialization has, as a downside, a legacy of waste, and the people of the world cannot enjoy the benefits of technology without dealing with the associated hazards.

The purpose of risk assessment is to support societal decision making. Risk assessment is the means by which democratic societies attempt to understand the adverse and unintended consequences of technology. Risk management is the use of risk assessment information to control or abate these consequences.

Ideally, both risk assessment and risk management will be conducted in a way that takes into account the interests of all stakeholders—this is no more than fair! In risk assessment, the central issue embodied in the ideal of is how we as a society take into account both the variation in human exposures to environmental hazards or stressors and the variation in human susceptibility to injury or illness.

RISK ASSESSMENT: DOES CONSISTENCY ACHIEVE THE GOAL OF FAIRNESS?

One way to be fair in risk assessment is consistency. Risk assessment sits at the uneasy interface of science and policy. Almost all decisions about risk assessment methods require considerations of issues in both policy and science. Science is constantly changing, whereas the pace of policy change at times seems glacial in comparison. Hence, there will always be the tension between the old and the new in risk assessment.

Some risk assessment practitioners have come to see this tension as a scientific culture war—a battle between those who would preserve the status quo, clinging

1

to old ways for consistency's sake, and those who heartily embrace new ideas and new information. Since the field of risk assessment began with the seventeenth-century mathematician Blaise Pascal's development of probability theory to ameliorate his winnings at games of chance,* there has been tension between those who view the best available science as a new and challenging opportunity and those who view change as an anathema. This conflict between new and old lies at the heart of modern environmental risk assessments. Ralph Waldo Emerson noted "a foolish consistency is the hobgoblin of little minds."[1] Consistency, however, is a way to create the perception of fairness.

If the goal of consistency were achieved by nothing more than using the same default values for exposure and the same toxicity criteria in every risk assessment, then this consistency would indeed be foolish and will never be fair. This makes the job of a risk assessor tough—one must understand not only the science but also the policy goals and use this knowledge of both in an honest and forthright manner. A commitment to rigorous intellectual honesty in the evaluation of the data and scientific knowledge used in a risk assessment allows one to move away from "foolish consistency" while still achieving fairness.

Scientific knowledge is constantly increasing, maybe 5% per year, maybe more.[2] Changes in policy occur more slowly, and thus, science will always be ahead of policy. For example, knowledge of the genetic code and the structure of DNA led not only to use of forensic DNA analysis but also to the growing field of genomics and its use in medicine. However, such information is relevant to differences in susceptibility to the health effects of environmental chemicals, and to date, most risk assessment practitioners have no way to incorporate such information into their considerations.

One of the things these genomic studies show is the remarkable ability of humans and other species to modulate gene expression in a subtle and context-dependent way and thus produce biologically appropriate responses to the ever-changing internal and external stimuli living organisms experience.[3,4] There also exists the same degree of variation in human behavior and resulting exposure characteristics as evidenced by the wide variation seen in time–activity studies in children.[5–7]

Given the range of human variability, how can one account for this range in the exposure and toxicity assessments in an honest way that is fair to all stakeholders? The amelioration of the scientific knowledge base underpinning risk assessment is inevitable—why would one not want to avail oneself of all this information?

Of course, some risk assessments are better than others and some risk-based decisions are better than others. As scientists and risk assessment practitioners, we use the tools provided by toxicologists, chemists, statisticians, and others to understand the exposures and effects of environmental stressors and account for human variability in both these aspects. Our efforts inform decision makers so that they can balance the competing interests of many

---

* Success at poker requires skills in both risk assessment and risk management.

stakeholders and hold the ideal of fairness paramount. In the simplest and best terms, risk assessors transform information into knowledge so that the decision makers can act with true wisdom.[8–10]

Risk assessment is a predictive activity—the practitioners attempt to predict potential consequences using the knowledge at hand and the principles of probability. Indeed, by their very nature, all risk assessments are predictive and, as such, are also probabilistic. For at least the last decade, the term "probabilistic risk assessment" (PRA) has been used to refer to risk evaluations that include a statistical and quantitative treatment of either variability or uncertainty or both; however, it is important for risk assessment practitioners not to lose sight of the inherently probabilistic nature of their activities.

The culture war of the old and familiar versus the new and innovative will likely continue to intensify. There will be many examples in this book that make clear that the cultural divide in risk assessment was evident throughout its history and will no doubt continue.

Some of this cultural divide is due to the difference between the goals of science versus the goals of government regulation. Regulation seeks resolution of competing agendas by affecting human behavior—the pressure brought to bear on regulators for decisions is very often unrelated to science; societal or political factors may play a greater role and potentially lead to peremptory, episodic, and ill-considered actions.

Science, on the other hand, investigates and attempts to explain observed phenomena in a cautious and incremental fashion. Risk assessment is the attempt to utilize science to inform these societal decisions—hence, risk assessment uses science but is a tool of regulation.

Scientists today stand at the threshold of a new biology—one that integrates genomics, proteomics, metabolomics, transcriptomics, computational methods, and systems biology in an attempt to reach a new understanding applicable to toxicology, risk assessment, medicine, and other human endeavors. This new science is recognized as the only feasible way to attempt to obtain toxicity information on the approximately 80,000 chemicals in commerce today.[11] However, an understanding of this new science and its predictive value for risk assessment remains elusive.[12]

The largest hurdle to progress in risk assessment remains the emphasis on consistency and the relatively slow pace with which change occurs in regulatory risk assessment.[13] One wonders if risk assessment practitioners in the late twenty-first century, 50–100 years hence, will characterize today's regulatory risk assessors as Luddites, much as we think of the naysayers who protested the building of the Liverpool–Manchester railroad in 1825:

> … the railway would prevent cows grazing and hens laying. The poisoned air from the locomotives would kill birds as they flew over them and render the preservation of pheasants and foxes no longer possible. There would no longer be any use for horses; and if the railways extended, the species would become extinguished, and oats and hay would be rendered unsalable commodities.[14]

How similar are the attitudes of today's regulators toward this new science? In the fullness of time, how will history judge today's society with regard to their open-mindedness and acceptance of new science? As a society, we cannot turn back the clock, no matter how much we might long for a simpler and less complex time. Science and technology have changed the world in both large and small ways, and risk assessment is the best decision tool to apportion the burdens of the technology we all enjoy in a democratic manner.

KNOWLEDGE VERSUS FEAR: THE PRECAUTIONARY PRINCIPLE AND UNINTENDED CONSEQUENCES

In 1980, under Public Law 96-528 passed by the US Congress, the National Research Council (NRC) of the National Academy of Sciences (NAS) undertook an effort to strengthen the reliability of and objectivity of the scientific assessment underlying the federal regulatory policies on carcinogens and other public health hazards. The report produced by the NRC was titled *Risk Assessment in the Federal Government: Managing the Process*.[8] When released in 1983, this book had a red cover and came to be known as the Red Book.

Carcinogens were a focus of the Red Book likely because of the fear of cancer ingrained in western society.[15] For many years, treatments for cancer were horrific and generally disfiguring—if they worked at all.[16] This fear is reflected in the adoption of the Delaney clause by the US Congress in 1958 that concludes that no food additive that has been shown to induce cancer in man or experimental animals can be considered safe.[17]

The precautionary principle was given voice to encourage policies that protect human health and the environment in the face of uncertain risks. It was first stated in the Rio Declaration of 1992:

> ... when an activity raises threats of harm to human health or the environment, precautionary measures should be taken even if some cause and effect relationships are not fully established scientifically.[18]

The dictum of better safe than sorry might sound like good advice, but acting before sufficient information is available to support the decision or failing to consider all the available information may actually make matters worse. The precautionary principle is intended to address scientific uncertainty and to act, when needed, despite this uncertainty. This approach is necessary because scientific evidence will always be incomplete.

However, many who apply the precautionary principle act out of fear and the perception that "something needs to be done right away." The actions intended by the precautionary principle should be in proportion to the scope of the problem. When actions are taken in haste, the application of the precautionary principle runs head on into the law of unintended consequences. Chapter 7 will provide much additional discussion.

The prominent role given to risk assessment by government agencies and international bodies over the past 20 years confirms that, as a policy, societal decisions are best served by the unbiased application of scientific knowledge. As a tool of these decisions, risk assessment occupies the interface between science and policy. In the next section, the history of risk assessment in the United States will be examined in detail.

HISTORY OF ENVIRONMENTAL RISK ASSESSMENT IN THE UNITED STATES

In 1969, the 91st US Congress enacted the National Environmental Policy Act (NEPA), and President Richard M. Nixon signed the act into law on January 1, 1970.

NEPA requires that every agency in the executive branch of the US government takes steps to implement the policies set forth in the act. NEPA created the Council on Environmental Quality in the Executive Office of the President with the responsibility of ensuring that other federal agencies met their obligations under NEPA.

RISK ASSESSMENT UNDER THE NATIONAL ENVIRONMENTAL POLICY ACT

The NEPA process involved evaluation of the environmental impact of any proposed action. Some actions, such as minor facility renovations or improvement of existing hiking trails, were categorically excluded from the process. Actions without a categorical exclusion (CE) would undergo an environmental assessment, the results of which would be reported in an environmental impact statement. Section 101 of the act conveys the ambition and desire for protection of the environment:

> The Congress, recognizing the profound impact of man's activity on the interrelations of all components of the natural environment, particularly the profound influences of population growth, high-density urbanization, industrial expansion, resource exploitation, and new and expanding technological advances and recognizing further the critical importance of restoring and maintaining environmental quality to the overall welfare and development of man, declares that it is the continuing policy of the Federal Government, in cooperation with State and local governments, and other concerned public and private organizations, to use all practicable means and measures, including financial and technical assistance, in a manner calculated to foster and promote the general welfare, to create and maintain conditions under which man and nature can exist in productive harmony, and fulfill the social, economic, and other requirements of present and future generations of Americans. (42 USC § 4331)[19]

Figure 1.1 shows a schematic for the NEPA process. Box 6 in the middle of the diagram refers to "environmental assessment." A stated policy goal of NEPA is to "attain the widest range of beneficial uses of the environment without degradation,

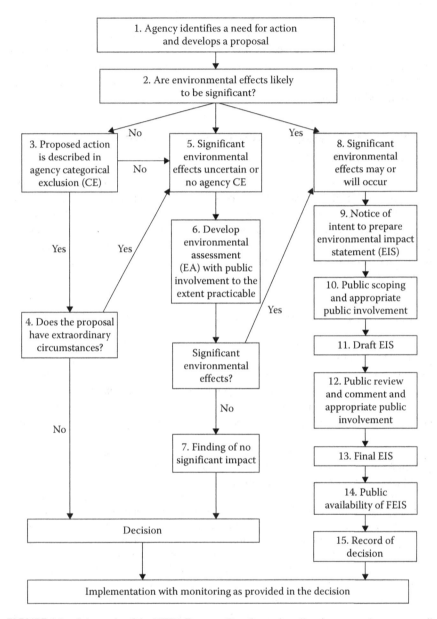

FIGURE 1.1 Schematic of the NEPA Process. Box 6 mentions "environmental assessment." This is the first mention of anything related to environmental risk assessment from the US government.

risk to health or safety, or other undesirable and unintended consequences."[19] This is likely the first explicit mention of environmental risk assessment by the US government. Hence, the Red Book, as the first risk assessment document of the US government, was necessary for implementation of NEPA.

EVENTS OF THE LATE 1960s FACILITATED THE PASSAGE OF NEPA

NEPA was passed by the senate in a unanimous vote on July 10, 1969, and passed by the House of Representatives by a vote of 372–15 on September 23, 1969. Clearly, NEPA received bipartisan support.

It is instructive to understand the American mindset in the late 1960s as the backdrop leading up to the passage of such a far-reaching act as NEPA. Indeed, it is highly unlikely that NEPA would be passed today, and the history of that time is worth considering as a backdrop to the passage of NEPA.

On March 18, 1968, presidential candidate Robert F. Kennedy spoke at the University of Kansas:*

> Too much and too long, we seem to have surrendered community excellence and community values in the mere accumulation of material things. Our gross national product ... if we should judge America by that—counts air pollution and cigarette advertising, ... It counts the destruction of our redwoods and the loss of our natural wonder in chaotic sprawl ... the gross national product does not allow for the health of our children, the quality of their education, or the joy of their play. It does not include the beauty of our poetry or the strength of our marriages, the intelligence of our public debate or the integrity of our public officials. It measures neither our wit nor our courage; neither our wisdom nor our learning; neither our compassion nor our devotion to our country; it measures everything, in short, except that which makes life worthwhile. And it tells us everything about America except why we are proud that we are Americans.

Two notable events in 1968 united political will and likely had bearing on the enactment of NEPA. These events were the following:

- April 4, 1968—Dr. Martin Luther King Jr. was assassinated in Memphis.
- June 6, 1968—Robert F. Kennedy was assassinated in Los Angeles.

On December 13, 1968, Garrett Hardin published his famous article "The Tragedy of the Commons" in *Science*.[20] Hardin wrote as follows:

> The rational man finds that his share of the cost of the wastes he discharges into the commons is less than the cost of purifying his wastes before releasing them. Since this is true for everyone, we are locked into a system of "fouling our own nest," so long as we behave only as independent, rational, free-enterprisers ... but the air and waters surrounding us cannot be readily fenced, and so the tragedy of

---

* This speech, from March 18, 1968, can be heard in its entirety on YouTube at http://www.youtube.com/watch?v=z7-G3PC_868&feature=related.

the commons as a cesspool must be prevented by different means, by coercive laws
or taxing devices that make it cheaper for the polluter to treat his pollutants than to
discharge them untreated.[20]

In 1969, two environmental disasters occurred. These also likely facilitated
the passage of NEPA:

- January 31, 1969—an offshore oil well near Santa Barbara, California,
  blew out, spilling 235,000 gal of oil that covered 30 miles of beach
  with tar.
- June 22, 1969—the Cuyahoga River in downtown Cleveland burst into
  flames five stories high from chemical and oil pollution.

With the passage and signing of NEPA on January 1, 1970, the practice of
environmental risk assessment was first codified into law.

HOW MUCH RISK IS ENOUGH?

The Occupational Safety and Health Act was also passed in 1970 and established
the Occupational Safety and Health Administration (OSHA). OSHA based their
early regulatory decisions on whether or not a hazard was identified—a qualitative criterion. In 1978, the American Petroleum Institute challenged OSHA's benzene lifetime permissible exposure limit (PEL). This case went all the way to the
Supreme Court. The court ruled that OSHA must establish that the chemical poses
a "significant" risk before establishing a standard. The court wrote as follows:

> Some risks are plainly acceptable and others are plainly unacceptable. If for exam-
> ple, the odds are one in a billion that a person will die from cancer by taking a
> drink of chlorinated water, the risk clearly could not be considered significant. On
> the other hand, if the odds are one in a thousand that regular inhalation of gasoline
> vapors that are 2 percent benzene will be fatal, a reasonable person might well con-
> sider the risk significant and take the appropriate steps to decrease or eliminate it.[21]

OSHA chose a risk level of 10^{-3} or "one in a thousand" as an appropriate standard for the workplace. The court identified acceptable risk as somewhere in the
million-fold risk range from "one in a billion" to "one in a thousand." OSHA
ended up choosing 10^{-3} as an acceptable risk, the upper end of the range stated
by the court.

In 1958, the US Congress passed the Delaney clause to the Food, Drug, and
Cosmetic Act of 1938. This clause banned the use in food of "any chemical
additive found to induce cancer in man, or, after tests, induce cancer in animals." In 1959, just after the passage of the Delaney clause, there occurred
the Thanksgiving Day cranberry scare because Arthur Sherwood Flemming,
then Secretary of Health, Education, and Welfare, announced publicly in early
November that aminotriazole, a weed killer that causes thyroid cancer in laboratory rats, was discovered in some grocery store cranberries. Flemming asked

the National Cancer Institute (NCI) to help establish a "safe" level of carcinogens in food. The NCI used a definition of safety of "1 in a 100,000,000" or 10^{-8} from a 1961 publication by Nathan Mantel, a biostatistician at NCI. The purpose of the article was to develop guidelines for the number of animals required to establish the safety of a chemical.[22]

The US Food and Drug Administration (USFDA) approved diethylstilbestrol (DES) in 1954 for use as a growing/finishing food additive for cattle. In 1971, a report appeared in the *New England Journal of Medicine* about the occurrence of vaginal tumors in young women from mothers who had received DES during pregnancy.[23] This discovery also increased the fear of carcinogens in food.

Because of the economic impact of the Thanksgiving Day cranberry scare, FDA felt it had the ability to define a "safe" level of carcinogens in food—in direct contradiction to the Delaney clause. FDA used Mantel's value of 10^{-8} in their proposed rule in the Federal Register in 1973 but increased this level to "one in a million" in the 1977 final rule. There was also considerable controversy leading up to FDA's ban of DES for use in cattle production in 1979.[24–26]

The proposed rule for the National Contingency Plan for Oil and Hazardous Substances identified a risk range of 10^{-7} to 10^{-4} as acceptable for Superfund cleanups. When the final rule was promulgated in 1990, the risk range was changed to 10^{-6} to 10^{-4}.

The US Environmental Protection Agency (USEPA) clarified the use of 10^{-4} as the upper end of the target risk range in a memorandum in 1991 titled *Role of the Baseline Risk Assessment in Remedy Selection*.[27] This guidance document identified 10^{-4} as a "soft bright line" as follows:

The upper boundary of the risk range is not a discrete line at $1 \times 10(-4)$, although EPA generally uses $1 \times 10(-4)$ in making risk management decisions.[27]

Although this memorandum attempted to introduce flexibility into the Superfund remedy selection process, risk managers may not always be aware of this flexibility.

In 1998, EPA's assistant administrator of OSWER, Tim Fields, released a memo titled *Approach for Addressing Dioxin in Soils at RCRA and CERCLA Sites*.[28] It stated as follows:

Based on presently available information, and using standard default assumptions for reasonable maximum exposure scenarios, the upper-bound lifetime excess cancer risk from residential exposure to a concentration of 1 ppb dioxin is approximately 2.5×10^{-4}, which is at the higher end of the range of excess cancer risks that are generally acceptable at Superfund sites. The calculated upper-bound excess cancer risk associated with a lifetime commercial/industrial exposure to 5 ppb, or the lower end of the range recommended for commercial/industrial soils, is approximately 1.3×10^{-4}, which is also within the CERCLA risk range.[28]

Hence, the upper end of the Comprehensive Environmental Response, Compensation, and Liability Act (CERCLA) risk range of 1×10^{-4} could be interpreted as somewhere between 10^{-4} and 10^{-3}. In essence, this represents a return to the Supreme Court's definition of acceptable risk as "one in a thousand" from the benzene decision.

ENVIRONMENTAL RISK ASSESSMENT PARADIGM AS DEFINED BY THE RED BOOK

The Red Book defines risk assessment as the use of the factual base to define the health effects of exposure of individuals and populations to hazardous materials and situations. Risk assessment contains the following four steps:

- *Hazard identification (HI)*—whether a chemical or other stressor can be causally linked to a particular adverse health outcome
- *Dose–response assessment*—the relationship between the magnitude of exposure and the likelihood of occurrence of the identified hazards
- *Exposure assessment*—the extent, frequency, and magnitude of human contact with the chemical or stressor
- *Risk characterization*—a description of the nature and magnitude of the risk associated with the situation being considered, including both qualitative and quantitative risk descriptors and the uncertainties attendant in these descriptors

Figure 1.2 shows this four-part scheme along with other factors that influence environmental decisions.

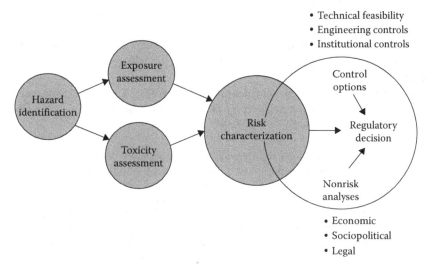

FIGURE 1.2 Schematic of the risk assessment paradigm from the Red Book with nonrisk decision factors also shown. As noted in the text, the Red Book highlighted the need for separation between risk assessment and risk management.

In addition to setting out the paradigm shown in Figure 1.2, the Red Book strongly advocated for the separation of risk assessment from risk management. The first recommendation made was as follows and occurs on page 151:

> Regulatory agencies should take steps to establish and maintain a clear conceptual distinction between assessment of risks and the consideration of risk management alternatives; that is, the scientific findings and policy judgments embodied in risk assessments should be explicitly distinguished from the political, economic, and technical considerations that influence the design and choice of regulatory strategies.[8]

Value judgments are needed to weigh the tradeoffs between the potential for adverse health consequences and economic, political, and social considerations. These judgments about nonscience issues are the proper function of government in a democratic society. The clear motivation for this recommendation was to place a firewall between these value judgments and the scientific information and science policy judgments that are the purview of the risk assessment. Above all, science requires integrity, and the writers of the Red Book foresaw and attempted to forestall the potential for economic or political factors of risk management to affect the scientific considerations of risk assessment.

The second recommendation on page 153 was as follows:

> Before an agency decides whether a substance should or should not be regulated as a health hazard, a detailed and comprehensive written risk assessment should be prepared and made publicly accessible. This written assessment should clearly distinguish between the scientific basis and the policy basis for the agency's conclusions.[8]

A written assessment would permit stakeholders to voice agreement or disagreement in an informed manner. The selected risk management alternative will be based on both science and policy considerations—it is highly appropriate for the regulated community or other stakeholders to disagree with the interpretation of the science, the policy considerations, or the value judgments. However, the scientific information, the data upon which the risk assessment is based, is generally not up for debate.

The third recommendation was for review of risk assessments by an independent scientific panel with members selected for their scientific and technical competence. The Red Book recommended that panel members be selected from the private and public sectors, universities, and government research agencies. Personnel from the agency conducting the risk assessment or employees of an entity with a substantial interest, economic or other, in the societal decision, should not be members of the scientific review panel.

The NRC went on to recommend that inference guidelines be developed to create a degree of uniformity in the risk assessment process. The Red Book specifically suggested that guidelines be developed for cancer risk assessment and for exposure assessment. In addition, inference guidelines should be periodically updated and revised. The writers of the Red Book were wise in that they were aware of the conflict between old and new and they recognized the need for both consistency and progress.

SCIENCE AND JUDGMENT IN RISK ASSESSMENT: THE BLUE BOOK

In 1970, the US Congress passed the Clean Air Act (CAA) regulating emissions from both mobile and stationary sources of air pollution. EPA was authorized to establish National Ambient Air Quality Standards (NAAQS) to protect public health and welfare. Before 1990, risk assessment was used to establish standards for six common classes of pollutants—sulfur oxides, particulate matter, carbon monoxide, nitrogen oxides, hydrocarbons, and photochemical oxidants such as ozone and formaldehyde. The 1990 CAA amendments authorized EPA to regulate the emissions of 189 toxic chemicals that were carcinogenic, mutagenic, or toxic to reproduction or development. Section 112 of the CAA required EPA to set emission standards for hazardous air pollutants to protect public health with "an ample margin of safety." The regulatory standards were not based on risk but rather the maximum achievable control technology (MACT).[29] Hence, EPA interpreted Section 112 to place technology-based regulation in a primary role and health-based risk assessment in a secondary role.

The National Resources Defense Council (NRDC) sued EPA in the District of Columbia Circuit Court of Appeals, seeking to compel the agency into a zero emissions policy for carcinogenic air pollutants. In the choice of technology-based regulation, EPA had adopted a generic method for determining whether the emissions of a specific pollutant would meet the bar of an "ample margin of safety." NRDC argued that this decision rendered all potential carcinogens as having an "ample margin of safety." In 1987, the court upheld the NRDC claim, indicating that the intent of Section 112 was protection of public health and that EPA's generic method was inadequate.

In the interim, EPA was active in implementing the recommendation in the Red Book for development of uniform inference guidelines. In 1986, EPA's Risk Assessment Forum (RAF) released the first version of the *Guidelines for Carcinogen Risk Assessment*, the *Guidelines for Mutagenicity Risk Assessment*, and the *Guidelines for Human Health Risk Assessment of Chemical Mixtures*.[30–32] In 1989, EPA's Office of Solid Waste and Emergency Response (OSWER) that regulates under the CERCLA, more commonly known as Superfund, produced a comprehensive guidance on the application of risk assessment to hazardous waste sites—*Risk Assessment Guidance for Superfund (RAGS). Part A*.[33] Although, in the years to come, parts B through F of RAGS were released, this 1989 document has become commonly known as "RAGS." The generic quantitative risk equations for carcinogens and non-carcinogens are shown in Box 1.1.

BOX 1.1   GENERIC RISK EQUATIONS FROM RAGS[32]

Carcinogenic Risk

$$\text{Risk} = \frac{\text{CSF} \times (\text{CR} \times \text{ED} \times \text{EF})}{(\text{BW} \times \text{AT})} \tag{1.1}$$

Noncancer HQ

$$\text{HQ} = \left(\frac{1}{\text{RfD}}\right) \times \frac{(\text{CR} \times \text{ED} \times \text{EF})}{(\text{BW} \times \text{AT})} \tag{1.2}$$

where
   CSF is the cancer slope factor
   RfD is the reference dose
   HQ is the hazard quotient
   CR is the contact rate
   ED is the exposure duration
   EF is the exposure frequency
   BW is the body weight
   AT is the averaging time

The *carcinogenic risk* is expressed as a unitless probability; the assumption is that the dose–response relationship for carcinogens is linear in the low-dose region and even one molecule of a substance poses some risk, albeit vanishingly small. The HQ is the ratio between the RfD and the average daily dose (ADD); an HQ value less than one indicates that it is unlikely even for sensitive populations to experience adverse health effects.

The 1990 amendments to the CAA rewrote Section 1212 to enhance the role of risk assessment. Rather than assuming that the MACT would result in acceptable level of risk, the 1990 amendments codified a tiered approach such that a risk assessment would always be performed; if the MACT standard results in a risk of greater than one in a million for the most highly exposed individual, then a residual risk standard would be also developed. Hence, risk assessment would play a central role in the CAA regulation.

From 1984 until 1992, EPA developed three versions of uniform guidelines on exposure. The *Final Guidelines for Exposure Assessment* was published in the Federal Register in 1992.[34] These guidelines were detailed and specific and distinguished between various types of dose terms (Box 1.2). The recognition of internal dose and delivered dose enabled the use in risk assessment of physiologically based pharmacokinetic (PBPK) models that were commonly used in

> ### BOX 1.2   DOSE TERMS USED IN EPA'S 1992 EXPOSURE GUIDELINES[33]
>
> *Exposure dose*—contact of a chemical with the outer boundary of a person, for example, skin, nose, mouth.
>
> *Potential dose*—amount of chemical contained in material ingested, air inhaled, or applied to the skin.
>
> *Applied dose*—amount of chemical in contact with the primary absorption boundaries, for example, skin, lungs, gastrointestinal tract, and available for absorption.
>
> *Internal dose*—amount of a chemical penetrating across an absorption barrier or exchange boundary via physical or biological processes.
>
> *Delivered dose*—amount of chemical available for interaction with a particular organ or cell.

pharmacology and drug development. These models are also known as absorption/ distribution/metabolism/excretion (ADME) models and will be discussed briefly at the end of this chapter and at greater length in Chapter 4.

Part of the 1990 CAA amendments directed EPA to engage the NAS to review EPA's methods for estimating carcinogenic potency of chemicals and methods for estimating exposure to both hypothetical and actual maximally exposed individuals. *Science and Judgment in Risk Assessment* was released by the NRC in 1994 and was known as the Blue Book.[9]

One of the major observations of the Blue Book was the desire of most people to "understand whether and how much their exposures to chemicals threaten their health and well-being."[9] The Blue Book also noted some common themes that cut across various aspects of risk assessment and suggested strategies for improvement. These common themes were the following:

- The use of default options or values
- Validation of data, models, and methods
- Information and data needs
- Accounting for uncertainty
- Dealing with variability
- Aggregation of risks

EPA's Use of Defaults

The Blue Book concluded that the practice of using default options was reasonable when there is doubt about the choice of values or models. Essentially, the NRC concluded that the use of defaults was a necessary evil. Regarding Section 112 of the CAA, the Blue Book noted that scientific disagreements fostered both concern and skepticism about risk assessment, and the use of defaults could potentially ameliorate this situation.

The Blue Book also indicated that the scientific or policy basis of each default should be clearly articulated. Lastly, the Blue Book indicated that a clear and transparent process for choosing to depart from default options includes full public discussion and peer participation by the scientific community.

VALIDATION OF MODELS, METHODS, AND DATA

The Blue Book called on EPA to establish the predictive accuracy of the methods and models used in risk assessment with greatest priority given to the scientific basis of the default options. Regarding exposure models and data, the Blue Book indicated EPA should consider both population mobility and time–activity relationships. Regarding the toxicity assessment, the Blue Book indicated that EPA should continue to use laboratory animal bioassay data as needed, but should not automatically assume that animal carcinogens are necessarily human carcinogens.

In an almost prescient fashion, the Blue Book discussed mode of action (MOA) without using the term. MOA has come to be central to the dose–response assessment. The term MOA was first used in the 1990s as a means of providing a structured approach to understanding the process of cancer induction in test animals and the relevance to this process to humans.[35] MOA is described in Table 1.1 and Figure 1.3 shows a diagram of the MOA for dioxin-induced liver tumors in rats. MOA information is often used to determine human relevance. The Blue Book indicated on page 141 that animal tumor data should not be used as the exclusive evidence to classify chemicals as human carcinogens if "the mechanisms operative in laboratory animals are unlikely to be operative in humans."[9]

INFORMATION AND DATA NEEDS

The Blue Book pointed out that EPA had not defined the types, quantities, and qualities of data needed for risk assessment. There should also be standards for the collection of environmental data to ensure that the data collected support the risk assessment to the greatest extent possible. Indeed, this recommendation led to the development of data quality objectives.

Because the Blue Book was written to address issues regarding the CAA, much of the document is relevant to air rather than other environmental media. The Toxic Release Inventory (TRI) Program is a database of material released during manufacturing and storage; basically, the amounts of chemicals in finished products are subtracted from the amounts purchased and held in inventory, and this difference is assumed to be released. If available, direct measurements of releases may also be used. The information was self-reported by the regulated entity and EPA had no way to check on the accuracy of the information.

TABLE 1.1
Common Concepts and Terms in Risk Assessment

Concept	Definition
Cancer risk	In RAGS, USEPA indicates that the presumption of a threshold for carcinogenic chemicals is inappropriate. Hence, any nonzero dose will provide some quantifiable prediction of the likelihood of cancer. The methodology underlying this type of risk assessment is to obtain a measure of the statistical upper bound of the slope within the low-dose region of the presumed dose–response relationship. The slope represents the linear relationship between risk and dose and is in the units of risk per dose.
Cancer slope factor (CSF)	RAGS defines this as a plausible upper-bound estimate of the probability of cancer per unit intake of a chemical over a lifetime. The most common units of the CSF is the reciprocal of mg chemical per kilogram body weight per day or $(mg/kg/d)^{-1}$.
Hazard quotient (HQ)	The ratio of a single substance exposure level over a specified time period (e.g., chronic) to an RfD for that substance derived from a similar exposure period.
Hazard index (HI)	The sum of more than one HQ for multiple substances and/or multiple exposure pathways. The HI is calculated separately for chronic, subchronic, and shorter-duration exposures.
RfD	An estimate (with uncertainty spanning perhaps an order of magnitude or greater) of a daily exposure level for the human population, including sensitive subpopulations, that is likely to be without an appreciable risk of deleterious effects during a lifetime.
MOA	A sequence of key events and processes, starting with interaction of an agent with a cell, proceeding through operational and anatomical changes, and resulting in the formation of cancer or other adverse effects.
Key event	An empirically observable causal precursor step to the adverse outcome that is itself a necessary element of the MOA. Key events are required events for the MOA, but often are not sufficient to induce the adverse outcome in the absence of other key events.
Associative event	Biological processes that are themselves not causal necessary key events for the MOA but are reliable indicators or markers for key events. Associative events can often be used as surrogate markers for a key event in a MOA evaluation or as indicators of exposure to a xenobiotic that has stimulated the molecular initiating event (MIE) or a key event.
Modulating factor	A biological factor that modulates the dose–response behavior or probability of inducing one or more key events or the adverse outcome.

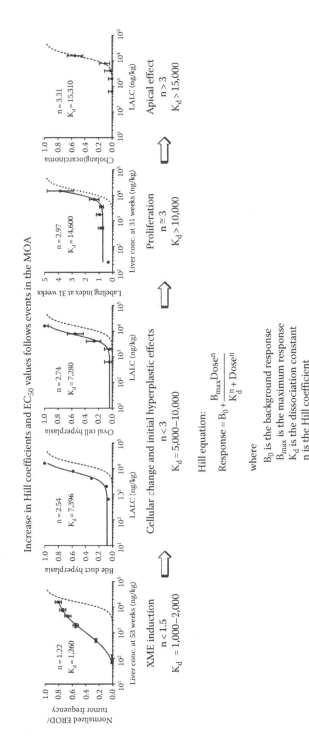

FIGURE 1.3 MOA for the production of cholangiocarcinoma in female Sprague–Dawley rats by 2,3,7,8-tetrachlorodibenzodioxin (TCDD). Tumor induction involves the key events of induction of xenobiotic metabolizing enzymes (XME), cellular and hyperplastic tissue changes, cell proliferation, and the apical event of tumors. Cholangiocarcinoma is a tumor that arises from liver stem cells likely residing near the bile ducts. The Hill equation shown at the bottom of the figure was used to model the dose–response of all key events and tumor occurrence. Higher values of the dissociation constant indicate an effect with a higher threshold. Higher values of the Hill coefficient indicate a more steeply rising dose–response. TCDD binds to and activates the aryl hydrocarbon receptor (AHR) producing changes in gene expression including induction of cytochrome p450 1A1 (CYP1A1) as shown in the leftmost panel. The later effects of hyperplasia and proliferation occur at higher thresholds and with a steeper dose–response. The apical effect of cholangiocarcinoma has the highest threshold and steepest response of all the effects shown. (Dotted line shown in all plots: Adapted from Simon et al., *Toxicol Sci.* 2009 September 23; 112(2):490–506.)

In 1992, Amoco, EPA, and the Commonwealth of Virginia agreed to conduct a multimedia assessment of releases at the Amoco refinery at Yorktown, Virginia.[36] One purpose of this study was to assess the accuracy of TRI data and the study revealed that the TRI data were not accurate. This fact is hardly surprising given that the preparation of TRI reports did not generate revenue for the regulated entity; since the regulators at EPA had no way to check, accuracy was sacrificed in favor of timely completion of the paper work for regulatory compliance.

ACCOUNTING FOR UNCERTAINTY

EPA did not account for the uncertainties inherent in risk assessment in either a qualitative or quantitative fashion. Hence, the Blue Book suggested that EPA develop guidelines for uncertainty analysis and the contributions of the various sources of uncertainty to the total uncertainty in a risk assessment.

Many EPA risk assessments were presented as a single point estimate of risk—the entire risk assessment was boiled down to a single number, hardly a nuanced or transparent presentation. In 1992, F. Henry (Hank) Habicht II, the deputy administrator of EPA from 1989 until 1992, wrote a memorandum, *Guidance on Risk Characterization for Risk Managers and Risk Assessors*.[37] Habicht wrote quite eloquently as follows:

> Specifically, although a great deal of careful analysis and scientific judgment goes into the development of EPA risk assessments, significant information is often omitted as the results of the assessment are passed along in the decision making process. Often, when risk information is presented to the ultimate decision-maker and to the public, the results have been boiled down to a point estimate of risk. Such 'short hand' approaches to risk assessment do not fully convey the range of information considered and used in developing the assessment. In short, informative risk characterization clarifies the scientific basis for EPA decisions, while numbers alone do not give a true picture of the assessment.[37]

The Blue Book included the Habicht memo in its entirety as Appendix B. This memo was the first official statement from EPA that the standard operating procedure for risk assessment failed to convey the full picture of risks, especially when the results of a complex and time-consuming assessment were transmitted to decision makers and the public as a single number.[38]

UNDERSTANDING AND DEALING WITH VARIABILITY

One size fits all? This is patently untrue for most aspects of humans with the possible exception of tube socks. It is especially untrue for ski boots. The idea of human variation is captured most eloquently of all in the quotation from Shakespeare's *Hamlet* at the beginning of this chapter.

Before 1994, EPA had chosen to regulate based on estimated risk to the maximally exposed individual—the worst-case scenario in which a hypothetical individual experienced a 70-year, 24 h/day exposure to the maximum estimate of the long-term average concentration. The Blue Book indicated that this could be used as a bounding estimate but was a poor choice for the basis of regulation.

Instead, the suggestion was to select an individual at the 90th percentile of exposure or above as a reasonable "high-end" estimate.

The Blue Book also recommended that EPA begin to use frequency distributions of both exposure and susceptibility to express the range of human heterogeneity and incorporate a range of values from these distributions to obtain a set of risk estimates more informative than a single point estimate.

AGGREGATION OF RISKS

The Blue Book also called for the consideration of how to account for separate but related causes of risk such as the occurrence of multiple chemicals from a single source. Aggregation of both exposures and effects were discussed. These concepts are still being debated today as questions arise about synergy of effects and whether multiple subthreshold exposures to chemicals operating via a common MOA or common adverse effect should be considered in terms of additivity of dose or additivity of effect.[39]

1997 FEDERAL COMMISSION REPORT

The Blue Book also advocated the formation of a joint Presidential/Congressional Commission on Risk Assessment and Risk Management. The report of this commission, *Framework for Environmental Health Risk Management*, dealt to a much greater extent with risk management and provided yet another diagram of the interface between risk assessment and risk management (Figure 1.4).[10] One of

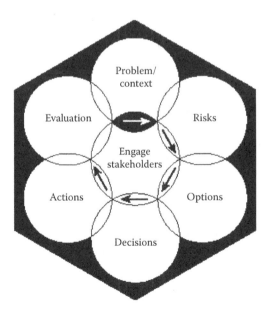

FIGURE 1.4 Risk assessment/risk management framework developed by the joint Congressional/Presidential Commission in 1997.

the major points made by the commission report was that inclusion of various stakeholders in the decision process generally led to better decisions and that government officials and other risk managers should take into account the economic, social, cultural, ethical, legal, and political ramifications of their decisions in addition to considerations of risk to human health or the environment.

In addition, the report specifically called for analysis of costs, benefits, and potential unintended consequences of environmental decisions. The report provided a controversial example. Assume a new regulation bans a commonly used but potentially carcinogenic pesticide resulting in a significant price increase of fruits and vegetables. Those who can afford the price increase will enjoy the benefit of reduced health risks, but others who cannot afford the higher prices will suffer poorer nutrition and increased cancer risk associated with a diet low in fruits and vegetables. Sadly, this example from the Blue Book came true only 2 years after its publication. Chapter 7 provides the details.

Statements in the report such as these forced EPA to consider the economic aspects of risk. The report was very clear that both risk–risk and risk–benefit tradeoffs must be considered. The report included the following potential adverse consequences:

- Reduced property values or loss of jobs
- Environmental justice issues, such as disregard for dietary needs, preferences or status of a particular group, or prioritizing cleanups in affluent areas
- Potential harm to the social fabric and life of a community by relocating people away from a highly contaminated area

How should risk issues be balanced against economic issues? In western society, life is considered priceless—this is, of course, why slavery and murder are illegal.

For risk–benefit analysis, economists have developed a measure of the worth of a human life called the value of a statistical life (VSL). The VSL is very different than the value of an actual life. The VSL can be estimated by statistical regression analysis of the wages of different occupations versus the risk of injury or death associated with the occupations. Another way to estimate the VSL is to ask people about their willingness to pay for a reduction in risk. EPA used the central estimate of $7.4M in 2006$ as the VSL for cost–benefit analysis.[40] There will be additional discussion of risk–benefit and risk–risk tradeoffs in a later chapter.

REALISM, COST, AND THE SEPARATION OF RISK ASSESSMENT AND RISK MANAGEMENT

The *Guidelines for Exposure Assessment*[34] document describes three tiers of exposure assessment. The first tier is a preliminary evaluation to produce bounding estimates. This preliminary evaluation described in the guidelines is similar to the majority of risk assessments conducted in that conservative assumptions are used to ensure that cleanups will be protective. The second tier of exposure assessment is the refinement of these preliminary risk descriptors by incorporation of site-specific considerations. The third tier of exposure assessment includes a probabilistic component and explicitly acknowledges human variability. The upper end of the distribution of risk should be characterized, and high-end estimates of individual risk, such as the hypothetical reasonable maximum exposure (RME) individual, should fall at the 90th percentile or above. Additionally, the exposure guidelines provide a detailed and cogent discussion of uncertainty assessment that concludes as follows:

> It is fundamental to exposure assessment that assessors have a clear distinction between the variability of exposures received by individuals in a population, and the uncertainty of the data and physical parameters used in calculating exposure.[34]

The exposure guidelines were a prescient document, and soon after their publication, EPA regional offices in Denver and Philadelphia issued guidance on the appropriate use of probabilistic methods in risk assessment.[41,42]

In 1995, administrator Carole Browner issued the far-reaching *Memorandum on EPA's Risk Characterization Program*.[13] This memo called for disclosure of the scientific analyses, uncertainties, assumptions, and science policy choices underlying the decisions made in the course of risk assessment and risk management. The memo endorsed the core values of transparency, clarity, consistency, and reasonableness (TCCR). The RAF's 1995 *Guidance on Risk Characterization*[38] that accompanied the Browner memo stated that it was imperative that risk assessors distinguish between variability and uncertainty. Similar to the Habicht memo, the *Guidance on Risk Characterization* also took the agency risk assessors to task for oversimplifying the results of their risk assessments:[38]

> Often risk assessors and managers simplify discussion of risk issues by speaking only of the numerical components of an assessment. ... However, since every assessment carries uncertainties, a simplified numerical presentation of risk is always incomplete and often misleading.[38]

In December 2000, the Science Policy Council of the USEPA issued the *Risk Characterization Handbook*.[43] This document also echoed the message of the Habicht memo, pointing out how risk characterization communicates the results of the risk assessment including key findings, uncertainties, strengths, and weaknesses of the analysis to decision makers and stakeholders in a conscious deliberate and transparent way. This document also emphasized the core values of TCCR.

USEPA ADDRESSES VARIABILITY AND UNCERTAINTY

In the spring of 1997, EPA deputy administrator Fred Hansen released a memorandum, *Policy for Use of Probabilistic Analysis in Risk Assessment*.[44] According to the policy statement of the memorandum, probabilistic analysis techniques, "given adequate supporting data and credible assumptions, can be viable statistical tools for analyzing variability and uncertainty in risk assessments." Along with this policy statement, the RAF released the *Guiding Principles for Monte Carlo Analysis*.[45]

COMPOUNDING CONSERVATISM

It is likely that PRA policy memo and the guiding principles document were released in response to calls for an increased role of science in EPA's decision making. These calls came from outside the agency. One of the most vocal critics of EPA's risk assessment methodology during the 1990s was Dr. David Burmaster of Alceon in Cambridge, MA. Sadly, Burmaster was a man ahead of his time[46] and suffered a great deal of frustration during the early 1990s when EPA risk assessors turned a deaf ear to his requests that they consider PRA.[47-49] Burmaster has a comprehensive knowledge of human biology and statistics. As early as 1993, Burmaster was clearly aware of the culture war between old and new discussed earlier in this chapter and its effect on the practice of risk assessment. He wrote in a perspective feature in the journal *Risk Analysis* as follows:

> An unfortunate trend – created in the name of policy consistency – has replaced the science in risk assessment with simplistic policy assumptions that have the effect of making risk assessments even more conservative. Too many risk assessors with advanced degrees are being forced to use (and to defend in public) contrived and biased methodologies to conform to EPA policy guidance. These methodologies cannot be defended as good science, and they subvert the Agency's stated risk management policies aimed at protecting the public health against reasonably expected future exposures....[49]

The degree of conservatism compounds dramatically for deterministic point estimates of risk constructed from upper percentiles of input parameter.[50]

One can actually do a very simple calculation to estimate the degree of conservatism in a risk assessment. An example is shown in Box 1.3, suggesting that the default methodology for cancer risk assessment is very conservative on a numerical basis.

There were other more powerful critics of USEPA at that time—notably John R. Graham, then head of the Harvard School of Public Health Center. Graham later became head of the White House Office of Management and Budget (OMB). In 2003, under Graham's leadership, the OMB issued guidelines strongly advocating the use of formal quantitative uncertainty analysis for regulatory decisions with economic effects of over $1B.[51] As will be seen in the next section, the document may have influenced the development of risk assessment methodology in China.

BOX 1.3   ESTIMATING THE PERCENTILE OF
CONSERVATISM IN A RISK ASSESSMENT

The basic equation we will use is

$$\text{Risk} = \text{Exposure} \times \text{Toxicity} \tag{1.3}$$

To estimate the percentile of conservatism in a risk assessment, we will use the following equation:

$$P_{\text{Conservatism}} = 1 - (1 - P_{\text{Exposure}}) \times (1 - P_{\text{Toxicity}}) \tag{1.4}$$

Looking back at Box 1.1, we can see that the CR, EF, and ED are exposure factors that are usually set at high-end values representing the 95th percentile. Hence, the percentile of conservatism for exposure would be

$$P_{\text{Exposure}} = 1 - (1 - 0.95) \times (1 - 0.95) \times (1 - 0.95) = 0.999875 \tag{1.5}$$

The toxicity term we will consider is the CSF. For CSFs, the 95th percentile lower confidence limit on the dose at the POD is currently used. Previously, the 95% upper confidence limit on the slope in the low-dose region was used. Hence, the conservatism of the toxicity term will be assumed to be 95%.

The overall percentile of conservatism can be estimated using the products of the percentiles for exposure and toxicity:

$$P_{\text{Conservatism}} = 1 - (1 - 0.999875) \times (1 - 0.95) = 0.9999375 \tag{1.6}$$

This calculated RME risk at greater than the 99.99th percentile cannot likely be distinguished from maximum risk in terms of the degree of conservatism.

RISK ASSESSMENT AS PRACTICED INTERNATIONALLY

The events in the United States that led to the passage of NEPA and the formation of the EPA were the first of their kind and influenced events in other countries. In this section, risk assessment methodology and its use in regulation will be considered in China, Europe, and Canada. The point is to show commonality of methods rather than to conduct an exhaustive comparison of the differences. The influence of the methodology developed in the United States will be apparent.

RISK ASSESSMENT IN CHINA MAY BLAZE A TRAIL FOR THE DEVELOPING WORLD

The average growth rate of China's GDP over the past 20 years has been 9.7%. With 22% of the world's population, environmental impacts in China have the potential to affect the rest of the world.[52] In 1990, the State Environmental Protection Administration (SEPA) of the People's Republic of China required the performance of environmental risk assessments for potential environmental pollution accidents. Before 2004, environmental risk assessments in China were conducted using guidance documents from other countries. In 2004, SEPA issued their *Technical Guidelines for Environmental Risk Assessment for Projects*.[53]

Much of the guidance currently being developed in China is copied from US guidance; for example, the Beijing Municipal Environmental Protection Bureau issued *Technical Guidelines for Environmental Site Assessment* and the Ministry of Environmental Protection (MEP) of the People's Republic of China, which replaced SEPA, issued *Guidelines for Risk Assessment of Contaminated Sites*. Both documents were based on USEPA's *Soil Screening Guidance*.[53–55]

Similar to the Red Book[8] and RAGS,[33] the MEP guidelines indicated five steps for environmental risk assessment:

- Damage identification
- Exposure assessment
- Toxicity assessment
- Risk characterization
- Expected value of soil remediation

The majority of the values for exposure assumptions, for example, incidental soil ingestion and inhalation rate, were obtained directly from USEPA guidance. What was dissimilar to USEPA guidance was the incorporation of risk management considerations into the risk assessment. The MEP thus has taken the same position as the US OMB in *Circular A-4*.[51]

One aspect of the toxicity assessment that will be discussed in a later chapter is the growing awareness that assumption of the linear no-threshold (LNT) hypothesis for chemical carcinogens is very likely incorrect.[56] The assumption stems from the acceptance of claims of an LNT dose–response hypothesis for radiation. This hypothesis was widely accepted in the 1950s and 1960s, but of late, the explosion of biological knowledge due to genomics, proteomics, computational, and systems biology strongly suggests that this hypothesis is incorrect.[57] In China, because of the desire of the government to maintain economic growth, the LNT hypothesis is being considered very carefully—China is at an early stage in the development of an environmental regulatory framework and wishes to base such a framework on the best available science.[53]

RISK ASSESSMENT IN THE EUROPEAN UNION

REACH stands for the *Registration, Evaluation, Authorisation and Restriction of Chemical Substances* under European Commission (EC) regulation 1907/2006 of the European Parliament. The European Union (EU) promulgated a far-reaching regulatory initiative in 2006. With the ever-increasing ability of analytical chemistry to measure ever lower levels of chemicals in the human body, the ubiquity of these chemicals in our bodies, albeit at vanishingly tiny levels, has increased public concern because of the perceived hazards associated with these chemicals.[58–62]

REACH has specified that the task of providing information has been placed onto industry. The European Chemicals Agency (ECHA) has developed detailed guidance on information requirements and chemical safety assessment from available data. The ECHA guidance provides 12 categories of information:

1. Physicochemical properties
2. Skin/eye irritation, corrosion, or respiratory irritation
3. Skin or respiratory sensitization
4. Acute toxicity
5. Repeated dose toxicity
6. Reproductive or developmental toxicity
7. Mutagenicity and carcinogenicity
8. Aquatic toxicity
9. Degradation and biodegradation
10. Bioconcentration and bioaccumulation
11. Effects on terrestrial organisms
12. Toxicokinetics

REACH soon realized that due to the number of untested chemicals in commerce (~80,000), the data requirements would be impossible to fulfill for each chemical. Hence, ECHA has led the way in developing guidance for the use of alternate toxicity testing methods such as quantitative structure–activity relationships (QSAR), in vitro testing, and read-across.[63–65]

For chemicals produced in amounts greater than 10 tons/year, REACH requires a chemical safety report (CSR).[66] The CSR consists of six elements:

1. Human health hazard assessment
2. Human health hazard assessment for physicochemical properties
3. Environmental hazard assessment
4. Persistent, bioaccumulative, and toxic (PBT) and very persistent, very bioaccumulative (vPvB) assessment
5. Exposure assessment
6. Risk characterization

The last two steps are performed only if the first four suggest the substance should be classified as dangerous.

Step 1, the human health hazard assessment, determines a "derived no-effect level" (DNEL) based on all relevant and available data. DNELs are developed for each route of exposure (oral, dermal, inhalation) and each observed adverse endpoint.

Step 2, the human health hazard assessment for physicochemical properties, considers flammability, explosivity, and oxidizing properties as well as the potential for safety hazards of this nature to occur.

Step 3, the environmental hazard assessment, determines the environmental concentration thought to have no impact, the "predicted no-effect concentration" (PNEC). PNECs are developed for water and sediment in marine and freshwater environments. If necessary, PNECs for air or food chain exposure may also be developed.

Step 4, the evaluation of persistence and bioaccumulation, involves data on biodegradability, octanol–water partition coefficient, and environmental toxicity.

If the substance is determined to be dangerous under EC Regulation 1272, then steps 5 and 6, exposure assessment and risk characterization, are conducted.[66]

These last two steps are iterative, and REACH permits a conservative exposure assessment to be replaced with a more realistic one if it can be shown that exposure control can be achieved.

RISK ASSESSMENT IN CANADA

Canada has taken a pragmatic approach to health risk assessment. Health Canada has recognized that the socioeconomic and physical environment, early childhood experiences and events, personal health practices, and biology are determinants of health. Other determinants of health include age, gender, income, education, literacy, and genetics.[67]

In 1990, Health and Welfare Canada published a preliminary framework for risk assessment and risk management.[68] In 1993, Prime Minister Kim Campbell split this department into two—Health Canada and Human Resources and Labor Canada. The next year, Health Canada published a revised framework, *Human Health Risk Assessment for Priority Substances.*[69] In 2000, Health Canada again revised the framework. Health Canada's *Decision-Making Framework for Identifying, Assessing and Managing Health Risks* is generally based on the 1993 revision and the US Presidential/Congressional Committee Framework (Figure 1.4).[10,70] The framework can be grouped into five phases:

1. Issue identification
2. Risk and benefit assessment
3. Identification and analysis of risk management options

4. Selection and implementation of an option
5. Monitoring and evaluation of outcomes

This framework also emphasized the need for stakeholder involvement at all stages of the process. This emphasis was the essence of the 2000 revision.

In 1999, a report titled *Risk, Innovation and Values: Examining the Tensions*[71] from the Treasury Board of Canada Secretariat explored the competing pressures on public sector risk managers regarding decisions with uncertainty outcomes. These competing pressures include the value of innovation versus the value that society places on a certain level of certainty. The report suggested that change was inevitable and risk managers should adopt innovation as a regular thought process. This would actually require a sea change for managers whose traditional way of thinking emphasized caution and for whom the whole notion of managing risk had become synonymous with avoiding risk.

Since 2004, Health Canada has released a plethora of guidance documents that deal with both preliminary and more advanced risk assessment methods, risk communication, and other aspects of risk assessment. These guidance documents can be found at http://healthcanada.gc.ca.

SANGAMO-WESTON SUPERFUND SITE: WHAT HAPPENS WHEN THINGS GO RIGHT!

From 1955 until 1977, the practice at a capacitor manufacturing plant in Pickens, South Carolina, was to discharge untreated wastewater into Towns Creek. Polychlorinated biphenyls (PCBs) used in the manufacturing process migrated downstream into Twelve Mile Creek, a major tributary of Lake Hartwell. Located on the border of Georgia and South Carolina, Lake Hartwell occupies 56,000 acres and provides for public recreation, flood control, and hydropower generation. The US Army Corps of Engineers impounded the lake between 1955 and 1963. Approximately 300,000 people visit the lake each year for recreation. Many of the lake visitors harvest and consume fish from the lake.[72]

The PCB-contaminated sediment from the Sangamo-Weston facility was prevented from reaching the lake for a time by the dams of three small hydro-electric plants on Twelve Mile Creek. Periodic flushing of the sediment in the impoundments behind the dams discharged the PCBs further downstream until approximately 730 acres of the lake bottom sediment in the Seneca River arm was contaminated. The PCBs entered the food chain and became concentrated in the fish living in Lake Hartwell. High levels of PCBs were detected in fish collected from Lake Hartwell in 1976.

Within the Superfund program, remediation is based on risks to the hypothetical receptor experiencing RME, defined as the highest exposure that is reasonably expected to occur at a site.[33,73] The National Oil and Hazardous Substance Pollution Contingency Plan (National Contingency Plan or NCP; 40 CFR 300) is the regulation under which the Superfund program operates. The preamble to the

NCP indicates that a major objective of the risk assessment is "to target chemical concentrations associated with levels of risk that will be adequately protective of human health for a particular site."[73,74]

Acceptable cancer risks defined by CERCLA are between 10^{-4} and 10^{-6} with a preference for the lower end of the risk range.[73]

Lake Hartwell is a popular inland fishing destination. As part of the CERCLA actions at Sangamo-Weston, a risk assessment was conducted for fish consumption. The cancer risk associated with the fish ingestion pathway for the RME receptor was as high as 1% or 10^{-2} for some areas of the lake. At its time in the early 1990s, the risk assessment was quite sophisticated and included data from a site-specific creel and fish consumption survey and probabilistic estimates of exposure and risk.[75]

Eight remedial alternatives were proposed for cleanup. These included institutional controls with public education and fish and sediment monitoring and continuance of an existing fish advisory. Details of the various remedial alternatives are shown in Table 1.2. Figure 1.5 shows the locations of the Twelve Mile Creek and Seneca River arms of the various alternatives.

Within the Superfund program, the proposed plan is a description of the site, the risks, and EPA's preferred alternative. EPA chose Alternative 2B, fisheries isolation, and presented the proposed plan at a public meeting in Clemson, SC, on April 19, 1994. Unwavering public opposition toward the fish fence was voiced at this meeting and in the public comments received—EPA concluded that the public consensus supported Alternative 2A—institutional controls.

One likely reason for EPA choosing Alternative 2B was cost; this remedy was chosen less than a year before the formation of the National Remedy Review Board (NRRB). Any Superfund remedy costing more than $25M must be reviewed by the NRRB to assure that the risk management decision is consistent with Superfund law, regulation, and guidance. The NRRB was formed in 1995 with the intention of controlling remedy costs and promoting consistent and cost-effective decisions.[76]

The public input to the risk management decision at Sangamo-Weston was an example of how the process could work when stakeholders are consulted and their input valued—exactly what the 1997 Federal Commission report advocated. This is likely the correct decision. One of the reasons that the effort was successful is the care and thought that went into the risk assessment and that the opinion of the public living in the Lake Hartwell area was taken into account.[72]

PERCEPTION IS REALITY: RISK COMMUNICATION AND STAKEHOLDER PARTICIPATION

Conducting a risk assessment rooted in the best available science and conducted with a scope and level of detail commensurate with the scope of the problem is a necessary part of decision making, but it is not all. Communicating these results to stakeholders in the decision is vital for successful implementation.

TABLE 1.2

Remedial Alternatives for the Sangamo-Weston Superfund Site

Remedial Alternative	Cost in 1994$	Description
Alternative 1—no action	$130,000	This alternative was evaluated to serve as the basis for comparison with other active cleanup alternatives. Under this no-action alternative, no further remedial actions for the contaminated sediments or fish at the site would be conducted. This alternative would not affect the existing health advisory issued by SCDHEC, who would be expected to continue the advisory until PCB concentrations in fish tissue decline to levels below 2 mg/kg (FDA tolerance level). The advisory currently warns against the consumption of fish from the Seneca River arm of Hartwell Lake above the Hwy 24 bridge and fish larger than 3 lb throughout the entire lake. The advisory would be modified if warranted by future trends regarding PCB levels in fish. Maintenance of the fish advisory is assumed to entail periodic replacement of existing signs that advise against fish consumption.
Alternative 2A— institutional controls	$3,208,000	This alternative would consist of four parts: (1) continuance of the existing fish advisory, (2) a public education program on fish handling and cooking techniques for reducing the intake of PCBs, (3) fish and sediment monitoring, and (4) regulation and periodic flushing of the sediment from behind the dams on Twelve Mile Creek.
Alternative 2B— fisheries isolation	$4,244,000	This alternative involved construction of a barrier or fish fence to prevent the movement of migratory species, striped bass, hybrid bass, and walleye, into the contaminated upper Seneca River arm of the lake. The barrier would be constructed from the water level down the lake bottom and would not impede boat traffic.
Alternative 3A—capping	$51,139,000	This alternative would isolate PCB-contaminated sediments by placing an 18 in. clean sediment cap over the areas with the highest contamination. The cap thickness was designed to minimize the impacts on sediment dwelling biota such as *Hexagenia* mayflies. The cap would be placed with a barge and hydraulic sand spreader.
Alternative 3B— sediment control structure	$53,591,000	This alternative proposed building a fixed-crest weir near the mouth of Twelve Mile Creek arm to maintain a constant pool elevation in this arm and prevent sediment erosion and transport.

(continued)

TABLE 1.2 (continued)
Remedial Alternatives for the Sangamo-Weston Superfund Site

Remedial Alternative	Cost in 1994$	Description
Alternative 3C— optimal capping/ sediment control structure	$34,049,000	In this alternative, a fixed-crest weir would be built further upstream on Twelve Mile Creek, and the sediment downstream of the weir would be capped.
Alternative 4— confined disposal facility	$46,909,000	This aggressive alternative would involve rerouting a 1600-foot section of Twelve Mile Creek and dredging sediment with PCB concentrations >1 mg/kg. The dredge spoils would be placed in a confined disposal facility.
Alternative 5—stabilization	$581,957,000	The upper portion of Twelve Mile Creek would be dewatered and the sediment excavated. The lower portion would be dredged. The dredge and excavation spoils would be stabilized with cement and placed in the confined disposal facility.

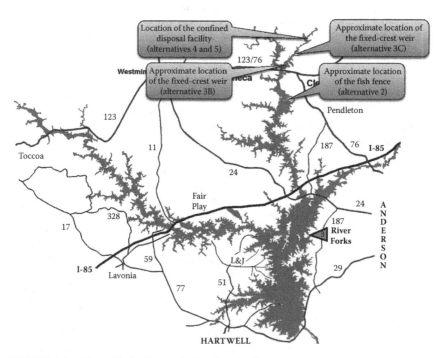

FIGURE 1.5   Map of Lake Hartwell showing the Sangamo-Weston Superfund remedial alternatives. See text for details.

Ideally, decision making in a democratic society involves all those who have a stake in the decision and minimizes the downside for those stakeholders to the greatest extent possible. Not all stakeholders can initially understand the many technical complexities of high-level risk assessments. Nonetheless, they have a valid interest in the outcome and should be treated as partners. What success at risk communication truly requires is the honest appreciation of and respect for a range of viewpoints. These viewpoints will, of course, reflect individual experience, education, and background. The essence of successful risk communication is respect for others who may have different outlook on the situation.

There are four aspects to risk communication:

- The message—how risky is the situation
- The source—who conveys the risk information
- The channel—the means of communication, that is, a public meeting, television, and radio
- The receiver—the public or other stakeholders in the decision

In 1992, EPA published *Seven Cardinal Rules of Risk Communication*, a pamphlet authored by Dr. Vincent J. Covello, the director of the Center for Risk Communication at Columbia University.[77] Covello has worked in this area for many years, most notably, in helping General Norman Schwarzkopf craft the public dialogue regarding the first Gulf War in 1990.

These seven cardinal rules are the following:

1. Accept and involve the public as a legitimate partner.
2. Plan carefully and evaluate your efforts.
3. Listen to the public's specific concerns.
4. Be honest, frank, and open.
5. Coordinate and collaborate with other credible sources.
6. Meet the needs of the media.
7. Speak clearly and with compassion.

While these cardinal rules are excellent general principles, environmental risk communication has become enormously more difficult since the 1990s due to the increasing risk-averse nature of western societies.

The information available to the public is often conflicting. In May 2010, the NCI released the *2008–2009 Report of the President's Cancer Panel*. The report suggested that the prevailing regulatory approach to environmental chemicals and cancer was reactionary rather than precautionary. The report also stated that the "true burden of environmentally induced cancer has been grossly underestimated."[78]

Dr. Michael J. Thun, vice president emeritus of Epidemiology and Surveillance Research of the American Cancer Society, immediately criticized the report. Dr. Thun stated that the report was unbalanced in its perspective for dismissing prevention efforts aimed at known causes of cancer such as tobacco, obesity, alcohol, infections, hormones, and sunlight. Dr. Thun also stated that the report did not represent the scientific consensus on environmentally induced cancer.[79] A number of other scientists have also suggested that this report is essentially incorrect.

The purpose here is not to discuss the veracity of the President's Cancer Panel Report, but rather to point out that this report along with sensationalistic news reporting such as Dr. Sanjay Gupta's series on CNN, *Toxic Towns USA*, tends to increase the outrage many people feel about environmental risks. Dr. Peter Sandman of Rutgers University, an expert in risk communication, has written extensively on outrage and how real communication about risk cannot begin until outrage is addressed.[80] Television journalism often tends to be prurient or alarmist to maintain or increase the viewing audience—at times, journalistic credibility is sacrificed for ratings.

What all this means is that the receiver—most often the public—will usually have some strongly held preconceptions about the nature and seriousness of environmental risks. An example of such a situation is provided in the following text.

PUBLIC PERCEPTION OF HEXAVALENT CHROMIUM: A CAUTIONARY TALE

The movie *Erin Brockovich*[81] was released in 2000 and chronicled the true story of the former beauty queen who became a legal aide and her participation in the toxic tort suit brought by the residents of Hinkley, California, against Pacific Gas and Electric (PG&E). This movie thrust the issue of Cr(VI) in drinking water into the public and political spotlight. This suit was settled in 1996 with PG&E paying a settlement of $333M to 600 residents.

As of the summer of 2013, the only school in Hinkley has closed because of falling enrollment. Property values have plummeted and the Community Advisory Committee for the town has yet to reach an agreement with PG&E about how to provide drinking water to the remaining residents. The San Bernadino Sun has characterized Hinkley as *a town that's fading away.*[82] In the fall of 2013, the California Department of Public Health issued a public health goal for Cr(VI) in drinking water of 10 parts per billion, 10 fold lower than the federal standard, that would cost an estimated $500M per year just for the Coachella Valley Water District just south of Hinkley.[83]

As a backdrop to this legal action, a controversy had been brewing about the data relating stomach cancer to hexavalent chromium exposure first reported in the *Chinese Journal of Preventive Medicine*.[84] The authors claimed that stomach cancer rates were elevated among those residing near the Jinzhou Iron Alloy Plant who consumed water containing up to 20 mg/L Cr(VI). At these concentrations, the water would be bright yellow. The collection of water samples and the estimates of cancer may have been influenced by government policies in the 1970s and 1980s when the data were collected. This influence is clearly evident in the last paragraph translated from a 1976 report from Dr. Zhang:

> Our studies have concluded that, guided by Chairman Mao's philosophy and thought, in work to prevent and treat tumors, with the consistent emphasis on prevention and further propaganda and popularization of scientific knowledge on the subject of cancer prevention and remedy, building the confidence that malignant tumors are not to be feared, that malignant tumors can be beaten, to build in-depth,

concrete, and broad-based epidemiological studies and research, to solve the epidemiological questions of malignant tumors, to strengthen the regular survey and treatment of common illnesses, recurrent maladies, and chronic illnesses, to make every effort to accomplish the goal of early detection and early treatment, in order to make the efforts of preventing and remedying tumors a service to the solidarity of the proletariat, is among the missions which we must make every effort to complete.[84]

The statistical comparison of cancer rates among those living near the plant with a comparison group was not clearly presented—neither in early reports nor in the 1987 publication.[84–86] These data were collected during the turmoil of the Cultural Revolution, and one can do no more than speculate about the political pressures on Dr. Zhang.

During 1995, Drs. Zhang and Li were in contact with scientists at McLaren-Hart/ChemRisk, and this contact may be the reason for the improved statistical analysis in the 1997 paper. In a letter dated June 9, 1995, Dr. Zhang wrote as follows:

> I've received the written draft you sent. After reading it, I quite agree with you. In the draft, one sentence reads "The death rate in each village does not show the positive correlation with the distance of the village to the source of pollution or the intensity of chromium pollution.[87]

The work was republished in 1997 in the *Journal of Occupational and Environmental Medicine* with the appropriate and clear statistical comparisons. In this updated study, no relationship could be demonstrated between distance from the plant (used as the measure of Cr(VI) concentration) and the rate of all cancers, either lung cancers or stomach cancers.[86]

The correspondence between Drs. Zhang and Li and scientists at McLaren-Hart/ChemRisk revealed a discussion of nuances of interpretation among scientific peers.[87] Nonetheless, the relationship between these Chinese scientists and the American consulting company soon became controversial with the publication of the letters as part of the PG&E trial record.

Peter Waldman of the *Wall Street Journal* reported on December 23, 2005, about the correspondence between Dr. Zhang and McLaren-Hart/ChemRisk, and his reporting further fueled this controversy.[88] Following Waldman's exposé, the Environmental Working Group (EWG) got involved in chromium. EWG, founded in 1993, has a stated mission to use public information to protect public health and the environment. EWG and the *Wall Street Journal* alerted Dr. Paul Brandt-Rauf, the editor in chief of the *Journal of Occupational and Environmental Medicine*, to this correspondence. In response, Brandt-Rauf retracted the paper in July 2006.[89–91] Dr. Zhang died around 2000 and thus could obviously not contest the retraction.

Six months after the retraction of the paper, Dr. Shukun Li stepped forward to say the paper was withdrawn unfairly and disputed the claim that she had agreed to the paper's retraction. Further, she demanded that the paper be republished. This never occurred. One possible reason is that Brandt-Rauf indicates that the communication with Dr. Li was conducted through a translator and some miscommunication may have occurred.[92]

The correspondence and relationship between Dr. Zhang and the McLaren-Hart scientists was characterized by Waldman as a cautionary tale about "what can happen when the line between advocacy and science blurs."[88] But perhaps Waldman had a stake here as well—was he building his career with a sensational story?

Truth is not absolute in science—a scientist may change his interpretation of conclusions based on new data or updated methods, and the scientific method is an iterative process used to learn about the world. Dr. Li has publicly stated that the *Wall Street Journal* article is false and demanded that it be retracted.[92]

Recently, the data of Zhang and Li from the 1987 paper were reanalyzed using the entire Liaoning province as the comparison group and showed a significantly increased rate for stomach cancer.[93] Further adding to the controversy, the scientists who worked at McLaren-Hart/ChemRisk also reanalyzed the data using the cancer rates from nearby agricultural villages that were not exposed to Cr(VI) and observed no statistical difference.[94] Hence, these data are hardly robust if they can be interpreted differently depending on the choice of which unexposed group serves as a control group. In short, these data have become like the elephant inspected by seven blind men—each man handles a different part of the beast and each one comes away with a different perception.

Perception Is Reality: Why the Movie *Erin Brockovich* Changed the Public's View of Chromium

The public perception of risk from hexavalent chromium stemming from the movie may have significantly undermined the process of scientific investigation and credible regulatory evaluation.[95] In one of the exercises at the end of this chapter, you will be asked to watch and comment on a video of a noted scientist testifying about hexavalent chromium before the US Senate Subcommittee on Environment and Public Works.

Many individuals fear chemical exposure, and this fear is completely understandable—there are instances of individuals being poisoned with disastrous consequences.[96] Some individuals are more sensitive to common odors such as perfume or tobacco smoke and may experience cognitive and emotional effects associated with these odors.[97–99] Such individuals also have a significantly higher lifetime prevalence of mood and anxiety disorders.[100–102]

This observation begs the question of to what extent the fear of chemicals is the result of the associated cognitive and emotional effects. Recently, a number of illness outbreaks have been attributed to psychogenic causes likely triggered by an odor.[103,104]

Perceptions of risk may also contribute to fear. A number of studies have noted an awareness bias in which individuals who perceive an environmental threat such as proximity to a landfill or industrial facility and who also worry about potential health effects associated with the perceived threat tend to report more ill health in the absence of any measurable medical or biological effect.[101]

The current public perception of hexavalent chromium as highly dangerous may actually be a mass psychogenic fear of chromium in the American public because of the movie *Erin Brockovich* and the reporting of Peter Waldman of the *Wall Street Journal*.

ASSUMPTION OF CAUSATION

The Scottish philosopher David Hume remarked that causation could not be empirically observed but rather is induced logically. This is a two-edged sword—one can never obtain final or absolute proof of causation or lack thereof. All that exists is an observed association between two phenomena. This is a difficulty for science that seeks to understand the world in terms of cause and effect based on observation and reasoning. Karl Popper, the twentieth-century philosopher and professor at the London School of Economics, attempted to address the issue of causation by arguing against the classical scientific method that uses observation and induction. Popper noted that theories resulting from induction can be tested only indirectly, by their implications. Even a large body of experimental or observational data that confirms a scientific theory does not constitute proof—yet a single instance of a counterexample can be decisive. The central idea of Popper's ideas about science is the asymmetry between verification and falsifiability—theories must be stated so that there exists the possibility of being falsified based on observation.

The philosophic difficulties with causation provided the intellectual background for the development of the considerations for causation of Sir Austin Bradford Hill.[104] These considerations have caught the fancy of many scientists because of their logic and simplicity.

Hill's "viewpoints" or "features to be considered" are difficult to apply in any field. They provide no easy formula or guidance.[107–109] Hill's considerations have been eloquently characterized as "guideposts on the road to common sense."[110] These considerations necessitate rigorous scientific thinking; unfortunately, the intellectually lazy will likely apply them as a checklist or set of criteria.[110,111] This is not what Hill intended, and he indicates that none of the nine "viewpoints" can be required as a *sine qua non* for causation.[106]

The Hill considerations have been adapted for many purposes, including the assessment of the MOA of reproductive and developmental toxins and chemical carcinogens.[112,113] USEPA's cancer guidelines specifically and correctly point out that the framework is a structure for organizing the available information.[113] Table 1.3 shows Hill's original considerations and the adaptations for use in assessing human relevance of a particular MOA in risk assessment.

TABLE 1.3
Hill's Considerations for Causation and Their Adaptation for Assessing Human Relevance of a MOA

Considerations for Causality[106]	Framework for Evaluating an Animal MOA[35,113,116,117]
Strength of association	*Postulated MOA*
The disease is associated with the exposure to a significant extent, as measured by valid statistical tests.	A description of the sequence of measured events starting with chemical administration and ending with the occurrence of the apical event.
Consistency	*Key events*
If a relationship is causal, we would expect to find it consistently in different studies, in different populations, and in a range of circumstances.	Clear descriptions of each of the key events that comprise the MOA.
Specificity	*Dose–response/biological gradient*
In most cases, it is impossible to attribute a specific effect to a single cause. Causality is most often multiple.	Dose–response relationships for each key event and comparisons of these DR relationships between key events and with the apical event.
Temporality	*Temporality*
The exposure must always precede the disease.	Sequence of key events of time leading to the occurrence of the apical event.
Dose–response or biological gradient	*Strength, consistency, and specificity of association*
If a dose–response relationship is present, it is a strong evidence for a causal relationship. However, as with specificity, the absence of a dose–response relationship does not rule out a causal relationship.	Assessment of relationships among key events, precursor or sentinel lesions and the apical effect.
Biological plausibility	*Biological plausibility and coherence*
The putative agent produces effects in a manner that is plausible, given the currently accepted understanding of biological processes, and a theoretical basis exists for making an association between an agent and a disease.	Determination of whether key events and their sequence are consistent with current biological thinking, with consideration of species specificity and the occurrence of the apical event.
Coherence	*Alternative MOAs*
The association should be compatible with existing theory and knowledge.	How likely are alternative MOAs compared to that proposed?
Experiment	*Conclusion about the MOA*
Can the condition be altered or prevented by an appropriate experimental regimen?	An overall indication of the level of confidence in the proposed MOA.
Analogy	*Uncertainties, inconsistencies, and data gaps*
What other factor or factors could produce the observed effects?	Identification and description of information deficiencies, inconsistencies, and contradictions in the overall data and proposals to ameliorate data gaps.

KEY CONCEPTS IN MODERN RISK ASSESSMENT

The last section of this chapter will present definitions and descriptions of key terms in modern risk assessment. This list is by no means complete, but the concepts described here are those that will be encountered again—both in this book and in the practice of risk assessment. Additional terms are shown in Table 1.2.

MODE OF ACTION

This term was first used in 1980[114] and defined in a regulatory context in 1998.[111] Since then the concept has been used to inform the dose–response assessment of both carcinogens and noncarcinogens, and a number of frameworks for characterizing MOA have been developed.[35,58,116–120]

The concept of MOA was developed in the context of a framework or structured approach to evaluate the overall weight of evidence (WOE) for a biologically plausible explanation of the manner in which a chemical produces an adverse effect. Key events in a carcinogenic MOA are measurable occurrences that occur before the apical endpoint. In essence, key events are biomarkers of effect. USEPA's 2005 *Guidelines for Carcinogen Risk Assessment* observes that the dose–response of biomarkers or precursor effects may be used in lieu of the apical effect for obtaining a point of departure (POD) for low-dose extrapolation.[113] MOA is also used to determine the human relevance of effects seen in animals.[118,119] An example of a MOA analysis is shown in Figure 1.3.

POINT OF DEPARTURE

The POD marks the quantitative value of a response and the dose associated with this response from which extrapolation to lower doses occurs. If quantitative dose–response modeling is not conducted, the POD will usually be the no observed adverse effect level (NOAEL) or lowest observed adverse effect level (LOAEL). For quantal or binomial data* such as from an animal cancer bioassay, 10% is the value of the POD most often chosen. For continuous data, there has been considerable debate about the choice of POD and whether this choice should be based on statistical or biological significance.

BIOMARKER

A biomarker is a physiological quantity measurable in humans in vivo. The best-known example is the alcohol breath test for sobriety. Because ethyl alcohol will volatilize from the blood to the air in the lungs, it can be measured in breath. Breath alcohol is not a direct measure of blood alcohol. Biomarkers can reflect either exposure or effect. The occurrence of arsenic in urine is a poor biomarker

---

* Quantal data are expressed as a proportion or percentage, that is, 4/50 animals were found to have hepatocellular adenomas. Continuous data are expressed as nonintegral measurement such as enzyme activity in mg substrate/g liver/h.

for chronic exposure to arsenic because it may reflect recent seafood consumption due to the prevalence of arsenobetaine in fish; however, arsenic in toenails is more representative of long-term exposure.[121]

Biomarkers of effect measure some physiological variable that is altered or affected by exposure. For example, both consumption of broccoli and exposure to the dioxin-like chemicals may increase the activity of the liver enzyme that metabolizes caffeine.[122,123]

BIOMONITORING EQUIVALENT

A biomonitoring equivalent (BE) is defined as the concentration or range of concentrations of a chemical or its metabolites in a biological medium (blood, urine, or other medium) that is consistent with an existing health-based exposure guidance value such as a reference dose (RfD) or tolerable or acceptable daily intake (TDI or ADI). There is an ongoing effort to express regulatory toxicity criteria (TDIs, ADIs, RfDs) in terms of BE values so that the increasing amount of human biomonitoring data can be understood in terms of the potential for adverse effects.

PHYSIOLOGICALLY BASED PHARMACOKINETIC MODELING

Mathematical modeling of the distribution of chemicals in the body is also known as physiologically based toxicokinetic modeling (PBTK). It is also known as ADME modeling. Simply, PBPK modeling is a way of dividing up the body into functional compartments into which chemicals may accumulate or be metabolized and excreted. One of the best-known PBPK models is very simple—it is the Widmark model used for retrograde extrapolation of measured blood or breath alcohol levels in forensic evaluation of potential drunk driving cases. The Widmark model assumes the body is a single compartment and that elimination of alcohol occurs by a zero-order kinetic process. What this means is that a constant amount of alcohol is metabolized and excreted per time unit.

EXERCISES FOR THOUGHT AND DISCUSSION

There are, of course, no right answers to the following questions. Nonetheless, these exercises can be thought-provoking, instructive, and often entertaining.

CURRENT DEBATE ABOUT CHEMICAL SAFETY

To help understand the current debate about chemical safety, please watch the videos at two websites. The first one is "The Story of Cosmetics" (http://www.storyofstuff.org/movies-all/story-of-cosmetics/).

This provides the perspective of a lay person who is fearful about chemicals present in the environment, food, and cosmetics. This video runs about 7 min.

The second one is "Oversight Hearing on the Environmental Protection Agency's Implementation of the Safe Drinking Water Act's Unregulated Drinking Water Contaminants Program" (http://www.epw.senate.gov/public/index.cfm?FuseAction=Hearings.Hearing&Hearing_ID=fc5a8756-802a-23ad-454a-b9eeb7bf1c36).

This is an archived webcast from the Senate committee on Environment and Public Works titled *Oversight Hearing on the Environmental Protection Agency's Implementation of the Safe Drinking Water Act's Unregulated Drinking Water Contaminants Program*.[124] Please start watching at about minute 123, and see the interaction between Dr. Steven Patierno and Senator Barbara Boxer.

Do you believe the concerns expressed in the first video are reasonable? Why or why not? If not, how would you go about edifying this individual? What did Dr. Patierno do right in his interaction with the senator from California and what did he do wrong?

RISK ASSESSMENT HISTORY: ROBERT F. KENNEDY'S SPEECH AT THE UNIVERSITY OF KANSAS

This speech, from March 18, 1968, can be heard in its entirety on YouTube (http://www.youtube.com/watch?v=z7-G3PC_868&feature=related) and provides excellent discussion material.

ANIMAL AND HUMAN CARCINOGENS

Not all chemicals that cause cancer in laboratory animals also cause cancer in other species, including humans. For example, phenobarbital has been used as a sedative in humans since 1912 and is also used to treat epilepsy in dogs. Phenobarbital is a potent carcinogen in rodents at doses that produce sedation in humans. Gold and colleagues have developed a compendium of animal and human carcinogens. This paper is available free of charge at http://www.ncbi.nlm.nih.gov/pubmed/11794380.[125] After reading this paper, how would you decide if a chemical that has been shown to be carcinogenic in animals would also be carcinogenic in humans? When is an animal bioassay likely to produce a false-positive or a false-negative result?

REFERENCES

1. Emerson RW. *Self-Reliance, and Other Essays*. New York: Dover Publications, 1993.
2. Larsen PO and von Ins M. The rate of growth in scientific publication and the decline in coverage provided by Science Citation Index. *Scientometrics*. 2010, September;84(3):575–603.
3. McDonnell DP and Wardell SE. The molecular mechanisms underlying the pharmacological actions of ER modulators: Implications for new drug discovery in breast cancer. *Curr Opin Pharmacol*. 2010, December;10(6):620–628.
4. Venet D, Rigoutsos I, Dumont JE, and Detours V. Most random gene expression signatures are significantly associated with breast cancer outcome. *PLoS Comput Biol*. 2011, October 10;7(10):e1002240.
5. Auyeung W, Canales RA, Beamer P, Ferguson AC, and Leckie JO. Young children's hand contact activities: An observational study via videotaping in primarily outdoor residential settings. *J Expo Sci Environ Epidemiol*. 2006, September;16(5):434–446.

6. Xue J, Zartarian V, Moya J, Freeman N, Beamer P, Black K et al. A meta-analysis of children's hand-to-mouth frequency data for estimating nondietary ingestion exposure. *Risk Anal.* 2007, April;27(2):411–420.

7. Xue J, Zartarian V, Tulve N, Moya J, Freeman N, Auyeung W, and Beamer P. A meta-analysis of children's object-to-mouth frequency data for estimating non-dietary ingestion exposure. *J Expo Sci Environ Epidemiol.* 2010, September;20(6):536–545.

8. National Research Council. *Risk Assessment in the Federal Government: Managing the Process.* Washington, DC: National Research Council, 1983.

9. National Research Council. *Science and Judgement in Risk Assessment.* Washington, DC: National Research Council, 1994.

10. The Presidential/Congressional Commission on Risk Assessment and Risk Management. Framework for environmental health risk management, Final Report, Volumes 1 and 2, 1997. http://www.riskworld.com/Nreports/1996/risk_rpt/Rr6me001.htm

11. National Research Council. *Toxicity Testing in the 21st Century: A Vision and a Strategy.* Washington, DC: The National Academies Press, 2007. http://www.nap.edu/catalog.php?record_id=11970

12. Thomas RS, Black M, Li L, Healy E, Chu T-M, Bao W et al. A comprehensive statistical analysis of predicting in vivo hazard using high-throughput in vitro screening. *Toxicol Sci.* 2012, August;128(2):398–417.

13. United States Environmental Protection Agency (USEPA). Memorandum from Carole Browner. *EPA Risk Characterization Program*, Washington, DC, March 21, 1995. http://www.epa.gov/oswer/riskassessment/pdf/1995_0521_risk_characterization_program.pdf

14. Ricci PF, Cox LA, and MacDonald TR. Precautionary principles: A jurisdiction-free framework for decision-making under risk. *Hum Exp Toxicol.* 2004, December;23(12):579–600.

15. Whelan EM. The U.S. government versus carcinogens. *CA Cancer J Clin.* 1978;28(4):239–240.

16. Mukherjee S. *The Emperor of All Maladies: A Biography of Cancer.* New York: Scribner, 2011.

17. Amdur MOM, Doull JJ, and Klassen CDC. *Cassarett and Doull's Toxicology the Basic Science of Poisons*, 4th edn. New York: Macmillam Publ.

18. Kriebel D, Tickner J, Epstein P, Lemons J, Levins R, Loechler EL et al. The precautionary principle in environmental science. *Environ Health Perspect.* 2001, September;109(9):871–876.

19. Council on Environmental Quality. *A Citizen's Guide to NEPA: Having Your Voice Heard.* Washington, DC, December, 2007. http://energy.gov/nepa/downloads/citizens-guide-nepa-having-your-voice-heard

20. Hardin G. The tragedy of the commons. *Science.* 1968, December 12;162(5364):1243–1248.

21. United States Occupational Health and Safety Administration (OSHA) (1997) Regulations (Preamble to Final Rules). 62 Fed. Reg. 1494. Washington, DC, January 10, 1997. https://www.osha.gov/pls/oshaweb/owadisp.show_document?p_table=PREAMBLES&p_id=1007 (accessed October 20, 2013).

22. Mantel N and Bryan WR. "Safety" testing of carcinogenic agents. *J Natl Cancer Inst.* 1961, August;27:455–470.

23. Herbst AL. Adenocarcinoma of the vagina. Association of maternal stilbestrol therapy with tumor appearance in young women. *N Engl J Med.* 1971, April 4;284(15):878–881.

24. Jukes TH. Diethylstilbestrol in beef production: What is the risk to consumers? *Prev Med.* 1976, September;5(3):438–453.
25. Wade N. DES: A case study of regulatory abdication. *Science.* 1972, July 7;177(4046):335–337.
26. Wade N. FDA invents more tales about DES. *Science.* 1972, August 8;177(4048):503.
27. United States Environmental Protection Agency (USEPA). Memorandum for Don Clay, Assistant Administrator OSWER, *Role of the Baseline Risk Assessment in Superfund Remedy Selection Decisions.* OSWER Directive 9355.0-30. Washington, DC, April 22, 1991. http://www.epa.gov/oswer/riskassessment/baseline.htm
28. United States Environmental Protection Agency (USEPA). Memorandum from Timothy Fields, Jr. Acting Administrator OSWER, *Approach for Addressing Dioxin in Soil at CERCLA and RCRA Sites.* OSWER Directive 9200.4-26. Washington, DC, April 13, 1998. http://www.epa.gov/superfund/policy/remedy/pdfs/92-00426-s.pdf
29. United States Environmental Protection Agency (USEPA). *The plain English Guide to the Clean Air Act.* Publication No. EPA-456/K-07-001. Research Triangle Park, NC, April, 2007. http://www.epa.gov/air/caa/peg/pdfs/peg.pdf
30. United States Environmental Protection Agency (USEPA). *Guidelines for Carcinogen Risk Assessment.* EPA/630/R-00/004. Washington, DC, September 24, 1986. http://www.epa.gov/cancerguidelines/guidelines-carcinogen-risk-assessment-1986.htm
31. United States Environmental Protection Agency (USEPA). *Guidelines for Mutagenicity Risk Assessment.* EPA/630/R-00/003. Washington, DC, September 24, 1986. http://www.epa.gov/raf/publications/guidelines-mutagenicityl-risk-assessment.htm
32. United States Environmental Protection Agency (USEPA). *Guidelines for the Health Risk Assessment of Chemical Mixtures.* EPA/630/R-98/002. Washington, DC, September 24, 1986. http://cfpub.epa.gov/ncea/cfm/recordisplay.cfm?deid=22567#Download
33. United States Environmental Protection Agency (USEPA). *Risk Assessment Guidance for Superfund: Vol. 1: Human Health Evaluation Manual (Part A) (Interim Final) (RAGS).* EPA/540/1-89/002. Washington, DC, December, 1989. http://www.epa.gov/oswer/riskassessment/ragsa/
34. United States Environmental Protection Agency (USEPA). *Guidelines for Exposure Assessment.* EPA/600/Z-92/001. Washington, DC, May 29, 1992. http://www.epa.gov/raf/publications/guidelines-for-exposure-assessment.htm
35. Sonich-Mullin C, Fielder R, Wiltse J, Baetcke K, Dempsey J, Fenner-Crisp P et al. IPCS conceptual framework for evaluating a mode of action for chemical carcino genesis. *Regul Toxicol Pharmacol.* 2001, October;34(2):146–152.
36. Klee H and Mahesh P. Amoco-USEPA pollution prevention project. Yorktown, VA. Project summary. Chicago, IL, June, 1992. http://heartland.org/sites/default/files/9100p39s.pdf
37. United States Environmental Protection Agency (USEPA). Memorandum from F. Henry Habicht, Deputy Administrator, *Guidance on Risk Characterization for Risk Managers and Risk Assessors*, February 26, 1992. http://www.epa.gov/oswer/riskassessment/pdf/habicht.pdf
38. United States Environmental Protection Agency (USEPA). *Guidance for Risk Characterization.* Science Policy Council. Washington, DC, February, 1995. http://www.epa.gov/spc/2riskchr.htm
39. Borgert CJ, Sargent EV, Casella G, Dietrich DR, McCarty LS, and Golden RJ. The human relevant potency threshold: Reducing uncertainty by human calibration of cumulative risk assessments. *Regul Toxicol Pharmacol.* 2012, March;62(2):313–328.

40. United States Environmental Protection Agency (USEPA). *Valuing Mortality Risk Reduction for Environmental Policy: A White Paper*. DRAFT. Washington, DC, December 10, 2010. http://yosemite.epa.gov/sab/sabproduct.nsf/02ad90b136fc21ef85256eba00436 459/34d7008fad7fa8ad8525750400712aeb/$file/white+paper+(dec.+2010).pdf

41. United States Environmental Protection Agency (USEPA). *Use of Monte Carlo Simulation in Risk Assessments*. EPA903-F-94-001. Philadelphia, PA, February, 1994. http://www.epa.gov/reg3hwmd/risk/human/info/guide1.htm

42. United States Environmental Protection Agency (USEPA). *Use of Monte Carlo Simulation in Performing Risk Assessments*. Denver, CO, 1995.

43. United States Environmental Protection Agency (USEPA). *Risk Characterization Handbook*. EPA 100-B-00-002. Science Policy Council. Washington, DC, December, 2000. http://www.epa.gov/spc/pdfs/rchandbk.pdf

44. United States Environmental Protection Agency (USEPA). *Policy for the Use of Probabilistic Analysis in Risk Assessment*. Washington, DC, May 15, 1997. http://www.epa.gov/stpc/pdfs/probpol.pdf

45. United States Environmental Protection Agency (USEPA). *Guiding Principles for Monte Carlo Analysis*. EPA/630/R-97/001. Risk Assessment Forum. Washington, DC, March, 1997. http://www.epa.gov/raf/publications/pdfs/montecar.pdf

46. Burmaster DE and Anderson PD. Principles of good practice for the use of Monte Carlo techniques in human health and ecological risk assessments. *Risk Anal*. 1994, August;14(4):477–481.

47. Burmaster DE and Lehr JH. It's time to make risk assessment a science. *Ground Water Monitoring Rev*. 1991;XI(3):5–15.

48. Burmaster DE and von Stackelberg K. Using Monte Carlo simulations in public health risk assessments: Estimating and presenting full distributions of risk. *J Expo Anal Environ Epidemiol*. 1991, October;1(4):491–512.

49. Burmaster DE and Harris RH. The magnitude of compounding conservatisms in superfund risk assessments. *Risk Anal*. 1993;13(2):131–134.

50. Cullen AC. Measures of compounding conservatism in probabilistic risk assessment. *Risk Anal*. 1994, August;14(4):389–393.

51. United States Office of Management and Budget (OMB). *Guidelines for Regulatory Analysis, Circular A-4*. Washington, DC, September 17, 2003. http://www.whitehouse.gov/omb/circulars_a004_a-4

52. Zhang K-M and Wen Z-G. Review and challenges of policies of environmental protection and sustainable development in China. *J Environ Manage*. 2008, September;88(4):1249–1261.

53. Meng X, Zhang Y, Zhao Y, Lou IC, and Gao J. Review of Chinese environmental risk assessment regulations and case studies. *Dose Response*. 2012;10(2):274–296.

54. United States Environmental Protection Agency (USEPA). *Soil Screening Guidance: Users' Guide*. EPA/540/R-96/018. OSWER Publication 9355.4-23. Washington, DC, July, 1996. http://www.epa.gov/superfund/health/conmedia/soil/pdfs/ssg496.pdf

55. United States Environmental Protection Agency (USEPA). *Supplemental Guidance for Developing Soil Screening Levels for Superfund Sites*. OSWER Publication 9355.4-24. Washington, DC, December, 2002. http://www.epa.gov/superfund/health/conmedia/soil/pdfs/ssg_main.pdf

56. Calabrese EJ. The road to linearity: Why linearity at low doses became the basis for carcinogen risk assessment. *Arch Toxicol*. 2009, March;83(3):203–225.

57. Cohen BL. The linear no-threshold theory of radiation carcinogenesis should be rejected. *J Am Phys Surg*. 2008;13(3):70–76.

58. Meek ME, Sonawane B, and Becker RA. Foreword: Biomonitoring equivalents special issue. *Regul Toxicol Pharmacol*. 2008, August;51(3 Suppl):S3.

59. Becker K, Conrad A, Kirsch N, Kolossa-Gehring M, Schulz C, Seiwert M, and Seifert B. German Environmental Survey (GerES): Human biomonitoring as a tool to identify exposure pathways. *Int J Hyg Environ Health*. 2007, May;210(3–4):267–269.
60. Aylward LL, Becker RA, Kirman CR, and Hays SM. Assessment of margin of exposure based on biomarkers in blood: An exploratory analysis. *Regul Toxicol Pharmacol*. 2011, October;61(1):44–52.
61. Becker RA, Hays SM, Robison S, and Aylward LL. Development of screening tools for the interpretation of chemical biomonitoring data. *J Toxicol*. 2012;2012:941082.
62. Calafat AM. The U.S. National Health and Nutrition Examination Survey and human exposure to environmental chemicals. *Int J Hyg Environ Health*. 2012, February;215(2):99–101.
63. Boogaard PJ, Hays SM, and Aylward LL. Human biomonitoring as a pragmatic tool to support health risk management of chemicals—Examples under the EU REACH programme. *Regul Toxicol Pharmacol*. 2011, February;59(1):125–132.
64. Liu H, Papa E, and Gramatica P. QSAR prediction of estrogen activity for a large set of diverse chemicals under the guidance of OECD principles. *Chem Res Toxicol*. 2006, November;19(11):1540–1548.
65. Vink SR, Mikkers J, Bouwman T, Marquart H, and Kroese ED. Use of read-across and tiered exposure assessment in risk assessment under REACH—A case study on a phase-in substance. *Regul Toxicol Pharmacol*. 2010, October;58(1):64–71.
66. Williams ES, Panko J, and Paustenbach DJ. The European Union's REACH regulation: A review of its history and requirements. *Crit Rev Toxicol*. 2009;39(7):553–575.
67. Federal, Provincial and Territorial Advisory Committee on Population Health (ACPH). Toward a health future: Second report on the health of Canadians. Ottawa, Ontario, Canada, September, 1999. http://publications.gc.ca/collections/Collection/H39-468-1999E.pdf.
68. Health and Welfare Canada. *Health Risk Determination: The Challenge of Health Protection*. Health and Welfare Canada, Health Protection Branch. Ottawa, Ontario, Canada. Risk management frameworks for human health and environmental risks. *J Toxicol Environ Health B Crit Rev*. 2003;6(6):569–720.
69. Health Canada. *Human Health Risk Assessment for Priority Substances*. Ottawa, Ontario, Canada, April 21, 1994. http://www.hc-sc.gc.ca/ewh-semt/alt_formats/hecs-sesc/pdf/pubs/contaminants/approach/approach-eng.pdf
70. Health Canada. *Health Canada Decision-making Framework for Identifying, Assessing, and Managing Health Risks*. Ottawa, Ontario, Canada, August 1, 2000. http://www.hc-sc.gc.ca/ahc-asc/alt_formats/hpfb-dgpsa/pdf/pubs/risk-risques-eng.pdf
71. Treasury Board of Canada. *Risk, Innovation and Values: Examining the Tensions*. Ottawa, Ontario, Canada, April 15, 1999. http://www.tbs-sct.gc.ca/pubs_pol/dcgpubs/riskmanagement/rm-riv01-eng.asp
72. United States Environmental Protection Agency (USEPA). Final record of decision for the Sangamo Weston/Lake Hartwell/PCB contamination. EPA ID: SCD003354412. EPA/ROD/R04-94/178. Atlanta, GA, June 28, 1994. http://www.epa.gov/superfund/sites/rods/fulltext/r0494178.pdf
73. United States Environmental Protection Agency (USEPA). *National Oil and Hazardous Substances Pollution Contingency Plan (NCP) Proposed Rule*. 53 Federal Register 51394. Washington, DC, December 12, 1988. http://www.epa.gov/superfund/policy/remedy/sfremedy/pdfs/ncppropream.pdf
74. United States Environmental Protection Agency (USEPA). *National Oil and Hazardous Substances Pollution Contingency Plan (NCP) Final Rule*. 55 Federal Register 8666. Washington, DC, December 12, 1994. http://www.epa.gov/superfund/policy/remedy/sfremedy/pdfs/ncpfinalrule61.pdf

75. Bales CW and Self RL. Evaluation of angler fish consumption after a health advisory on lake Hartwell, Georgia and South Carolina. *Proc Annu Conf Southeast Assoc Fish and Wildl Agencies.* 1993;47:650–656.

76. United States Environmental Protection Agency (USEPA). Memorandum from Elliott Laws, Assistant Administrator for OSWER, *Formation of National Superfund Remedy Review Board.* Washington, DC, November 28, 1995. http://www.epa.gov/superfund/programs/nrrb/11-28-95.htm

77. United States Environmental Protection Agency (USEPA). *Seven Cardinal Rules of Risk Communication.* OPA-87-020. Washington, DC, April, 1988. http://www.epa.gov/publicinvolvement/pdf/risk.pdf

78. National Cancer Institute. *Reducing Environmental Cancer Risk: What We Can Do Now* 2008–2009 Annual Report of the President's Cancer Panel. Bethesda, MD, April, 2010. http://deainfo.nci.nih.gov/advisory/pcp/annualReports/pcp08-09rpt/PCP_Report_08-09_508.pdf

79. American Cancer Society (ACS). Cancer and the environment. ACS Pressroom Blog May 6. Atlanta, GA, 2010. http://acspressroom.wordpress.com/2010/05/06/cancer-and-the-environment/

80. Lachlan K and Spence PR. Communicating risks: Examining hazard and outrage in multiple contexts. *Risk Anal.* 2010, December;30(12):1872–1886.

81. Soderbergh, S. (2000) *Erin Brockovich.* Starring Julia Roberts and Albert Finney; Written by Susannah Grant.

82. Steinberg, J. (2013) *Hinkley: A town that's fading away*, San Bernadino Sun, July 8, 2013. http://www.sbsun.com/article/zz/20130709/NEWS/130709252 (accessed October 20, 2013).

83. James, I. (2013) Calif. chromium-6 limit sets off major debate in valley. MyDesert. October 11, 2013. http://www.mydesert.com/article/20131011/NEWS07/310110022/ (accessed October 20, 2013).

84. Zhang JD. [Epidemiological Analysis of Malignant Tumors in Rural Areas in the Suburbs of Jinzhou.] Section 6 of Liaoning provincial cancer death rate, report to Chinese Government. 1976 (Certified translation, original Chinese document and English translation provided by Dr. Brent Kerger of Cardno ChemRisk).

85. Zhang JD and Li XL. [Chromium pollution of soil and water in Jinzhou]. *Zhonghua Yu Fang Yi Xue Za Zhi.* 1987, September;21(5):262–264.

86. Zhang JD and Li S. Cancer mortality in a Chinese population exposed to hexavalent chromium in water. *J Occup Environ Med.* 1997, April;39(4):315–319.

87. Zhang JD. Letter from Dr. J.D. Zhang to Mr. Tony Ye of McLaren-Hart/ChemRisk and Draft of paper by Zhang, J.D. and Li, X.L. titled "Study of the relationship between the chromium pollution in jingzhou suburban region and the cancer death rate (in English). November 18, 1995 (Certified translation, original Chinese document and English translation provided by Dr. Brent Kerger of Cardno ChemRisk).

88. Waldman P. Toxic traces: New questions about old chemicals; second opinion: Study tied pollutant to cancer; then consultants got hold of it; 'clarification' of Chinese Study absolved chromium-6; did author really write it?; Echo of Erin Brockovich case. *Wall Street J.* 2005, December 12.

89. Brandt-Rauf P. Editorial retraction. *J Occup Environ Med.* 2006, July;48(7):749.

90. Phillips ML. Journal retracts chromium study. *The Scientist.* 2006, June 6. http://www.the-scientist.com/news/display/23590/

91. Waldman P. Publication to retract an influential water study. *Wall Street J.* 2006, June 6. http://online.wsj.com/article/SB114921401312569489.html

92. Phillips ML. Chromium paper retracted unfairly, author says. *The Scientist.* 2006, December 12. http://www.the-scientist.com/news/home/38457/

93. Beaumont JJ, Sedman RM, Reynolds SD, Sherman CD, Li L-H, Howd RA et al. Cancer mortality in a Chinese population exposed to hexavalent chromium in drinking water. *Epidemiology.* 2008, January;19(1):12–23.
94. Kerger BD, Butler WJ, Paustenbach DJ, Zhang J, and Li S. Cancer mortality in Chinese populations surrounding an alloy plant with chromium smelting operations. *J Toxicol Environ Health A.* 2009;72(5):329–344.
95. Steinpress MG and Ward AC. The scientific process and Hollywood: The case of hexavalent chromium. *Ground Water.* 2001, May;39(3):321–322.
96. Saurat J-H, Kaya G, Saxer-Sekulic N, Pardo B, Becker M, Fontao L et al. The cutaneous lesions of dioxin exposure: Lessons from the poisoning of Victor Yushchenko. *Toxicol Sci.* 2012, January;125(1):310–317.
97. Bell IR, Schwartz GE, Peterson JM, and Amend D. Self-reported illness from chemical odors in young adults without clinical syndromes or occupational exposures. *Arch Environ Health.* 1993;48(1):6–13.
98. Bell IR, Miller CS, Schwartz GE, Peterson JM, and Amend D. Neuropsychiatric and somatic characteristics of young adults with and without self-reported chemical odor intolerance and chemical sensitivity. *Arch Environ Health.* 1996;51(1):9–21.
99. Simon GE, Daniell W, Stockbridge H, Claypoole K, and Rosenstock L. Immunologic, psychological, and neuropsychological factors in multiple chemical sensitivity. A controlled study. *Ann Intern Med.* 1993, July 7;119(2):97–103.
100. Staudenmayer H, Binkley KE, Leznoff A, and Phillips S. Idiopathic environmental intolerance: Part 1: A causation analysis applying Bradford Hill's criteria to the toxicogenic theory. *Toxicol Rev.* 2003;22(4):235–246.
101. Staudenmayer H, Binkley KE, Leznoff A, and Phillips S. Idiopathic environmental intolerance: Part 2: A causation analysis applying Bradford Hill's criteria to the psychogenic theory. *Toxicol Rev.* 2003;22(4):247–261.
102. Tarlo SM, Poonai N, Binkley K, Antony MM, and Swinson RP. Responses to panic induction procedures in subjects with multiple chemical sensitivity/idiopathic environmental intolerance: Understanding the relationship with panic disorder. *Environ Health Perspect.* 2002, August;110(Suppl 4):4669–4671.
103. Jones TF, Craig AS, Hoy D, Gunter EW, Ashley DL, Barr DB et al. Mass psychogenic illness attributed to toxic exposure at a high school. *N Engl J Med.* 2000, January 1;342(2):96–100.
104. Page LA, Keshishian C, Leonardi G, Murray V, Rubin GJ, and Wessely S. Frequency and predictors of mass psychogenic illness. *Epidemiology.* 2010, September;21(5):744–747
105. Moffatt S, Mulloli TP, Bhopal R, Foy C, and Phillimore P. An exploration of awareness bias in two environmental epidemiology studies. *Epidemiology.* 2000, March;11(2):199–208.
106. Hill AB. The environment and disease: Association or causation? *Proc R Soc Med.* 1965, May;58:295–300.
107. Höfler M. Causal inference based on counterfactuals. *BMC Med Res Methodol.* 2005;5:28.
108. Höfler M. The Bradford Hill considerations on causality: A counterfactual perspective. *Emerg Themes Epidemiol.* 2005, November 11;2:11.
109. Höfler M. Getting causal considerations back on the right track. *Emerg Themes Epidemiol.* 2006;38.
110. Phillips CV and Goodman KJ. Causal criteria and counterfactuals; nothing more (or less) than scientific common sense. *Emerg Themes Epidemiol.* 2006;3:5.
111. Phillips CV and Goodman KJ. The missed lessons of Sir Austin Bradford Hill. *Epidemiol Perspect Innov.* 2004;1(1):3.

112. Moore JA, Daston GP, Faustman E, Golub MS, Hart WL, Hughes C et al. An evaluative process for assessing human reproductive and developmental toxicity of agents. *Reprod Toxicol*. 1995;9(1):61–95.

113. United States Environmental Protection Agency (USEPA) USEPA. Risk Assessment Forum. *Guidelines for Carcinogen Risk Assessment*. EPA/630/P-03/001F. March, 2005. http://www.epa.gov/raf/publications/pdfs/CANCER_GUIDELINES_FINAL_3-25-05.PDF

114. Stara JF, Kello D, and Durkin P. Human health hazards associated with chemical contamination of aquatic environment. *Environ Health Perspect*. 1980, February;34:145–158.

115. Dellarco VL and Wiltse JA. US Environmental Protection Agency's revised guidelines for Carcinogen Risk Assessment: Incorporating mode of action data. *Mutat Res*. 1998, September 9;405(2):273–277.

116. Seed J, Carney EW, Corley RA, Crofton KM, DeSesso JM, Foster PM et al. Overview: Using mode of action and life stage information to evaluate the human relevance of animal toxicity data. *Crit Rev Toxicol*. 2005;35(8–9):664–672.

117. Meek ME, Bucher JR, Cohen SM, Dellarco V, Hill RN, Lehman-McKeeman LD et al. A framework for human relevance analysis of information on carcinogenic modes of action. *Crit Rev Toxicol*. 2003;33(6):591–653.

118. Boobis AR, Cohen SM, Dellarco V, McGregor D, Meek ME, Vickers C et al. IPCS framework for analyzing the relevance of a cancer mode of action for humans. *Crit Rev Toxicol*. 2006, November;36(10):781–792.

119. Boobis AR, Doe JE, Heinrich-Hirsch B, Meek ME, Munn S, Ruchirawat M et al. IPCS framework for analyzing the relevance of a noncancer mode of action for humans. *Crit Rev Toxicol*. 2008;38(2):87–96.

120. Boobis AR, Daston GP, Preston RJ, and Olin SS. Application of key events analysis to chemical carcinogens and noncarcinogens. *Crit Rev Food Sci Nutr*. 2009, September;49(8):690–707.

121. Rivera-Núñez Z, Meliker JR, Meeker JD, Slotnick MJ, and Nriagu JO. Urinary arsenic species, toenail arsenic, and arsenic intake estimates in a Michigan population with low levels of arsenic in drinking water. *J Expo Sci Environ Epidemiol*. 2012;22(2):182–190.

122. Abraham K, Geusau A, Tosun Y, Helge H, Bauer S, and Brockmoller J. Severe 2,3,7,8-tetrachlorodibenzo-p-dioxin (TCDD) intoxication: Insights into the measurement of hepatic cytochrome P450 1A2 induction. *Clin Pharmacol Ther*. 2002;72(2):163–174.

123. Lampe JW, King IB, Li S, Grate MT, Barale KV, Chen C et al. Brassica vegetables increase and apiaceous vegetables decrease cytochrome P450 1A2 activity in humans: Changes in caffeine metabolite ratios in response to controlled vegetable diets. *Carcinogenesis*. 2000, June;21(6):1157–1162.

124. United States Senate Committee on Environment and Public Works (2012) Full Committee hearing entitled, *Oversight Hearing on the Environmental Protection Agency's Implementation of the Safe Drinking Water Act's Unregulated Drinking Water Contaminants Program*. Tuesday, Washington, DC, July 12, 2011. http://www.epw.senate.gov/public/index.cfm?FuseAction=Hearings.Hearing&Hearing_ID=fc5a8756-802a-23ad-454a-b9eeb7bf1c36 (accessed October 20, 2013).

125. Gold LS, Manley NB, Slone TH, and Ward JM. Compendium of chemical carcinogens by target organ: results of chronic bioassays in rats, mice, hamsters, dogs, and monkeys. *Toxicol Pathol*. 2001;29(6):639–652.

FURTHER READINGS

Bernstein PL. *Against the Gods: The Remarkable Story of Risk.* New York: Wiley, 1996, 1998.

Pohjola MV and Tuomisto JT. Openness in participation, assessment, and policy making upon issues of environment and environmental health: A review of literature and recent project results. *Environ Health.* 2011;10:58, doi:10.1186/1476–069X-10–58. http://www.ehjournal.net/content/10/1/58

Ruckelshaus W. Stopping the Pendulum. *Environ Toxicol Chem.* 1996;15(3):229–232.

United States Environmental Protection Agency. *Risk Assessment Guidance for Superfund: Volume 1: Human Health Evaluation Manual (Part A) (RAGS).* Risk Assessment Guidance for Superfund: Volume 1: Human Health Evaluation Manual (Part A) (RAGS). 1989; http://www.epa.gov/oswer/riskassessment/ragsa/

2 Perception, Planning and Scoping, Problem Formulation, and Hazard Identification
All Parts of Risk Assessment

> Whereas many persons live in great fear and apprehension of some of the more formidable and notorious diseases, I shall set down how many died of each: that the respective numbers, being compared with the total ... those persons may better understand the hazard they are in.
>
> **John Graunt**
> *Natural and Political Observations Made upon the Bills of Mortality, 1662*

Risk means different things to different people. If you watched the "The Story of Cosmetics" video from the first exercise at the end of Chapter 1, you should now realize that the woman in this video was confusing hazard identification (HI) with risk assessment. For environmental risk, it is important to realize that both toxicity and exposure contribute to risk; both are necessary for risk. Hence, low but detectable levels of chemicals known to be toxic may be present in the environment but would not present a concern because the exposure and resulting risk are very low.

WHAT IS RISK AND HOW CAN WE ESTIMATE RISK?

Risk perception also changes the way risk is viewed. Risks may be voluntary or involuntary. The classic example of this distinction is the individual attending a public meeting about a nearby hazardous waste site who lights up a cigarette and asks: "So what are you going to do to prevent me from getting cancer from these nasty environmental chemicals? And, by the way, why did you have this meeting on Saturday? That's when I go skydiving." This somewhat facetious example is provided to illustrate the difference between the perceptions of voluntary and involuntary risk. This individual chose to smoke and jump from airplanes—both voluntary risks—but did not want to experience risk brought on by the actions of others.

RISK OF BEING STRUCK BY LIGHTNING

Risk means different things to different people. Insurance actuaries, gamblers, toxicologists, statisticians, epidemiologists, and laypeople all use this word in different contexts. The idea of risk arose in the field of statistics; as a result, there are a number of ways in which numerical values for risk can be expressed. In the following discussion, the risk of being struck by lightning will be used as an example.

Frequentist or Actuarial Risk

The risk of being struck by lightning in the United States can easily be calculated based on historical values.[1] The average number of reported deaths per year from lightning during the period from 2001 to 2010 was 39. The average number of reported injuries over the same period was 241. From the 2000 census, the US population was 281,421,906. From the 2010 census, the US population was 308,745,538. We will assume the average for the period from 2000 to 2010 is represented by the average of these two values or 295,083,722. Hence, the risk of being struck by lightning can be calculated as follows:

$$\frac{(\text{Deaths} + \text{Injuries})}{\text{Population at risk}} = \frac{(39 + 241)}{295,083,722} = 9.5\text{E} - 07 \qquad (2.1)$$

Hence, the risk of being struck by lightning is around one in a million. This estimate is based on the assumption that the frequency of past events reflects the likelihood of similar events occurring in the future. This is known as frequentist or actuarial risk. Insurance companies use actuarial risk in setting rates. Most accidental death policies do cover death by lightning strike, so insurance policies consider being struck by lightning and dying to be a low-risk event—a good bet, so to speak.

There are several uncertainties inherent in this lightning risk estimate—the number of reported deaths and injuries may not be accurate. The accuracy of the population estimate will generally have a much lower effect because this number is so large and occurs in the denominator.

Predicted Risk: Using a Model

There is another way to estimate the risk of being struck by lightning—to use a prediction model. For example, one could estimate three quantities on a yearly basis:

- Average number of lightning ground strikes per year
- Average number of people at risk during a single lightning event
- Probability of a lightning strike hitting a person

The model would be a simple multiplication of these factors:

$$\frac{\text{Strikes}}{\text{Year}} \times \frac{\text{People at risk}}{\text{Strike}} \times \text{Probability of a strike} = \text{Risk of being struck by lightning}$$

$$(2.2)$$

This simple prediction model seems quite logical until one attempts to obtain data on these quantities. There is a wide geographic variation in the number of strikes per year. For example, Singapore has one of the highest rates of lightning activity in the world.[2]

For this example of a prediction model, we will consider "Lightning Alley," the area between Tampa and Orlando in central Florida that sees more lightning than any other area in the United States with as many as 50 strikes per square mile per year,[2] corresponding to 20 strikes per km². If one assumes that the Orlando metropolitan area is a target, this area is home to 2M people and has an area of 260 km², one can determine the risk of being struck by lightning in Orlando.[3]

Multiplying 260 km² by 20 strikes/km², one can calculate 5200 strikes per year in the Orlando metropolitan area. If one further assumes that lightning has an equal probability of striking any part of the area with similar frequency* and that a single strike affects an area of 100 m², then the lightning strikes affect 0.2% of the area each year. Ten percent of the population would likely never experience this risk because they spend little to no time outside during storms. Hence, the population at risk would be 1.8M not 2M—although the value of 2M could still be used in the denominator as representing the entire potentially affected population depending on the assumption and type of risk estimate desired. Further, if the population, similar to the assumed occurrence of lightning, is evenly distributed over the metropolitan area[†] and each individual has a 1% chance of being outside during a storm,[‡] then the prediction model would be as follows:

$$\text{Population at risk} \times \frac{\text{Likelihood of being outside}}{\text{Entire population}}$$

$$= \text{Risk of being struck by lightning in Orlando} \qquad (2.3)$$

$$\text{Population at risk} = 0.2\% \times 90\% \times 2,000,000 = 3600 \qquad (2.4)$$

$$\text{Susceptibility} = \text{Likelihood of being outside during a storm} = 1\% \qquad (2.5)$$

$$\text{Entire population} = 2,000,000 \qquad (2.6)$$

$$3600 \times \frac{1\%}{2,000,000} = 1.8\text{E}-05 \qquad (2.7)$$

---

* "They" do say that lightning never strikes twice in the same place.
† Both these assumptions are obviously incorrect!
‡ This value is based on professional judgment, which is code or risk-speak, meaning that no one has any idea what the true value is.

This model produces a risk estimate that is about 20-fold higher than that for the entire United States. Perhaps this difference reflects the increased occurrence of lightning in Orlando. There is considerable uncertainty about the assumptions used to calculate the risk estimate specific to Orlando, but there is also uncertainty about whether the simple calculation that yielded a risk of one in a million for the entire United States is applicable to Orlando.

In 2011, eight people were injured by a lightning strike occurring at SeaWorld's Discovery Cove water park in Orlando. Three of these people were guests at SeaWorld and the others were employees.[4] The fact that eight people were injured in a single lightning strike suggests that the population density in the Orlando area may be a factor in increasing the risk. However, the fact that Orlando is a tourist destination means that the estimated number of people at risk, that is, the denominator, may be higher than the number of residents. In 2011, a total of 55M people visited Orlando.[3] Would visitors who likely spend more of their time outside be more susceptible to lightning? The number of visitors is much higher than the resident population and would likely affect calculation of both the numerator and denominator. Just how the presence of visitors would affect the risk estimate is not entirely clear because of the many assumptions used in the model.

Perceived Risk

The news report from SeaWorld suggests still another way to determine risk—perception or subjective judgment. Following this news report, many people would no doubt be concerned about the risk of lightning in Orlando; with time and fading memory, this perception would also likely fade.

If one were concerned about the risk of being struck by lightning, such news might suggest that the higher of the two very uncertain estimates is more correct. This is not necessarily true. While the uncertainties in these estimates were discussed, albeit briefly, it should be abundantly clear that one could have little to no confidence in the accuracy of either estimate.

Extrapolating from the estimate of one in a million from the entire United States to Orlando seems to produce an underestimate of the risk because of the greater density of lightning in Orlando. The use of the Orlando-specific values for resident population density along with the assumption that the population is evenly distributed over this area is also clearly incorrect. The lightning strike at SeaWorld injuring eight indicates that people tend to cluster. A tourist attraction such as SeaWorld will tend to produce such clusters. Will these clusters increase or decrease risk? Well, it all depends where people are when lightning strikes.

The point of this discussion has been to use the risk from a familiar event to illustrate just how one can characterize risk. All of the estimates of the risk of being struck by lightning are incorrect, but they do illustrate the types of risk estimates and the difficulty of having sufficient confidence in a risk estimate to undertake a risk management action.

DESIGNING RISK ASSESSMENTS: PLANNING AND SCOPING VERSUS PROBLEM FORMULATION

Risk management decisions necessarily proceed from the initial steps of (1) planning and scoping activities and (2) problem formulation. There will be practical implications stemming from the results of both steps about which adverse outcomes to investigate, which data are most relevant to the problem, and which risk management options should be considered. Stakeholders will likely have different goals and likely a range of divergent perspectives on the problem. These divergent perspectives will necessarily lead to different problem formulations and thus different decisions.

For example, a response to concern about the risk of being struck by lightning when visiting SeaWorld would be to shut down this popular attraction. Stakeholders in this decision would, of course, be the owners and patrons of the facility as well as those fearful of lightning. Obviously, the owners would be opposed to closure—as would those children who had been looking forward to their summer vacation.

In these initial risk assessment activities, the wishes of all parties need to be considered. Inclusion of a range of stakeholders in planning and scoping is important both to be fair and to avoid missing aspects of the problem of which a more limited group of stakeholders might not be cognizant.

Early in the development of ecological risk assessment guidance, EPA separated the initial steps of risk assessment into (1) planning and scoping and (2) problem formulation.[5–8]

Planning and scoping involves discussions between risk managers and stakeholders with risk assessors playing a supporting role. The result of planning and scoping activities is ideally a broad conceptual statement of the problem, options for its solution, and any trade-offs that need to be considered.

Problem formulation involves discussions between risk managers and risk assessors to develop the detailed design for the assessment, including technical and scientific considerations. The result of problem formulation would ideally reflect that broad conceptual statement developed in planning and scoping.

The NEPA did not discuss problem formulation per se; however, NEPA outlines a scoping process for environmental impact statements that includes elements of problem formulation.[9]

The first use of the term "problem formulation" and first discussion regarding environmental risk assessment was presented in EPA's 1992 *Framework for Ecological Risk Assessment*.[5] Although the definition presented was limited to ecological risk assessment, the definition provided was applicable to any type of risk assessment:

> Problem formulation is the first phase of ... and establishes the goals, breadth, and focus of the assessment. It is a systematic planning step that identifies the major factors to be considered in a particular assessment, and it is linked to the regulatory and policy context of the assessment.[5]

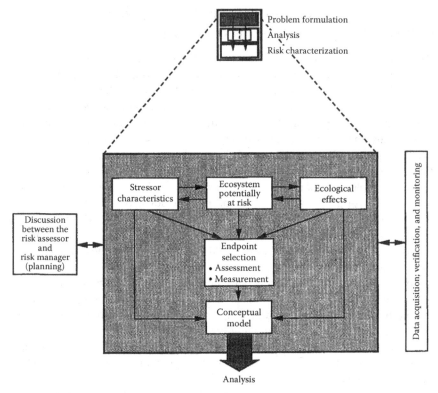

FIGURE 2.1 Problem formulation in ecological risk assessment. (From United States Environmental Protection Agency (USEPA), Risk Assessment Forum, *Framework for Ecological Risk Assessment*, EPA/630/R-92/001, Washington, DC, February 1992, http://www.epa.gov/raf/publications/pdfs/FRMWRK_ERA.PDF.)

The centerpiece of problem formulation is a conceptual model. This model contains a set of working hypotheses of how stressors might affect human health or aspects of the natural environment. Hence, problem formulation would determine the nature and extent of data collection efforts to support the risk assessment. Figure 2.1 shows EPA's diagrammatic representation of problem formulation in ecological risk assessment.[5]

In the 1996 report *Understanding Risk: Informing Decisions In A Democratic Society* from the NAS, problem formulation is the first step in the analytic–deliberative process that results in a risk characterization.[10] This was the first document related to human health risk assessment that recognized the importance of careful up-front planning and the need to formalize problem formulation as part of risk assessment.

NEED FOR PROBLEM FORMULATION IS NOT LIMITED TO ECOLOGICAL RISK ASSESSMENT

The 1997 Federal Commission Report indicated that problem definition and context should be the first step in a risk assessment.[11] In human health risk assessment, the general activity of planning seems to have been given short shrift by EPA until well past the year 2000. The discussion of "scoping" provided by EPA in RAGS is extremely general and brief.[12] As noted, in ecological risk assessment, planning and scoping is distinguished from problem formulation, and the latter is defined very specifically. Problem formulation is a process for generating and evaluating preliminary hypotheses about why ecological effects may occur or have occurred. The process includes plans for developing a conceptual model, collecting and analyzing data, and characterizing risk.[7,13]

Other organizations and disciplines also use the process of problem formulation. It is the focus of activities in operations research, and problem formulation as a formal defined activity is used in evaluation of medical treatments, public health interventions, and military planning.[14–18]

For example, problem formulation is an inherent part of the North Atlantic Treaty Organization's (NATO) *Code of Best Practice for Command and Control Assessment* (COBP).[19] The principles and methods described by NATO for the assessment for risk associated with war and operations other than war (OOTW) also apply to environmental risk assessment. A multidisciplinary and eloquent description of problem formulation is provided in the COBP:

> Explicit problem formulation must precede construction of concepts for analysis or method selection. This is not a trivial exercise …
>
> Problem formulation must not only provide problem segments amenable to analysis, but also a clear and valid mechanism for meaningful synthesis to provide coherent knowledge about the original, larger problem. The formulated problem is thus an abstraction of the real problem … that can be interpreted in terms of decisions and actions.
>
> Problem formulation must be broad and iterative in nature, accepting the minimum of a priori constraints and using methods to encourage creative and multi-disciplinary thinking, such as proposing a number of hypotheses for the expression of the problem.
>
> Practical constraints such as data availability, study resources (including time), and limitations of tools should be treated as modifiers of the problem formulation rather than initial drivers. Such constraint may, in the end, drive the feasible solutions, but it is important to recognise this as a compromise rather than an ideal. *Proper problem formulation takes substantial time and effort!*
>
> It is important that problem formulation address risk from multiple perspectives. In addition to sensitivity analysis of the dependent variables, risk analysis techniques should be used to directly explore options to mitigate risk.[19]

The iterative nature of problem formulation becomes clear from the diagram of the problem formulation process used by NATO (Figure 2.2).

RECOGNITION OF THE IMPORTANCE OF PROBLEM FORMULATION FOR HUMAN HEALTH RISK ASSESSMENT

Although several EPA guidance documents indicated that problem formulation as well as planning and scoping should become a part of human health risk assessment, in practice, this does not always happen.[5–8]

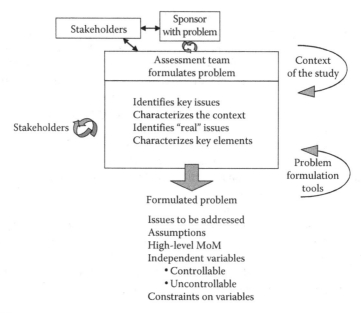

FIGURE 2.2 NATO overall study plan for problem formulation.

In 2009, the NRC released the report *Science and Decisions: Advancing Risk Assessment*.[20] This report is known as the Silver Book because of the color of its cover. Its purpose was to provide the EPA a plan for improving the risk assessment/risk management process in the short term of 2–5 years and the long term of 20 years. The report recommended a three-phase approach to risk assessment— phase I would be problem formulation (Figure 2.3). The report made several recommendations to EPA and indicated that implementing these recommendations would constitute a significant transformation of the culture of risk assessment and decision-making within EPA.

Risk assessment is both a process and a product in that the process creates the product. As such, a major challenge in risk assessment is to design a process that includes considerations of technical quality to create a product that is useful to a community of consumers who may have disparate and, at times, conflicting concerns. The Silver Book uses the term "design" to imply the adoption of viewpoint of user-friendliness to develop both a transparent and science-based assessment process and decision support tool useful to all stakeholders. Problem formulation is a design activity that occurs or should occur early in the risk assessment process and involves understanding and weighing the risk management objectives, the recognition for statutory and state-of-knowledge constraints, and explicit acknowledgment of the need for trade-offs.[20]

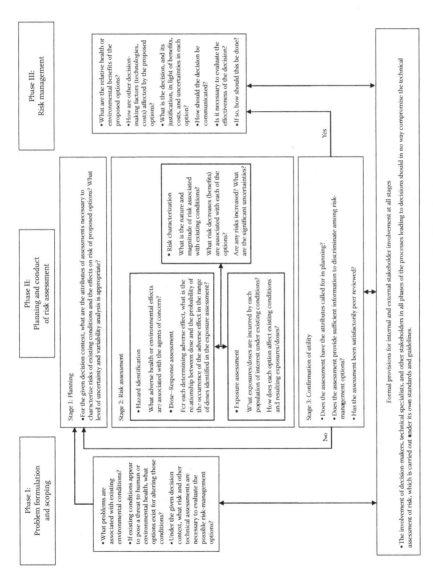

FIGURE 2.3   The role of problem formulation in the overall process defined by the Silver Book.

The 1983 Red Book discussed in Chapter 1 was strongly supportive of the need for a conceptual distinction between risk assessment and risk management so that the scientific/technical process of risk assessment was free from political or economic influence.[21] The recognition of this conceptual distinction by both risk assessors and risk managers is critical during problem formulation.

The product of a problem formulation likely the most useful to the largest number of stakeholders is the conceptual model. Figure 2.4 shows a conceptual model for an air pollution risk assessment. The major elements of a conceptual model that provide information for the risk assessment process about what data to obtain and how to interpret and analyze these data are as follows:

- Sources
- Stressors or pollutants
- Exposure pathways and/or exposure media
- Routes of exposure
- Exposed populations
- Endpoints/outcomes of concern
- Metrics used for decision support

The usefulness of conceptual models and problem formulation led the NRC to recommend that EPA formalize these as part of the human health risk assessment process.[20]

EPA faces significant challenges in the implementation of problem formulation and other Silver Book recommendations.[22] One of the NRC committee members even suggested, quite eloquently, scrapping the divide between risk assessment and risk management because it interferes with effective problem formulation.[23]

> The basic dogma holds that risk assessment must precede risk management. But there is an opposite and perhaps better way: the opening question should not be "How bad is the problem?" but "How good are the solutions we might apply to the problem?"[23]

EPA is currently leading a multiagency collaborative effort called the NextGen program—presumably to address these challenges. The agencies involved include the National Institute of Environmental Health Sciences, the National Center for Advancing Translational Sciences, the Centers for Disease Control (CDC), the National Center for Toxicological Research of the FDA, the Department of Defense, and one state agency, California's EPA.

Although a very recent publication by EPA staffers on NextGen references the Silver Book, no specific details of how recommendations therein, including those related to either planning and scoping or problem formulation, will be implemented or addressed.[24] Hence, the future role of problem formulation remains to be determined.

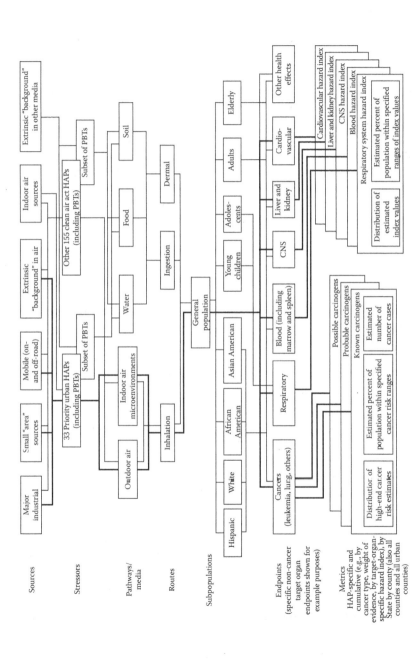

FIGURE 2.4 Example of a conceptual model for air pollution risk assessment. (From National Research Council (NRC), *Science and Decisions: Advancing Risk Assessment*, Washington, DC: The National Academies Press, 2009, http://www.nap.edu/catalog.php?record_id=12209. With permission.)

HAZARD IDENTIFICATION VERSUS RISK CHARACTERIZATION

At the beginning of this chapter, we considered the risk of being struck by lightning. The point of that discussion was to use the risk from a familiar event to illustrate ways one can characterize risk. Both quantitative estimates of the risk of being struck by lightning were wrong—nonetheless, they illustrate two types of risk estimates.

How might one determine in which risk estimate one has more confidence? One way is to examine the weight of the scientific evidence supporting each estimate.

Relating these lightning risks to HI, all this information could serve as the basis of identifying lightning as a hazard—and, of course, not just in Orlando. The single incident of the lightning strike and resulting injuries at SeaWorld are sufficient to identify lightning as a hazard even if the risk cannot be accurately predicted. This is the essence of the difference between HI and risk characterization.

The task of HI is complete when one can say: "We know it's bad, but we don't know how bad." Conversely, risk characterization requires one to provide an estimate of just how bad it is.

It is critical to realize that life is not risk-free—there is no such thing as zero risk. Hence, the task of HI is based to a large degree on ascertaining when conditions have the potential for unreasonable risk. Chapter 1 dealt extensively with the definition of unreasonable risk, and "unreasonable" is most always defined by societal consensus, most often in the form of an elected government.

Obviously, there has to be some information upon which to base the claim of hazard. Weight of evidence (WOE) (discussed in depth in the succeeding text) is an important consideration.

For example, high doses of the artificial sweetener saccharin have been shown to be carcinogenic in rats evidenced by a dose-related increase in urinary bladder neoplasms. High doses of the sweetener aspartame produced an increase in brain tumors in rats.[25] Shouldn't we be concerned about these two carcinogenic chemicals that are used ubiquitously as sweeteners and consumed every day by millions of people? Why are these substances approved for human consumption? Shouldn't they be banned? Such a ban would certainly be in the spirit of the Delaney Clause?

Additional studies on both substances suggest that the mechanism by which saccharin causes cancer in rats is not relevant to humans and that the experimentally observed carcinogenicity of aspartame could not be confirmed in later experiments.[26–28]

The point is that HI and hazard prediction, that is, risk assessment, are not the same. In passing the Delaney Clause banning all suspected carcinogens from food packaging, the US Congress confused HI with risk assessment, just like the woman in the story of stuff video at the end of Chapter 1.

WHAT IS HAZARD IDENTIFICATION?

Likely the first attempt at developing information on the hazard of various diseases was that of John Graunt, a London merchant who was born in 1620. In 1603, the city of London began keeping records of births and deaths. There were several reasons—that year one of the worst outbreaks of plague occurred and information about the population of London was needed for tax revenues. In 1663, Graunt published the frequency of various causes of death experienced by Londoners in 1663 in *Natural and Political Observations Made upon the Bills of Mortality*, quoted at the start of this chapter.[29]

HI, as defined by the Red Book, is the procedure for determining whether exposure to a chemical can increase the incidence of an adverse health outcome. HI characterizes the nature and weight of the evidence and thus involves consideration of causation. Hence, there is almost never a simple yes–no answer supported by definitive data. The evidence for HI may depend on epidemiologic or other human studies, on results from laboratory animal testing, in vitro studies of cells of both human and animal origin, quantitative structure–activity estimates, and other in silico predictive methods.

By itself, HI can provide the conclusion that a chemical presents little or no risk to human health or the environment and, thus, is not of regulatory concern. However, a chemical may be deemed potentially hazardous, and then the other three steps of the risk assessment process—exposure assessment, dose–response assessment, and risk characterization—would be undertaken.

Following the Red Book in 1986, a consortium of US government agencies essentially codified the use of the term "hazard" as separate from "risk."[30] Figure 1.1 showed the four parts of risk assessment detailed in the Red Book and elsewhere, and, for risk to occur, there must be both a hazard and sufficient exposure to this hazard.[21]

The Red Book pointed out that the regulatory actions stemming from the Delaney Clause were based on HI only, not on a complete risk assessment. In fact, the Delaney Clause precluded the performance of a risk assessment for known or suspected carcinogens.[21] To be perfectly clear, the Delaney Clause, inappropriate or not, was an exercise of the precautionary principle with a disproportionate response.

The Blue Book defines HI as "the identification of the contaminants that are suspected to pose health hazards, quantification of the concentrations at which they are present in the environment, a description of the specific forms of toxicity (neurotoxicity, carcinogenicity, etc.) that can be caused by the contaminants of concern, and an evaluation of the conditions under which these forms of toxicity might be expressed in exposed humans."[31] Data for HI are typically obtained from environmental monitoring data as well as epidemiologic and animal studies. Like the Red Book, the Blue Book was also careful to distinguish HI from both dose–response assessment and risk characterization.[31]

The 1997 report of the Presidential/Congressional Commission on risk assessment and risk management titled *Framework for Environmental Health Risk Management* mentioned but did not discuss HI.[11] This report was focused on the interface between risk assessment and risk management. The Federal Commission Report notes that hazard is an intrinsic property of a substance or a situation and gives some useful examples. The report indicates benzene does not cause lung cancer, but can cause leukemia. A garter snake bite may be harmless, but a rattlesnake bite can kill if untreated.

In the framework for regulatory decision-making in the report, four data sources for HI of potential carcinogens were identified—epidemiology, lifetime rodent bioassays, in vitro tests, and structure–activity considerations. These data sources are discussed in the following text, and the situation today is similar to that in 1997 except that much more is known about the strengths and limitations of these data sources.

UNCERTAINTY CLASSIFICATIONS USED IN HAZARD IDENTIFICATION

In 1977, the International Agency for Research on Cancer (IARC) produced guidelines giving five classifications of evidence that a chemical might be carcinogenic in humans. IARC pointed out that "for practical purposes" and because of the lack of scientific evidence of a correlation between animal and human carcinogenicity, the pragmatic position was to assume that animal carcinogens could also be human carcinogens.[32] The five classes of evidence defined by IARC were as follows:

- *Sufficient evidence of carcinogenicity*—increased incidence of malignant tumors: (1) in multiple species or strains; (2) in multiple experiments (preferably with different routes of administration or using different dose levels); or (3) to an unusual degree with regard to incidence, type of tumor, or tumor site or age at onset
- *Limited evidence of carcinogenicity*—data suggesting a carcinogenic effect but are limited because (1) the studies involve a single species, strain, or experiment; (2) the studies use inadequate dosage levels, inadequate duration of exposure, inadequate follow-up or survival, low statistical power because of methods or number of animal; or (3) the neoplasms were likely spontaneously occurring or difficult to classify as malignant
- *Inadequate evidence of carcinogenicity*—inability to interpret the evidence as showing either the presence or absence of a carcinogenic effect because of major qualitative or quantitative limitations
- *Negative evidence of carcinogenicity*—within the limits of the tests
- *No data on carcinogenicity*

In 2006, IARC published a preamble to their monographs. The preamble presented groups related to the WOE of human carcinogenicity[33]:

- Group 1—carcinogenic to humans
- Group 2A—probably carcinogenic to humans
- Group 2B—possibly carcinogenic to humans
- Group 3—not classifiable as to its human carcinogenicity
- Group 4—probably not carcinogenic to humans

EPA's 1986 *Guidelines for Carcinogen Risk Assessment* indicated that six types of information should be included when considering HI for potential carcinogens[34]:

- Physical–chemical properties and routes and patterns of exposure
- Structure–activity relationships
- Metabolic and pharmacokinetic properties
- Toxicological effects

- Short-term tests, including in vitro tests
- Long-term animal studies
- Human studies

EPA indicated that the WOE for human carcinogenicity was expressed in a letter grouping. These groups were as follows:

- Group A—carcinogenic to humans
- Group B—probably carcinogenic to humans
- Group C—possibly carcinogenic to humans
- Group D—not classifiable as to human carcinogenicity
- Group E—evidence of noncarcinogenicity in humans

A number of state agencies and other groups inappropriately adopted these WOE summary criteria as hard-and-fast guidelines. These groupings were the result of attempting to provide an overall picture of nuanced and often contradictory evidence. Hence, their use as "bright" lines is inappropriate. In 2005, these groups were updated with the revision to the cancer guidelines[35] as follows:

- Carcinogenic to humans
- Likely to be carcinogenic to humans
- Suggestive evidence of carcinogenic potential
- Inadequate information to assess carcinogenic potential
- Not likely to be carcinogenic to humans

The NRC published *Science and Decisions: Advancing Risk Assessment* in 2009. This report does not break any new ground in HI but points out that any quantitative analysis of uncertainty in HI would require sophisticated methods to synthesize information from multiple scientific domains and multiple scales of information.[36] One important insight provided here was the recognition that uncertainty associated with HI was related to lack of information about critical cause–effect relationships. This uncertainty cannot easily be quantified in a way that would yield a CI or statement of probability. Hence, to some extent, HI ends up being as much an exercise in cognitive science, understanding of bias and heuristics, and epistemology as it is an analysis based in toxicology or risk assessment.[37–40]

WEIGHT OF EVIDENCE

Although this term was used in the Red Book, no specific definition was provided. Despite this, the Red Book advanced a long list of questions regarding how risk assessors should weigh various types of evidence, especially when considering carcinogenicity. Although WOE plays a central role in risk assessment, a specific definition has yet to emerge.

Uncertainty is the unwelcome but constant handmaiden of science. The decision process for HI to which the term "weight of evidence" is applied has so far been an unstructured combination of scientific evidence, interpretation of that evidence, and so-called expert judgment—vulnerable to the biases of those involved.

The term "weight of evidence" may possess three meanings:

- *Metaphorical*, where WOE refers to a collection of studies and an unspecified approach to examining these studies in the aggregate
- *Methodological*, where WOE refers to interpretive methodologies used in meta-analysis, systematic review, or application of considerations of causation
- *Theoretical*, where WOE has a specific meaning for pattern recognition in cognitive psychology and still another theoretical meaning in the legal system[41]

Psychologically, humans tend to "weigh" evidence incorrectly, and heuristic evaluations of evidence lead to over- or underconfidence in the conclusions (see Box 2.1). Physicians have been most obviously guilty of such errors in judgment. To address this issue in the practice of medicine where decisions based on scientific evidence are routine, Dr. Archie Cochrane, a student of Sir Austin Bradford Hill, published a monograph in 1971 titled *Effectiveness and Efficiency* strongly critical of subjective decision-making processes of most physicians because of the inherent biases and inconsistencies. Cochrane was a strong and vocal champion of randomized clinical trials as the best evidence for medical decisions.[42]

BOX 2.1   WEIGHT OF EVIDENCE VERSUS STRENGTH OF EVIDENCE

It is vital in assessing the confidence or degree of belief in a given hypothesis to distinguish between the weight of the evidence, its predictive validity, and the strength of the evidence—its level of extremeness. Human nature tends to make one overconfident when strength is high and weight is low and underconfident when strength is low even if the weight is high. In other words, humans tend to rely on anecdotal evidence. Scientists, being human, must exercise discipline to conduct an appropriate WOE analysis that a particular chemical is linked to cancer or some other adverse effect.

Because of human nature, the formation of a belief that a particular chemical causes a particular effect may depend on a single study in which the effect was easily notable or a large proportion was affected. This belief may persist even though the predictive validity of the study was weak. Predictive validity depends on study design, sample size, background rates, and other factors. Hence, even experienced scientists may tend to use heuristics when a "wow" factor is present.[43,44]

Hypothesis-based WOE evaluation is an increasingly used tool to understand and put into context the growing body of data on the effects of chemicals.[45] One particularly difficult issue is the distinction between effects that

BOX 2.1 (continued)   WEIGHT OF EVIDENCE
VERSUS STRENGTH OF EVIDENCE

are adaptive and effects that are adverse.[46] Two factors have given new urgency to the task of understanding just which effects are adverse and which are not—first, the ever-increasing ability of chemists to measure lower and lower levels of chemicals in human tissues as well as other biomarkers, and second, the extreme sensitivity of many in vitro assays that use biologically engineered systems with unknown relevance to humans. WOE evaluations in toxicology are now based on some understanding of the MOA, as discussed in Chapter 1. WOE frameworks are expected to be a way to avoid bias when interpreting the plethora of available toxicological data.[47]

Decisions based on so-called expert judgment have been called "authority-based" rather than evidence-based.[48] The Cochrane Collaboration develops systematic reviews of medical and scientific evidence for health-care interventions.[49] This collaboration was developed precisely because physicians tended to base their health-care decisions on heuristics with the associated biases and overconfidence. The Cochrane Collaboration is the foundation of evidence-based medicine and provides systematic reviews of health-care interventions so that providers can use evidence to inform their medical decisions.

Recently, the evidence-based toxicology consortium at the Center for Alternatives to Animal Testing (CAAT) at Johns Hopkins University is attempting to adopt the methods of evidence-based medicine to toxicology to determine the best use of human, animal, and in vitro data to inform regulatory decisions.[50–52]

One motivation for using in vitro testing is a concern for animal welfare and the growing realization that high-dose experiments in animals are unlikely to yield high quality information about low-dose effects in humans.[53]

High-throughput in vitro testing is used routinely in the pharmaceutical industry to identify potential drug candidates. EPA and others are attempting to use these same methods and assays to determine the risks of the many untested chemicals in commerce today. To date, it appears that high-throughput in vitro testing can identify certain positive assay results as no more than qualitative risk factors.[54] What this means is that the results of in vitro testing cannot, at present, be used for quantitative predictions of risk, but may possibly be a useful tool for HI.

EPIDEMIOLOGIC STUDIES FOR HAZARD IDENTIFICATION

From a scientific point of view, well-conducted epidemiologic studies provide the most convincing evidence for identifying chemicals or other agents as hazards or not. Although a well-conducted epidemiologic study provides evidence of the greatest weight linking a chemical to an adverse effect in humans, such studies are few.

In most epidemiologic studies, the statistical power is low, the latency between exposure and disease occurrence is long, and there will always be confounders such as exposure to multiple chemicals and genetic variation in susceptibility.

The most uncertain aspect of observational epidemiology is generally the exposure characterization. Exposure in epidemiologic studies is a time-varying complex quantity for which a summary must be developed before any relationship to health outcomes can be determined. This summary of exposure is most often cumulative over time, for example, pack years for cigarette smoking.

The two general types of exposure measures that can be used in epidemiologic studies are biomonitoring and historical reconstruction. These are discussed in the succeeding text in relation to HI.

Biomonitoring and the Use of Biomarkers

For some chemicals, there may exist biomonitoring data or biomarkers, and these can greatly increase the accuracy of exposure characterization. Biomarkers may represent exposure or effect. Biomarkers of exposure are not viewed in the same way as biomarkers of effect. These are discussed in more detail in the succeeding text with an example.

In the United States, the CDC conducts the National Health and Nutrition Examination Survey (NHANES) with the National Center for Health Statistics (NCHS). NHANES data are available at the CDC's website and include results from questionnaires, physical measurements such as height and weight, and measurements of hematological parameters and levels of various chemicals in blood. NHANES uses statistical sampling methods to ensure that the data are representative of the entire US population.[55] For example, NHANES includes measurements of more than 200 environmental chemicals, including mercury in hair and blood, lead in blood, as well as dioxin-like chemicals, perfluorinated compounds, and volatile organic compounds in blood.

Thus, for many chemicals, specific biomarkers exist. When such a biomarker is used to verify exposure, epidemiologic studies become much more usable because much of the uncertainty associated with exposure is eliminated.

A variety of human tissues or fluids have been used as sources of biomonitoring or biomarker data. These tissues or fluids include whole blood or serum, urine, adipose tissue, hair, breast milk, saliva or sputum, semen, and exhaled air. The increasing ability of analytical chemists to measure extremely low concentrations of a variety of chemicals in human tissues requires that these data be presented in the proper context.[56] The presence of a chemical in the body is only indicative of the need for understanding what this finding means—as such, the finding of chemicals present in the body constitutes HI only.

Dr. Harvey Clewell, Melvin Andersen, and Jerry Campbell of the Hamner Institute, pioneers of the use of PBTK modeling (Box 2.4) in risk assessment, formed a band called the Belladonna Blues Band. Dr. Clewell had written a funny song called "Bad Blood Blues" that the band performed at the 2007 Society of Toxicology meeting that can be viewed on YouTube (http://www.youtube.com/watch?v=5b7HsDRWPkE). This song humorously illustrates the need for context when interpreting biomonitoring data.

Biomarkers of Exposure and Biomarkers of Effect

A number of endogenous and exogenous chemicals react with DNA to form adducts. The use of DNA adducts as biomarkers of exposure is problematic because a huge amount of DNA damage occurs due to endogenous substances and oxidative stress.[57,58] Endogenous DNA damage is often indistinguishable from that due to chemical exposure, and this greatly complicates the assessment of DNA adducts as biomarkers of exposure. The mere presence of adducts cannot be indicative of a potential deleterious effect of a chemical.

There remains confusion within the risk assessment community regarding the difference between genotoxicity and mutagenicity. A mutation is a heritable change in the DNA sequence that leads to a phenotypic effect, whereas DNA damage reflects genotoxicity only. DNA damage can be repaired, and even if this damage is not repaired, biological regulatory mechanisms will likely cause the cell to undergo apoptosis or programmed cell death.[58]

Mutations in certain genes are associated with cancer. A number of acquired characteristics or "hallmarks" constitute the phenotypic changes associated with metastatic transformation.[59] The presence of DNA adducts, as a potential source of mutations occurring via faulty DNA repair, is associated with cancer, but without identification of the source of these adducts or consideration of adducts within the overall cancer process, their use for HI cannot be supported.[60]

One may also measure biomarkers of effect. The most widespread use of this type of biomarker measurement is the level of liver enzymes. Pharmaceutical companies routinely use this sort of testing (see Box 2.2). However, there may be opportunities and motivation to discover chemical-specific biomarkers. For example, 1,3-butadiene used in rubber manufacture is metabolized to a highly reactive di-epoxide that reacts with both DNA and proteins; hemoglobin adducts have been shown to be a reliable biomarker of butadiene exposure.[61–64] Chromosomal aberrations and the occurrence of mutations in specific genes in lymphocytes have been used as biomarkers of effect for 1,3-butadiene.[57,62,65] (See Box 2.2.)

BOX 2.2   MEASURING BIOMARKERS OF EFFECT IN HUMANS

Cytochrome P450 (CYP) proteins comprise an enzyme superfamily present in virtually all organisms. The enzymes are also called the mixed-function oxidases (MFOs) and are located in the endoplasmic reticulum of hepatocytes (liver cells). CYPs biotransform many substances in food and drugs as part of phase I metabolism.[66] CYPs are classified according to their substrates—CYP2E1 metabolizes alcohol and trichloroethylene; CYP1A2 metabolizes caffeine and phenacetin.

(continued)

BOX 2.2 (continued)   MEASURING
BIOMARKERS OF EFFECT IN HUMANS

Induction of CYPs can occur via activation of nuclear receptors (NRs). NRs are a family of highly evolutionarily conserved transcription factors that bind a variety of ligands and then bind to DNA to alter gene expression. NRs regulate a constellation of biological processes, including development, hematopoiesis, and metabolism.[67] Classical NRs include the estrogen receptor, androgen receptor and thyroid hormone receptor. Other NRs such as the constitutive androstane receptor (CAR), the pregnane X receptor (PXR), and the aryl hydrocarbon receptor (AHR) bind ligands generally occurring in food or drugs and can alter metabolism by increasing CYP gene expression.[68]

A number of in vivo measurements of the activity of drug-metabolizing enzymes in humans are commonly used to assess liver function.[69] All these methods use a common drug that is metabolized by one or more of the CYPs. The most commonly used drug is caffeine, which is metabolized by CYP1A2, and metabolites of caffeine occurring in urine or blood may be used to assess CYP1A2 activity. The most accurate measure is obtained when using caffeine labeled with ^{13}C on the 3-position and then measuring $^{13}CO_2$ in exhaled breath.

Smoking, consumption of meat, consumption of brassica vegetables (broccoli), and exposure to dioxin-like chemicals all induce CYP1A2 and result in an increase in caffeine metabolism that can be measured in vivo.[70–73]

The occurrence of mutations may also serve as a biomarker of effect. The activity of X-linked gene for hypoxanthine-guanine phosphoribosyl-transferase (HPRT) can be used to select thioguanine-resistant mutants in peripheral human lymphocytes, and mutant clones can be expanded with the use of a mitogen.[74] The PIG-A gene codes an enzyme subunit involved in the biosynthesis of glycosylphosphatidylinositol. This molecule serves to tether a number of proteins to the surface of the red blood cell. Hence, erythrocytes deficient in PIG-A can easily be measured with flow cytometry.[75]

However, even biomarker-based studies may be difficult to interpret, and care in interpretation is warranted. For a long time, measurements of many persistent organic pollutants in human serum have been normalized to serum lipid concentrations. Hence, these lipid-adjusted measurements would be reported on as "pg/g lipid" or similar. Blood lipid concentrations may be measured gravimetrically or estimated with various formulae with considerable variation between these methods.[76,77]

Recently, the association of type 2 diabetes with a biomarker of exposure, dioxin in serum lipid, was investigated and shown to be the result of so-called reverse causation. For a number of years and in a number of studies, a positive association has been observed between lipid-adjusted measurements of serum dioxin and the occurrence of type 2 diabetes. This association was notable in the Ranch Hand cohort—herbicide workers in the US military during the Vietnam War; however, it was not observed in mortality studies of populations with occupational or environmental exposure. Recently, this association was examined among members of the Ranch Hand cohort, and the increases in serum dioxin concentrations were shown to be associated as well with hyperlipidemia, obesity, and poor diabetes control. In other words, the increases in serum lipids associated with diabetes also increased the levels of the lipid-soluble dioxins in serum rather dioxin-producing diabetes. The only reason this demonstration of reverse causation was possible was the existence of 20 years of longitudinal data on the Ranch Hand cohort.[78]

HISTORICAL RECONSTRUCTION OF EXPOSURE

A number of methods are available to epidemiologists for exposure reconstruction. These include the following:

- Self-reported occupational histories and exposures
- Occupational histories based on employer records
- Job-exposure and task-exposure matrices
- Expert assessment and the use of occupational/industrial hygiene measures

These various types of measurements have strengths and weaknesses and should be assessed on a case-by-case basis. For example, an individual may indicate that he/she was employed as a pipe fitter for 10 years and then may also note that after 6 years, the company used a different plastic solvent for pipes. This sort of information may not be included in the employer records.[79]

A job-exposure matrix (JEM) is often used as measure of exposure.[80] Exposure misclassification is a significant problem with JEMs due to variability of exposure within job titles. If exposure differences can be determined for specific tasks, a task-exposure matrix may be used instead.[81]

Expert assessment of occupational exposures is very similar to the exposure assessment in environmental risk assessment in that measurements of concentrations in abiotic media such dust, soil, or air are combined with estimates of the worker CR with these media. There may be between-worker and within-worker variability in exposure, and these variations may differ depending on the exposure medium.[82]

The goal of epidemiologic exposure reconstruction is to assign group average exposure levels that can be used to compute individual cumulative exposures. A sufficient quantity of individual exposure data is required. Often, these data are lacking or sparse. Ways around this data gap are as follows:

- Project results back in time from current exposure data.
- Recreate the actual physical processes that were assumed to result in exposure and take measurements.
- Use worker recall of exposures.
- Mathematical or statistical modeling for process reconstruction and exposure analysis.

False Positives in Epidemiologic Studies

A number of epidemiologists lament the search for and reporting of weak associations in observational epidemiology, and a major problem with the use of epidemiologic studies for HI and other aspects of risk assessment as well is the frequent occurrence of false-positive results.

For example, in 1993, a case-control study in New York city reported a link between breast cancer risk and serum levels of DDE, the major metabolite of the insecticide, DDT.[83] However, seven studies that followed and a cumulative meta-analysis failed to confirm the original findings.[84]

A larger analysis of studies reporting a positive association found that the main determinants of false-positive findings were the absence of a specific a priori hypothesis, small magnitude of association, failure to adjust for smoking, and the absence of a dose–response relationship.[85]

Another example is the relationship of pancreatic cancer to coffee consumption. In 1981, a small but significant positive association was noted.[86] By the end of the 1980s, subsequent studies failed to confirm this association, and in 2011, a pooled analysis that included over 850,000 individuals failed to show an association.[87] In 2012, a study of more than 33,000 men and more than 18,000 women demonstrated that when adjusted for smoking, inverse associations were observed for deaths due to heart disease, respiratory disease, stroke, injuries and accidents, diabetes, and infections, but not for deaths due to cancer. The association between coffee consumption and reduced total mortality showed a significant trend (about which many coffee drinkers were no doubt happy). Nonetheless, these authors were careful to state that the constellation of evidence was insufficient to view the association as causal.[88]

Use of Quantiles Reduces Statistical Power

It is tempting in many epidemiologic studies that employ continuous variables to split the entire cohort into groups based on quantiles of the continuous variable. Most often, tertiles, quartiles, or quintiles are used. Quantiles can be thought of as potentially low-, medium-, and high-risk groups and thus are intuitively appealing. In addition, quantiles afford the possibility of using generalized linear models to search for trends in the data.

There are a number of problems associated with the use of quantiles. Obviously, risk will vary between individuals within a given quantile to an unknown degree. On top of this is the ever-present potential for exposure misclassification. What happens in many cases is that the numbers of individuals in each quantile become so low that the study loses sufficient statistical power to justify its conclusions. The last problem with quantiles is that the cut points selected are most often based on the continuous variable and chosen for statistical convenience as opposed to biological relevance. Implicitly, individuals within a single quantile are assumed to be homogenous, and the choice of cut points has the potential to produce both false positives and false negatives.[89]

In Chapter 1, a discussion of the Hill considerations for causation was provided. This same sort of thinking is needed for examination of epidemiologic

**BOX 2.3   MULTIPLE COMPARISONS AND
THE FALSE-POSITIVE ERROR RATE**

When multiple comparisons are made between several treatment groups and a control group, the multiple comparisons can greatly elevate the chance of a type 1 error, that is, concluding that a difference exists when in fact it does not, that is, incorrectly rejecting the null hypothesis of no difference.

If the p-value of the overall error rate, known as a familywise type 1 error rate, is not adjusted, this p-value will be too high with increased likelihood of a type 1 error. The familywise type 1 error rate is calculated as follows:

$$\alpha_{FWE} = 1 - (1 - \alpha)^k \qquad (2.8)$$

where
α_{FWE} is the familywise error probability
α is the prespecified significance level for the individual comparisons
k is the number of comparisons

For three comparisons, such as would occur with three dose groups/quantiles and a control group, the commonly selected value $\alpha = 0.05$, the familywise error rate would be 0.1426. Thus, in order not to commit a type 1 error, the prespecified value of α for individual comparisons must be lowered.

There is a consensus that adjustment for multiple comparisons is a necessary procedure because of the likelihood of an inflated type 1 error rate without adjustment.[91–93]

There are a number of correction methods for adjusting the type 1 error rate. The best known of these is the Bonferroni method. The Bonferroni correction partitions the nominal level of significance in the equal components, α_B. For the case of three comparisons, α_B would be 0.05/3 or 0.0167 and α_{FWE} (Equation 2.8) would be 0.04917, very close to 0.05.

studies that use quantiles. Often, each quantile is separately tested against the control group. Such multiple comparisons increase the chance of a false positive. Box 2.3 provides a discussion of type 1 or false-positive error rate and appropriate p-value corrections for multiple comparisons.

EXAMPLE OF HAZARD IDENTIFICATION FROM EPIDEMIOLOGY

Indoor dampness and exposure to mold has been known or suspected of causing health problems since the beginning of humankind. Chapter 13 in the *Book of Leviticus* in the Bible provides instructions on detecting and remediating dampness, mold, and mildew within dwellings—even back then, mold was considered the cause of a number of diseases, including leprosy.

More recently, in 2004, the Institute of Medicine (IOM) of the National Academy of Science reviewed the evidence on health effects related to dampness and, in contrast to the Bible, found no causal associations. From the 45 studies reviewed, the IOM did find sufficient evidence of association between both indoor dampness and the presence of mold and coughing, wheezing, upper respiratory tract symptoms, and exacerbation of asthma symptoms.

An additional 354 studies published before 2009 and not included in the IOM review were the subject of a review in 2011. Evidence from epidemiologic studies and meta-analyses showed indoor dampness or mold to be associated consistently with increased asthma development and exacerbation, diagnosis of asthma, dyspnea, wheezing, cough, respiratory infections, bronchitis, allergic rhinitis, eczema, and upper respiratory tract symptoms. Associations were found in both allergic and nonallergic individuals. Evidence strongly suggested causation of asthma exacerbation in children. The evidence for causation was both the strong and consistent association of dampness and mold with worse asthmas symptoms and the complementary finding that remediation of mold and dampness led to dramatic reductions in asthma exacerbation.[90] The removal of a specific factor associated with a reduction in the effect or response provides a powerful counterfactual demonstration of the likelihood that the factor is causal to the effect.

The evidence for causation that could not be observed was any trend between quantitative measures of either dampness or the presence of mold—hence, there was no biological gradient or dose–response. No doubt with an eye on problem formulation, the authors conclude, appropriately or not, with consideration of risk management, pointing out that targeting of resources toward reduction of indoor dampness would likely be more effective for lowering the global burden of respiratory disease than would research to determine safe levels of dampness.[90]

What is instructive and possibly unique about this study is that the sole conclusion that could be drawn did not go beyond HI. Nonetheless, in the case of indoor dampness, nothing more than HI was needed to inform risk management. It seems this study confirmed what was written in the Bible and that Leviticus indeed had the first and last word on the assessment of risk from indoor dampness.

ANIMAL BIOASSAYS AS THE BASIS FOR HAZARD IDENTIFICATION

In 1775, Percivall Pott, a surgeon at St. Bartholomew's Hospital in London, recognized the relationship between cancer and occupation, observing that scrotal cancer was highly prevalent in chimney sweeps. In 1915, malignant tumors were produced by applying coal tar to the ears of rabbits. The same set of chemicals, polycyclic aromatic hydrocarbons (PAHs), are present in both soot and coal tar and were thus shown to be the potential causative agent.[94]

Often the results from animal bioassays confirm prior epidemiologic observations; however, predictions of human carcinogenicity from animal results are not as reliable.[95] Notwithstanding, animal bioassays have been used for both HI and human dose–response assessment for over 40 years. In the United States, the EPA, the FDA, the NCI, and the National Toxicology Program (NTP) have supported the use of animal bioassays. A 2-year bioassay in rats requires a minimum of 50 animals per dose group—more, if interim sacrifices are required—and may cost about $3M. Fifty animals per dose group per gender are needed to provide sufficient statistical power to observe a 10% risk of cancer.[96]

In addition to the uncertainty of qualitative species extrapolation—simply assuming that animal carcinogens are also human carcinogens—additional uncertainty is inherent when the dose at the 10% POD is extrapolated down to a one-in-a-million risk for compliance with target regulatory risk levels.

When the predictive ability of rat bioassays to predict mouse carcinogens and vice versa was examined, there was a sufficient lack of overlap to raise serious concerns about the predictive ability of a single rodent bioassay to predict human carcinogenicity.[97] Out of 392 tested chemicals, 76% of rat carcinogens also caused cancer in mice and 70% of mouse carcinogens were positive in rat bioassays.[98]

The IARC was a pioneer of HI of human carcinogens and pointed out that both long-term animal bioassays and short-term in vitro tests could be used to identify possible human carcinogens. As already noted, IARC developed one of the first WOE schemes with the goal of standardizing the evaluation of carcinogenic activity from human and animal studies. Sufficient evidence of carcinogenicity is provided by positive results in two species or in two or more independent studies in the same test species. Limited evidence of carcinogenicity is used when positive results are observed in a single bioassay only.[99,100]

Later studies indicated that any observed correlations tended to be highly dependent on study design. In fact, a mismatch was observed between chemicals predicted to be carcinogenic by IARC and those predicted by the NTP.[101]

The demonstration that a chemical is carcinogenic in both humans and animals with concordance of tumor site and tumor type constitutes a conclusive identification of hazard. Vinyl chloride used to manufacture PVC pipes and siding for houses causes liver hemangiosarcoma in both humans and rodents.[102] However, such evidence is hardly the norm. By far, animal experimentation without additional evidence has identified more chemicals as having the potential to produce adverse effects in humans than has any other source of information. Obviously, this source of uncertainty consumes both time and resources.

IN VITRO TESTING AS THE BASIS FOR HAZARD IDENTIFICATION

The Blue Book stated explicitly that laboratory animals were not human beings and that this obvious fact was a clear disadvantage of animal studies. Another disadvantage was the relatively high cost of animal studies containing enough animals to detect an effect of interest. In addition, extrapolation of effects observed in animal studies requires both interspecies extrapolation and extrapolation from high bioassay test doses to lower environmental doses. Even in the early 1990s, the scientific community was well aware of the uncertainties inherent in animal testing.[31,103]

The vision and framework articulated in NRC's 2007 report, *Toxicity testing in the 21st century: A vision and a strategy,* is that toxic effects result from the departure from homeostasis due to dysregulation of multiple biological pathways or systems.[104] This vision stems from recent advances in toxicogenomics, bioinformatics, systems biology, and computational toxicology that show considerable promise in changing the manner in which toxicity evaluations are performed. The aspiration in this "vision" is to transform hazard evaluation and risk assessment from a system that uses high-dose whole animal bioassays to one based primarily on computational profiling linked to in vitro methods that evaluate changes in biological processes using cells, cell lines, or cellular components, preferably of human origin. The "strategy" refers to the path forward for increasing use of in vitro high-throughput assays in lieu of animal bioassays for risk assessment and, when animal testing is necessary, to conduct only those tests that are absolutely needed.

There are three major motives for this effort. The first is the growing recognition that high-dose animal experiments are unlikely to provide useful information about low-dose effects in humans. The second is the consideration of animal welfare and the desire to reduce the use of animals in regulatory toxicity testing. The third is that conducting traditional animal toxicity testing for the 80,000 or so untested chemicals in commerce would be prohibitively expensive and would take longer, by decades, than the regulatory need for this information.

Toxicity Pathways, Adverse Outcome Pathways, and MOA

Toxicity pathways (TPs) are cellular response pathways that, when sufficiently perturbed, are expected to result in adverse health effects. Low exposures could cause small perturbations that do not lead to any alterations in normal biological functions. Higher exposures could lead to adaptive responses as the organism responds to maintain homeostasis. These adaptive responses do not necessarily compromise cellular or organ functions. When exposures are sufficiently large, then the magnitude of the perturbations in the response pathway would cause significant cell injuries and adverse effects.

The scope of TPs helps define the broader construct of adverse outcome pathways (AOPs) as representing "existing knowledge concerning the linkage between the MIE and an adverse outcome at the individual or population level." By definition, an AOP spans multiple levels of biological organization. The MIE is defined as the initial point of chemical–biological interaction within an organism, and an AOP links an MIE to an adverse outcome.[105]

Hence, the difference between a TP and an AOP is that TPs are identical to cellular response pathways and the level or type of activity in these pathways is what constitutes toxicity. On the other hand, an AOP represents a plausible set of connections leading from the MIE to the adverse effect, and thus, activity in the AOP is obligatorily connected to an adverse outcome. The connection between the initial event and the adverse outcome may not be obligatory, and the suggestion has been made to change the term MIE to initial molecular event (IME).[106]

IMPLEMENTING TOXICITY TESTING IN THE TWENTY-FIRST CENTURY (TT21C)

The ToxCast™ program of the EPA is an effort to incorporate the use of in vitro assays into risk assessment. ToxCast consists of a battery of both commercial and publically developed in vitro assays. One distinct and obvious advantage is that these assays can be robotically automated to generate data very quickly. For example, the robotic assay at the National Chemical Genomics Center can develop concentration–response information on about 10,000 chemicals in a single day.

Many of the assays used in ToxCast and other Tox21 efforts have been taken directly from the pharmaceutical field. For pharmaceutical purposes, the screening of chemicals with potent biological activities is a routine aspect of the drug candidate development process. In contrast, commodity chemicals are selected and designed because of their physicochemical properties with the goal of improving the performance of a specific product, and such chemicals typically possess orders of magnitude less biological activity than pharmaceutical agents.

This difference between commodity chemicals and drug candidates begs the question of whether the new in vitro testing paradigm is capable of delivering meaningful information to inform risk-based decision-making. For these assays to be useful, it is necessary to anchor each assay in terms of its biological context within a TP as well as what the in vitro results mean in terms of real-world exposures.

In 2003, the European 7th Amendment to the Cosmetics Directive banned the use of animal testing for acute toxicity. This ban was followed in 2009 by another on in vivo genotoxicity testing in animals.[107] On March 11, 2013, the European Commission banned animal testing for cosmetics within the EU territory.[108] This ban has significant ramifications that continue ripple throughout the European chemical industry.

In response, risk assessment approaches are being developed that use data from in vitro testing and prediction models with the goal of interpreting the predictions in terms of realistic exposure levels obtained from biomonitoring.[109] Three key components of this approach are envisioned—the development of exposure-driven risk assessment approaches, new targeted in vitro methods and predictive models, and understanding the applicability of new technologies to risk assessment.[110]

CAN IN VITRO ASSAYS COVER ALL TOXICITY PATHWAYS?

This remains to be seen. Whether the suite of ToxCast assays or those being developed in Europe cover the entire range of TPs in humans is not known. How many TPs are there? How many assays are needed for adequate coverage of each pathway?

When asked about the number of TPs, Dr. Melvin Andersen of the Hamner Institute answers, tongue in cheek, "132" and then qualifies his answer by adding, "as a toxicologist, I am used to working with false accuracy." Thomas Hartung, director of the CAAT at Johns Hopkins University, opines "as the number of cellular targets and metabolic pathways is finite, the number of PoT (Toxicity Pathways) should be, too. Evolution cannot have left too many Achilles heels given the number of chemicals surrounding us and the astonishingly large number of healthy years we enjoy on average."[111] Although there are evolutionary and energetic constraints on the complexity of human biology,[112] the question of coverage of the entire domain of TPs remains unknown, and the question of whether a sufficient number of pathways are represented remains unanswered.

Having a taxonomy of adverse effects is one means of addressing the question of the domain of applicability of the various assays and what constitutes reasonable and relevant integration and use of the assay results.[113] In response, the American Society for Cellular and Computational Toxicology (ASCCT) and the International QSAR Foundation have developed an "Effectopedia" to attempt to catalog the current list of AOPs (http://sourceforge.net/projects/effectopedia/).

In Vitro Assays May Be Useful for Hazard Identification Only

In order to interpret the results of in vitro assays, a prediction model is needed to determine how the concentrations used in the assay correspond to human exposures. Hence, confidence in the accuracy of this prediction model is necessary to be able to use in vitro assays in risk assessment. Recently, the difficulties in the interpretation of in vitro toxicity evaluations demonstrated in comprehensive analysis of the predictive performance of more than 600 in vitro assays across 60 in vivo endpoints using 84 different statistical classification methods.[54] In addition, the predictive power of these models was compared with that of QSAR and other in silico models (see succeeding text). The predictive power of the in vitro assays was not significantly different than that of in silico methods. Hence, the assays are currently seen as a survey of MIEs, and the responses of some assays or combinations of assays appear to be positive or negative "risk" factors for toxicity. From this study, it is clear that it may be possible to use the in vitro testing for HI—but not yet.

IN SILICO PREDICTION MODELS AS THE SOLE BASIS FOR HAZARD IDENTIFICATION

A structure–activity relationship relates features of chemical structure to a property, effect, or biological activity associated with that chemical. This can be done in either a qualitative or quantitative fashion. The underlying idea is that the structure of a chemical determines its physical and chemical properties and reactivities, which, in turn, specify biological/toxicological properties.

In 1869, Alexander Crum Brown, a graduate of the University of Edinburgh Medical School and the University of London, was the first to explore structure–activity relationships with his demonstration that discovered the first structure–activity link by showing that a number of alkaloids, including the convulsant strychnine, could be converted into muscle relaxants by converting them to quaternary amines.[114] The reason was finally understood more than 50 years later with the work of Henry Dale and Otto Loewi who identified acetylcholine, another quaternary amine, as the transmitter substance at a several different types of synapses including the neuromuscular junction.[115] The work of Arthur Cushny early in the twentieth century demonstrated that usually only a single member of a pair of optical enantiomers possessed biological activity.[116] Around 1900, Meyer and Overton independently advanced the theory that anesthetic potency was related to lipophilicity and developed a quantitative relationship between the solubility of a drug in olive oil versus that in water.[117] All these early discoveries led to the recognition that the biological activity of a chemical could be predicted from its chemical structure and associated physical–chemical properties.

QUANTITATIVE STRUCTURE–ACTIVITY RELATIONSHIPS

Some of the chemical features used in QSAR include the octanol–water partition coefficient as a measure of lipophilicity, molecular weight or molar refractivity as a measure of size, and polarity as a measure of reactivity. Initially, QSAR methods were used for prediction of aquatic toxicity in ecological risk assessment.[118] QSAR is used for human health risk assessment as well and is often combined with PBTK modeling. Hence, QSAR is used to estimate the toxicodynamic properties of a chemical, and PBTK is used to estimate the toxicokinetic properties.[119] (See Box 2.4.)

Of necessity, QSAR requires a prediction model for the biological activity. Often, this is a statistical regression of the predicted value versus the predictor value.[120] One of the early uses of QSAR was EPA's attempt to predict dermal permeability of chemicals from water. Measurements of the dermal permeability coefficient K_p were available for 90 chemicals and EPA used a regression model

---

BOX 2.4   TOXICOKINETICS AND TOXICODYNAMICS

Toxicokinetics refers to the distribution of chemicals in the body. Processes considered in toxicokinetics are absorption, distribution, metabolism and excretion (ADME). Often, quantitative descriptions of these processes are incorporated into a mathematical model called a physiologically-based toxicokinetic (PBTK) model. These are also known as physiologically-based pharmacokinetic (PBPK) models.

Toxicodynamics refers to the processes by which chemicals produce effects in the body. For example, in rodents, phenobarbital binds to CAR and alters gene expression. This binding leads to induction of enzymes, increased cell proliferation, and liver tumors. In humans, although phenobarbital binds to CAR, the hyperplastic effects are absent. Mathematical models of toxicodynamic effects may also be developed.

When a PBTK model and a toxicodynamic model are combined, the resulting model is known as a biologically based dose–response (BBDR) model.

In Chapter 4 on dose–response assessment, it will become abundantly clear just how consideration of both toxicokinetics and toxicodynamics plays an important role in risk assessment.

to estimate K_p for other chemicals. The independent variables were the octanol–water partition coefficient and the molecular weight. Unfortunately, this regression method did not work for high molecular weight highly lipophilic chemicals, and an effective prediction domain was established inside which the regression was applicable.[121]

CONCLUSIONS

Because of the expected ban on animal testing in Europe, much attention is being paid to the development and validation of in vitro and in silico models for predicting toxicity. At present, there remain thousands of untested chemicals used in commerce. EPA's ToxCast effort to address this large number of chemicals was developed around assays adapted from the pharmaceutical industry; similar efforts are ongoing in Europe.[122] However, confidence around the predictions from these results has not yet been achieved.[54,123] In addition, much human biomonitoring data have become the focus of attention in both Europe and North America. Understanding these data with an appropriate screening tool could provide another method of HI.[120,124,125]

Often, better results are obtained when in vitro data are combined with other information. Read-across is a technique that uses toxicity data from chemicals with structures similar to the one under consideration.[126] These structurally similar chemicals are then used to restrict the domain of application of QSAR results.[127] An associated area of interest is the grouping of chemicals and the identification of analogs. Analog identification may be based on presence within a congeneric series, similar functional groups, overall chemical similarity, mechanism of action, or 3D structure.[128–132]

Frameworks are being developed for integrating information from a variety of sources into HI and screening.[133] Hybrid modeling in which biological results and chemical structural properties are pooled has begun to show improvement in the accuracy of prediction over in silico models alone.[134–136]

The second decade of the twenty-first century is truly an exciting time to be studying environmental risk assessment. With the recognition that techniques and methods used routinely in the pharmaceutical industry could revolutionize toxicology came the challenge of implementing these techniques. Confidence in the predictive power of these methods has yet to be achieved. Hence, the extent to which these nontesting methods are considered in planning and scoping, in problem formulation, and in HI has yet to be determined.

EXERCISES FOR THOUGHT AND DISCUSSION

UNDERSTANDING STATISTICAL POWER

With a small or highly variable effect, a relatively large sample size is needed to make inferences with confidence, especially when effects are of small magnitude or highly variable. Statistical power is the likelihood that a study is able to detect the presence of an effect when the effect is indeed present. Statistical power is affected by the size of the effect and the sample size of the study.

Considerable variation in semen quality parameters exists in humans. This variation is due to both between- and within-person variability. The 2010 World Health Organization (WHO) *Laboratory Manual for the Examination and Processing of Human Semen*, Fifth Edition, flatly states that it is impossible to characterize semen quality from evaluation of a single sample.[137] Length of abstinence is a major determinant of this variation.[138–142] A commonly measured semen quality parameter is sperm concentration, the number of spermatozoa, usually in millions, in a milliliter of semen. Sperm concentration is related to time to pregnancy and is a predictor of conception.[143] Sperm concentration is measured by counting individual sperm using a hemocytometer grid.[144]

Using a relatively simple power calculation, we will examine a recent report of the effect of chemical exposure on sperm concentration.[145] The paper is available without charge at http://ehp03.niehs.nih.gov/article/fetchArticle.action?articleUR I=info%3Adoi%2F10.1289%2Fehp.10399.

In this paper, two groups of young men, one group exposed to 2,3,7,8-TCDD as boys and the other not exposed, were examined for reproductive parameters, including sperm concentration. It is important to note that the WHO manual provides the normal range of sperm concentration as 15–213M per mL. The group of 71 exposed individuals had a geometric mean value of 53.6M and the control group of 82 individuals had a geometric mean of 72.5M. The geometric standard deviations (GSDs) were 2.46 and 2.29, respectively. Please note that both samples were in the normal range. Statistical analysis of semen concentrations is most often conducted in logarithmic space, which is why the GM and GSD values are provided.

You can download an Excel file called "power.xls" from the website at http://www.crcpress.com/product/isbn/9781466598294. Alternatively, those of you proficient in Excel can set this up on your own.

The spreadsheet works by having random numbers coded with the formula = RAND() in the hidden columns A and B. Column C has 71 values that represent the exposed group, and column D has 82 values that represent the control group.

The value you will be looking at is in cell G9, labeled "p-value." What you will do is obtain 10 different alternate realizations of these two datasets and perform a student's t-test on the differences. The proportion of the times the p-value is less than $\alpha = 0.05$ is a measure of the ability to demonstrate that these two samples are indeed different.

Enter a value in cell J1. This will be a dummy value just to get the random number generator in Excel to turn over. Note whether the p-value is less than 0.05. Repeat nine more times. If the p-value was less than 0.05 on 2/10 trials, the power of the statistical test is 20%. Generally, one wants a power of 80% to show that the number of samples was adequate.

Based on your simple investigation of power, discuss what this means for your confidence in the results presented in the paper.[145]

DISCUSSION OF THE DIFFERENCES IN PROBLEM FORMULATION BETWEEN NATO AND THE SILVER BOOK

Figures 2.2 and 2.3 show these two problem formulation diagrams. The NATO diagram was developed essentially by individuals trained in operations research, whereas the Silver Book diagram was developed by statisticians, toxicologists, and epidemiologists. How are these similar? How are they different?

EXPLORING QSAR: PART 1

The Flynn dataset of dermal permeability values can be downloaded as an Excel file from the CRC Press website (http://www.crcpress.com/product/isbn/9781466598294). Read Appendix A of EPA's Supplemental Guidance for Dermal Risk Assessment[121] and try to develop a predictive regression for the dermal permeability coefficient. You should be able to explore this method using the LINEST function in Excel.

EXPLORING QSAR: PART 2

There exist a number of public domain tools for in silico prediction of toxicity. The links are shown in the following text. Download as many of these as you care to and give them a spin. Try them with chemicals of known toxicity to see if they can predict effects observed in humans. To obtain data on human health effects, use EPA's Integrated Risk Information System (IRIS) database at http://www.epa.gov/iris/index.html.

The links are as follows:

- Oncologic (EPA)—http://www.epa.gov/oppt/sf/pubs/oncologic.htm
- OpenTox—http://www.opentox.org
- Organisation for Economic Co-operation and Development (OECD) QSAR Toolbox—http://www.qsartoolbox.org
- OECD QSAR Project—http://www.oecd.org/env/existingchemicals/qsar
- OCHEM—http://www.ochem.eu
- Chembench—http://chembench.mml.unc.edu
- Toxmatch—http://ihcp.jrc.ec.europa.eu/our_labs/computational_toxicology/qsar_tools/toxmatch
- AIM (EPA)—http://www.epa.gov/oppt/sf/tools/aim.htm

REFERENCES

1. National Weather Service. Lightning Safety. n.d. http://www.lightningsafety.noaa.gov/medical.htm (accessed July 21, 2012).
2. Meteorological Service Singapore n.d. *Educational Tour*: Lightning. http://www.weather.gov.sg/wip/web/home/lightning (accessed October 20, 2013).
3. Visit Orlando: The official source for visit Orlando information and research. n.d. http://corporate.visitorlando.com/research-and-statistics/orlando-visitor-statistics/ (accessed July 21, 2012).

4. Anonymous. Sea World Lightning Strike Injures 8 People at Water Park in Orlando, Florida. *The Huffington Post.* 2008. http://www.huffingtonpost.com/2011/08/16/seaworld-lightning-strike-orlando-florida_n_928891.html (accessed July 21, 2012).

5. United States Environmental Protection Agency (USEPA). Risk Assessment Forum. *Framework for Ecological Risk Assessment.* EPA/630/R-92/001. Washington, DC, February 1992. http://www.epa.gov/raf/publications/pdfs/FRMWRK_ERA.PDF

6. United States Environmental Protection Agency (USEPA). Science Policy Council. *Guidance on Cumulative Risk Assessment, Part 1 Planning and Scoping.* Washington, DC, July 3, 1997. http://www.epa.gov/spc/2cumrisk.htm

7. United States Environmental Protection Agency (USEPA). *Guidelines for Ecological Risk Assessment.* EPA/630/R-95/002F. Washington, DC, April, 1998. http://www.epa.gov/raf/publications/guidelines-ecological-risk-assessment.htm

8. United States Environmental Protection Agency (USEPA). Risk Assessment Forum. *Framework for Cumulative Risk Assessment.* EPA/630/P-02/001F. Washington, DC, May, 2003. http://www.epa.gov/raf/publications/framework-cra.htm

9. Council on Environmental Quality (CEQ). *A Citizen's Guide to NEPA: Having Your Voice Heard.* Washington, DC, December, 2007. http://energy.gov/nepa/downloads/citizens-guide-nepa-having-your-voice-heard

10. National Research Council (NRC). *Understanding Risk: Informing Decisions In A Democratic Society.* Washington, DC: The National Academies Press, 1996. http://www.nap.edu/catalog.php?record_id=5138

11. The Presidential/Congressional Commission on Risk Assessment and Risk Management. *Framework for Environmental Health Risk Management,* Final Report, Volumes 1 and 2, 1997. http://www.riskworld.com/Nreports/1996/risk_rpt/Rr6me001.htm

12. United States Environmental Protection Agency (USEPA). *Risk Assessment Guidance for Superfund: Vol. 1: Human Health Evaluation Manual (Part A) (Interim Final) (RAGS).* EPA/540/1-89/002. Washington, DC, December 1989. http://www.epa.gov/oswer/riskassessment/ragsa/

13. United States Environmental Protection Agency (USEPA). *Guidelines For Exposure Assessment.* EPA/600/Z-92/001. Washington, DC, May 29, 1992. http://www.epa.gov/raf/publications/guidelines-for-exposure-assessment.htm

14. O'Brien PE. Bariatric surgery: Mechanisms, indications and outcomes. *J Gastroenterol Hepatol.* 2010, August;25(8):1358–1365.

15. Spinella PC, Dunne J, Beilman GJ, O'Connell RJ, Borgman MA, Cap AP, and Rentas F. Constant challenges and evolution of US military transfusion medicine and blood operations in combat. *Transfusion.* 2012, May;52(5):1146–1153.

16. Basáñez M-G, McCarthy JS, French MD, Yang G-J, Walker M, Gambhir M et al. A research agenda for helminth diseases of humans: Modelling for control and elimination. *PLoS Negl Trop Dis.* 2012;6(4):e1548.

17. Buckman JEJ, Sundin J, Greene T, Fear NT, Dandeker C, Greenberg N, and Wessely S. The impact of deployment length on the health and well-being of military personnel: A systematic review of the literature. *Occup Environ Med.* 2011, January;68(1):69–76.

18. Hauschild VD, Watson A, and Bock R. Decontamination and management of human remains following incidents of hazardous chemical release. *Am J Disaster Med.* 2012;7(1):5–29.

19. North Atlantic Treaty Organization (NATO). *NATO Code Of Best Practice For Command and Control Assessment.* RTO-TR-08. SAS-026. ISBN 92-837-1116-5. January, 2004, Updated 2008. http://www.cso.nato.int/pubs/rdp.asp?RDP=RTO-TR-081

20. National Research Council (NRC). *Science and Decisions: Advancing Risk Assessment.* Washington, DC: The National Academies Press, 2009. http://www.nap.edu/catalog.php?record_id=12209

21. National Research Council (NRC). *Risk Assessment in the Federal Government: Managing the Process.* Washington, DC: The National Academies Press, 1983. http://www.nap.edu/catalog.php?record_id=366

22. Rodricks J and Levy J. Science and decisions: Advancing toxicology to advance risk assessment. *Toxicol Sci.* 2012, August;131:1–8.

23. Finkel AM. "Solution-focused risk assessment": A proposal for the fusion of environmental analysis and action. *Hum Ecol Risk Assess.* 2011, July;17(4):754–787.

24. Cote I, Anastas PT, Birnbaum LS, Clark RM, Dix DJ, Edwards SW, and Preuss PW. Advancing the next generation of health risk assessment. *Environ Health Perspect.* 2012, August 8;120:1499–1502.

25. Ishii H. Incidence of brain tumors in rats fed aspartame. *Toxicol Lett.* 1981, March;7(6):433–437.

26. National Toxicology Program (NTP). *NTP Report on Carcinogens: Background Document for Saccharin. Final.* Research Triangle Park, NC, March 1999. http://ntp.niehs.nih.gov/ntp/newhomeroc/other_background/Saccharin1_3Apps_508.pdf

27. Cohen SM. Cell proliferation and carcinogenesis. *Drug Metab Rev.* 1998, May;30(2):339–357.

28. Takayama S, Sieber SM, Adamson RH, Thorgeirsson UP, Dalgard DW, Arnold LL et al. Long-term feeding of sodium saccharin to nonhuman primates: Implications for urinary tract cancer. *J Natl Cancer Inst.* 1998, January 1;90(1):19–25.

29. Bernstein PL. *Against the Gods: The Remarkable Story of Risk.* New York: John Wiley & Sons, Inc., 1998.

30. U.S. Interagency Staff Group on Carcinogens. Chemical carcinogens: A review of the science and its associated principles. *Environ Health Perspect.* 1986, August;67:201–282.

31. National Research Council (NRC). *Science and Judgment in Risk Assessment.* Washington, DC: The National Academies Press, 1994. http://www.nap.edu/catalog.php?record_id=2125

32. International Agency for Research on Cancer (IARC). *IARC Monographs on the Evaluation of the Carcinogenic Risks to Humans. Cancer and Industrial Processes Associated with Cancer in Humans. IARC Monographs,* Vols. 1–20. IARC Monographs Supplement #1. ISBN 9789283214021.Lyon, France, September, 1979. http://apps.who.int/bookorders/anglais/detart1.jsp?sesslan=1&codlan=1&codcol=72&codcch=1001

33. International Agency for Research on Cancer (IARC). *IARC Monographs on the Evaluation of Carcinogenic Risks to Humans: Preamble.* Lyon, France, 2006. http://monographs.iarc.fr/ENG/Preamble/CurrentPreamble.pdf

34. United States Environmental Protection Agency (USEPA). *Guidelines for Carcinogen Risk Assessment.* EPA/630/R-00/004. Washington, DC, September 24, 1986. http://www.epa.gov/cancerguidelines/guidelines-carcinogen-risk-assessment-1986.htm

35. United States Environmental Protection Agency (USEPA). Risk Assessment Forum. *Guidelines for Carcinogen Risk Assessment.* EPA/630/P-03/001F. March, 2005. http://www.epa.gov/raf/publications/pdfs/CANCER_GUIDELINES_FINAL_3-25-05.PDF

36. Lange T, Vansteelandt S, and Bekaert M. A simple unified approach for estimating natural direct and indirect effects. *Am J Epidemiol.* 2012, August 8;176(3):190–195.

37. Kepecs A and Mainen ZF. A computational framework for the study of confidence in humans and animals. *Philos Trans R Soc Lond B Biol Sci.* 2012, May 5;367(1594):1322–1337.

38. Zadeh LA. From computing with numbers to computing with words. From manipulation of measurements to manipulation of perceptions. *Ann N Y Acad Sci.* 2001, April;929:221–252.

39. Morizot B. Chance: From metaphysical principle to explanatory concept. The idea of uncertainty in a natural history of knowledge. *Prog Biophys Mol Biol.* 2012, May 5;110:54–60.

40. Menzies P. The causal structure of mechanisms. *Stud Hist Philos Biol Biomed Sci.* 2012, June 6;43:796–805.

41. Weed DL. Weight of evidence: A review of concept and methods. *Risk Anal.* 2005, December;25(6):1545–1557.

42. Shah HM and Chung KC. Archie Cochrane and his vision for evidence-based medicine. *Plast Reconstr Surg.* 2009, September;124(3):982–988.

43. Griffin D and Tversky A. The weighing of evidence and the determinants of confidence. *Cogn Psychol.* 1992;24:411–435.

44. Tversky A and Kahneman D. Judgment under uncertainty: Heuristics and biases. *Science.* 1974, September 9;185(4157):1124–1131.

45. Prueitt RL, Goodman JE, Bailey LA, and Rhomberg LR. Hypothesis-based weight-of-evidence evaluation of the neurodevelopmental effects of chlorpyrifos. *Crit Rev Toxicol.* 2011, November;41(10):822–903.

46. Keller DA, Juberg DR, Catlin N, Farland WH, Hess FG, Wolf DC, and Doerrer NG. Identification and characterization of adverse effects in 21st century toxicology. *Toxicol Sci.* 2012, April;126(2):291–297.

47. Borgert CJ, Mihaich EM, Ortego LS, Bentley KS, Holmes CM, Levine SL, and Becker RA. Hypothesis-driven weight of evidence framework for evaluating data within the US EPA's Endocrine Disruptor Screening Program. *Regul Toxicol Pharmacol.* 2011, November;61(2):185–191.

48. Guzelian PS, Victoroff MS, Halmes NC, James RC, and Guzelian CP. Evidence-based toxicology: A comprehensive framework for causation. *Hum Exp Toxicol.* 2005, April;24(4):161–201.

49. Chalmers I. The Cochrane collaboration: Preparing, maintaining, and disseminating systematic reviews of the effects of health care. *Ann N Y Acad Sci.* 1993, December 12;703:156–163; discussion 163–165.

50. Hartung T. Evidence-based toxicology—The toolbox of validation for the 21st century? *ALTEX.* 2010;27(4):253–263.

51. Hartung T. Food for thought… on evidence-based toxicology. *ALTEX.* 2009;26(2):75–82.

52. Guzelian PS, Victoroff MS, Halmes C, and James RC. Clear path: Towards an evidence-based toxicology (EBT). *Hum Exp Toxicol.* 2009, February;28(2–3):71–79.

53. Stephens ML, Barrow C, Andersen ME, Boekelheide K, Carmichael PL, Holsapple MP, and Lafranconi M. Accelerating the development of 21st-century toxicology: Outcome of a human toxicology project consortium workshop. *Toxicol Sci.* 2012, February;125(2):327–334.

54. Thomas RS, Black M, Li L, Healy E, Chu T-M, Bao W et al. A comprehensive statistical analysis of predicting in vivo hazard using high-throughput in vitro screening. *Toxicol Sci.* 2012, May 5;128:398–417.

55. United States Centers for Disease Control (CDC). *National Health and Nutrition Examination Survey (NHANES).* 2008. http://www.cdc.gov/nchs/nhanes.htm

56. Paustenbach D and Galbraith D. Biomonitoring and biomarkers: Exposure assessment will never be the same. *Environ Health Perspect.* 2006, August; 114(8):1143–1149.

57. Swenberg JA, Bordeerat NK, Boysen G, Carro S, Georgieva NI, Nakamura J et al. 1,3-Butadiene: Biomarkers and application to risk assessment. *Chem Biol Interact.* 2011, June 6;192(1–2):150–154.

58. Sablina AA, Budanov AV, Ilyinskaya GV, Agapova LS, Kravchenko JE, and Chumakov PM. The antioxidant function of the p53 tumor suppressor. *Nat Med.* 2005, December;11(12):1306–1313.
59. Hanahan D and Weinberg RA. Hallmarks of cancer: The next generation. *Cell.* 2011, March 3;144(5):646–674.
60. Jarabek AM, Pottenger LH, Andrews LS, Casciano D, Embry MR, Kim JH et al. Creating context for the use of DNA adduct data in cancer risk assessment: I. Data organization. *Crit Rev Toxicol.* 2009;39(8):659–678.
61. Preston RJ. Cancer risk assessment for 1,3-butadiene: Data integration opportunities. *Chem Biol Interact.* 2007, March 3;166(1–3):150–155.
62. Albertini RJ, Sram RJ, Vacek PM, Lynch J, Wright M, Nicklas JA et al. Biomarkers for assessing occupational exposures to 1,3-butadiene. *Chem Biol Interact.* 2001, June 6;135–136:429–453.
63. Albertini RJ, Srám RJ, Vacek PM, Lynch J, Nicklas JA, van Sittert NJ et al. Biomarkers in Czech workers exposed to 1,3-butadiene: A transitional epidemiologic study. *Res Rep Health Eff Inst.* 2003, June;(116):1–141; discussion 143–162.
64. Boogaard PJ, van Sittert NJ, and Megens HJ. Urinary metabolites and haemoglobin adducts as biomarkers of exposure to 1,3-butadiene: A basis for 1,3-butadiene cancer risk assessment. *Chem Biol Interact.* 2001, June 6;135–136:695–701.
65. Albertini RJ, Sram RJ, Vacek PM, Lynch J, Rossner P, Nicklas JA et al. Molecular epidemiological studies in 1,3-butadiene exposed Czech workers: Female-male comparisons. *Chem Biol Interact.* 2007, March 3;166(1–3):63–77.
66. Werck-Reichhart D and Feyereisen R. Cytochromes P450: A success story. *Genome Biol.* 2000;1(6):REVIEWS3003.
67. Chute JP, Ross JR, and McDonnell DP. Minireview: Nuclear receptors, hematopoiesis, and stem cells. *Mol Endocrinol.* 2010, January;24(1):1–10.
68. Sladek FM. What are nuclear receptor ligands? *Mol Cell Endocrinol.* 2011, March 3;334(1–2):3–13.
69. Streetman DS, Bertino JSJ, and Nafziger AN. Phenotyping of drug-metabolizing enzymes in adults: A review of in-vivo cytochrome P450 phenotyping probes. *Pharmacogenetics.* 2000, April;10(3):187–216.
70. Sinha R, Rothman N, Brown ED, Mark SD, Hoover RN, Caporaso NE et al. Pan-fried meat containing high levels of heterocyclic aromatic amines but low levels of polycyclic aromatic hydrocarbons induces cytochrome P4501A2 activity in humans. *Cancer Res.* 1994, December 12;54(23):6154–6159.
71. Abraham K, Geusau A, Tosun Y, Helge H, Bauer S, and Brockmoller J. Severe 2,3,7,8-tetrachlorodibenzo-p-dioxin (TCDD) intoxication: Insights into the measurement of hepatic cytochrome P450 1A2 induction. *Clin Pharmacol Ther.* 2002;72(2):163–174.
72. Kotake AN, Schoeller DA, Lambert GH, Baker AL, Schaffer DD, and Josephs H. The caffeine CO2 breath test: Dose response and route of N-demethylation in smokers and nonsmokers. *Clin Pharmacol Ther.* 1982, August;32(2):261–269.
73. Lampe JW, King IB, Li S, Grate MT, Barale KV, Chen C et al. Brassica vegetables increase and apiaceous vegetables decrease cytochrome P450 1A2 activity in humans: Changes in caffeine metabolite ratios in response to controlled vegetable diets. *Carcinogenesis.* 2000, June;21(6):1157–1162.
74. Noori P, Hou S, Jones IM, Thomas CB, and Lambert B. A comparison of somatic mutational spectra in healthy study populations from Russia, Sweden and USA. *Carcinogenesis.* 2005, June;26(6):1138–1151.
75. Peruzzi B, Araten DJ, Notaro R, and Luzzatto L. The use of PIG-A as a sentinel gene for the study of the somatic mutation rate and of mutagenic agents in vivo. *Mutat Res.* 2010;705(1):3–10.

76. Mills WJ, Nienow C, Sweetman GLM, Cox R, Tondeur Y, Webber JP, and Leblanc A. Lipids analysis is a significant, often unrecognized source of uncertainty in POPS results for human blood. *Organohalogen Compd.* 2007;69:1158–1161.

77. Bernert JT, Turner WE, Patterson DGJ, and Needham LL. Calculation of serum "total lipid" concentrations for the adjustment of persistent organohalogen toxicant measurements in human samples. *Chemosphere.* 2007;68(5):824–831.

78. Kerger BD, Scott PK, Pavuk M, Gough M, and Paustenbach DJ. Re-analysis of Ranch Hand study supports reverse causation hypothesis between dioxin and diabetes. *Crit Rev Toxicol.* 2012, June 6;42:669–687.

79. Teschke K, Olshan AF, Daniels JL, De Roos AJ, Parks CG, Schulz M, and Vaughan TL. Occupational exposure assessment in case-control studies: Opportunities for improvement. *Occup Environ Med.* 2002, September;59(9):575–593; discussion 594.

80. Mannetje A and Kromhout H. The use of occupation and industry classifications in general population studies. *Int J Epidemiol.* 2003, June;32(3):419–428.

81. Benke G, Sim M, Fritschi L, and Aldred G. Beyond the job exposure matrix (JEM): The task exposure matrix (TEM). *Ann Occup Hyg.* 2000, September; 44(6):475–482.

82. Symanski E, Maberti S, and Chan W. A meta-analytic approach for characterizing the within-worker and between-worker sources of variation in occupational exposure. *Ann Occup Hyg.* 2006, June;50(4):343–357.

83. Wolff MS, Toniolo PG, Lee EW, Rivera M, and Dubin N. Blood levels of organochlorine residues and risk of breast cancer. *J Natl Cancer Inst.* 1993, April 4;85(8):648–652.

84. Boffetta P, McLaughlin JK, La Vecchia C, Tarone RE, Lipworth L, and Blot WJ. False-positive results in cancer epidemiology: A plea for epistemological modesty. *J Natl Cancer Inst.* 2008, July 7;100(14):988–995.

85. Swaen GG, Teggeler O, and van Amelsvoort LG. False positive outcomes and design characteristics in occupational cancer epidemiology studies. *Int J Epidemiol.* 2001, October;30(5):948–954.

86. MacMahon B, Yen S, Trichopoulos D, Warren K, and Nardi G. Coffee and cancer of the pancreas. *N Engl J Med.* 1981, March 3;304(11):630–633.

87. Genkinger JM, Li R, Spiegelman D, Anderson KE, Albanes D, Bergkvist L et al. Coffee, tea, and sugar-sweetened carbonated soft drink intake and pancreatic cancer risk: A pooled analysis of 14 cohort studies. *Cancer Epidemiol Biomarkers Prev.* 2012, February;21(2):305–318.

88. Freedman ND, Park Y, Abnet CC, Hollenbeck AR, and Sinha R. Association of coffee drinking with total and cause-specific mortality. *N Engl J Med.* 2012, May 5;366(20):1891–1904.

89. Bennette C and Vickers A. Against quantiles: Categorization of continuous variables in epidemiologic research, and its discontents. *BMC Med Res Methodol.* 2012;12:21.

90. Mendell MJ, Mirer AG, Cheung K, Tong M, and Douwes J. Respiratory and allergic health effects of dampness, mold, and dampness-related agents: A review of the epidemiologic evidence. *Environ Health Perspect.* 2011, June;119(6):748–756.

91. Dmitrienko A, Offen W, Wang O, and Xiao D. Gatekeeping procedures in dose-response clinical trials based on the Dunnett test. *Pharm Stat.* 2006;5(1):19–28.

92. Leon AC, Heo M, Teres JJ, and Morikawa T. Statistical power of multiplicity adjustment strategies for correlated binary endpoints. *Stat Med.* 2007, April 4;26(8):1712–1723.

93. James S. Approximate multinormal probabilities applied to correlated multiple endpoints in clinical trials. *Stat Med.* 1991, July;10(7):1123–1135.

94. Pastoor T and Stevens J. Historical perspective of the cancer bioassay. *Scand J Work Environ Health.* 2005;31(Suppl 11):29–1140; discussion 119–122.

95. Wilbourn JD, Haroun L, Vainio H, and Montesano R. Identification of chemicals carcinogenic to man. *Toxicol Pathol.* 1984;12(4):397–399.
96. Haseman JK. Statistical issues in the design, analysis and interpretation of animal carcinogenicity studies. *Environ Health Perspect.* 1984, December;583:85–92.
97. Purchase IF. Inter-species comparisons of carcinogenicity. *Br J Cancer.* 1980, March;41(3):454–468.
98. Gold LS, Bernstein L, Magaw R, and Slone TH. Interspecies extrapolation in carcinogenesis: Prediction between rats and mice. *Environ Health Perspect.* 1989, May;812:11–19.
99. Wilbourn J, Haroun L, Heseltine E, Kaldor J, Partensky C, and Vainio H. Response of experimental animals to human carcinogens: An analysis based upon the IARC Monographs programme. *Carcinogenesis.* 1986, November;7(11):1853–1863.
100. Vainio H and Wilbourn J. Identification of carcinogens within the IARC monograph program. *Scand J Work Environ Health.* 1992;18(Suppl):164–173.
101. Fung VA, Barrett JC, and Huff J. The carcinogenesis bioassay in perspective: Application in identifying human cancer hazards. *Environ Health Perspect.* 1995;103(7–8):680–683.
102. United States Environmental Protection Agency (USEPA). *Toxicological Review of Vinyl Chloride (CAS No. 75–01–4) In Support of Summary Information on the Integrated Risk Information System.* EPA/635R-00/004. Arlington, VA, May, 2000. http://www.epa.gov/iris/toxreviews/1001tr.pdf
103. National Research Council (NRC). *Issues In Risk Assessment.* Washington, DC: The National Academies Press, 1993. http://www.nap.edu/catalog.php?record_id=2078
104. National Research Council (NRC). *Toxicity Testing in the 21st Century: A Vision and A Strategy.* Washington, DC: The National Academies Press, 2007. http://www.nap.edu/catalog.php?record_id=11970
105. Ankley GT, Bennett RS, Erickson RJ, Hoff DJ, Hornung MW, Johnson RD et al. Adverse outcome pathways: A conceptual framework to support ecotoxicology research and risk assessment. *Environ Toxicol Chem.* 2010, March;29(3):730–741.
106. Patlewicz G, Simon T, Goyak K, Phillips RD, Craig Rowlands J, Seidel S, and Becker RA. Use and validation of HT/HC assays to support 21(st) century toxicity evaluations. *Regul Toxicol Pharmacol.* 2013, January 3;65:259–268.
107. Pauwels M and Rogiers V. Human health safety evaluation of cosmetics in the EU: A legally imposed challenge to science. *Toxicol Appl Pharmacol.* 2010, March 3;243(2):260–274.
108. Bhattacharya S, Zhang Q, Carmichael PL, Boekelheide K, and Andersen ME. Toxicity testing in the 21st century: Defining new risk assessment approaches based on perturbation of intracellular toxicity pathways. *PLoS One.* 2011;6(6):e20887.
109. Fentem J, Chamberlain M, and Sangster B. The feasibility of replacing animal testing for assessing consumer safety: A suggested future direction. *Altern Lab Anim.* 2004, December;32(6):617–623.
110. Westmoreland C, Carmichael P, Dent M, Fentem J, MacKay C, Maxwell G et al. Assuring safety without animal testing: Unilever's ongoing research programme to deliver novel ways to assure consumer safety. *ALTEX.* 2010;27(3):61–65.
111. Hartung T and McBride M. Food for thought… on mapping the human toxome. *ALTEX.* 2011;28(2):83–93.
112. Mayr E. *The Growth of Biological Thought: Diversity, Evolution and Inheritance.* Cambridge, MA: Harvard University Press, 1982.
113. Boekelheide K and Campion SN. Toxicity testing in the 21st century: Using the new toxicity testing paradigm to create a taxonomy of adverse effects. *Toxicol Sci.* 2010, March;114(1):20–24.

114. Crum Brown A and Fraser TR. On the connection between chemical constitution and physiological action. I. On the physiological action of the salts of the ammonium bases, derived from strychnia, brucia, thebaica, codeia, morphia, and nicotia. *Trans R Soc Edinburgh.* 1869;25:151–203.

115. Dale H. Pharmacology and nerve-endings (Walter Ernest Dixon Memorial Lecture: Section of Therapeutics and Pharmacology). *Proc R Soc Med.* 1935, January;28(3):319–332.

116. Cushny AR. The action of optical isomers: III. Adrenalin. *J Physiol.* 1908, June 6;37(2):130–138.

117. Kaufman RD. Biophysical mechanisms of anesthetic action: Historical perspective and review of current concepts. *Anesthesiology.* 1977, January;46(1):49–62.

118. Bradbury SP. Quantitative structure-activity relationships and ecological risk assessment: An overview of predictive aquatic toxicology research. *Toxicol Lett.* 1995, September;79(1–3):229–237.

119. Blaauboer BJ. The integration of data on physico-chemical properties, in vitro-derived toxicity data and physiologically based kinetic and dynamic as modelling a tool in hazard and risk assessment. A commentary. *Toxicol Lett.* 2003, February 2;138(1–2):161–171.

120. Sobus JR, Tan Y-M, Pleil JD, and Sheldon LS. A biomonitoring framework to support exposure and risk assessments. *Sci Total Environ.* 2011, October 10;409(22):4875–4884.

121. United States Environmental Protection Agency (USEPA). *Risk Assessment Guidance for Superfund: Volume 1: Human Health Evaluation Manual (Part E, Supplemental Guidance for Dermal Risk Assessment) Final.* EPA/540/R/99/005. OSWER 9285.7-02EP. PB99-963312. Washington, DC, July, 2004. http://www.epa.gov/oswer/riskassessment/ragse/index.htm

122. National Institute for Public Health and the Environment (RIVM) the Netherlands. *Opinion on the Usefulness of in vitro Data for Human Risk Assessment: Suggestions for Better Use of Non-testing Approaches.* Letter report 320016002/2009. Bilthoven, The Netherlands, June 6, 2009. http://www.rivm.nl/bibliotheek/rapporten/320016002.pdf

123. Benigni R, Bossa C, Giuliani A, and Tcheremenskaia O. Exploring in vitro/in vivo correlation: Lessons learned from analyzing phase I results of the US EPA's ToxCast Project. *J Environ Sci Health C Environ Carcinog Ecotoxicol Rev.* 2010, October;28(4):272–286.

124. Becker RA, Hays SM, Robison S, and Aylward LL. Development of screening tools for the interpretation of chemical biomonitoring data. *J Toxicol.* 2012;2012:941082.

125. Boogaard PJ, Hays SM, and Aylward LL. Human biomonitoring as a pragmatic tool to support health risk management of chemicals—Examples under the EU REACH programme. *Regul Toxicol Pharmacol.* 2011, February;59(1):125–132.

126. European Centre for Ecotoxicology and Toxicology of Chemicals (ECETOC). High information content technologies in support of read-across in chemical risk assessment. Technical Report No. 109. Dec 2010.

127. Cronin MTD, Enoch SJ, Hewitt M, and Madden JC. Formation of mechanistic categories and local models to facilitate the prediction of toxicity. *ALTEX.* 2011;28(1):45–49.

128. Aladjov H, Todorov M, Schmieder P, Serafimova R, Mekenyan O, and Veith G. Strategic selection of chemicals for testing. Part I. Functionalities and performance of basic selection methods. *SAR QSAR Environ Res.* 2009;20(1–2):159–183.

129. Enoch SJ, Ellison CM, Schultz TW, and Cronin MTD. A review of the electrophilic reaction chemistry involved in covalent protein binding relevant to toxicity. *Crit Rev Toxicol.* 2011, October;41(9):783–802.

130. Enoch SJ, Cronin MTD, Schultz TW, and Madden JC. Quantitative and mechanistic read across for predicting the skin sensitization potential of alkenes acting via Michael addition. *Chem Res Toxicol.* 2008, February;21(2):513–520.

131. Pavan M and Worth AP. Publicly-accessible QSAR software tools developed by the Joint Research Centre. *SAR QSAR Environ Res.* 2008;19(7–8):785–799.

132. Sakuratani Y, Sato S, Nishikawa S, Yamada J, Maekawa A, and Hayashi M. Category analysis of the substituted anilines studied in a 28-day repeat-dose toxicity test conducted on rats: Correlation between toxicity and chemical structure. *SAR QSAR Environ Res.* 2008;19(7–8):681–696.

133. Doull J, Borzelleca JF, Becker R, Daston G, DeSesso J, Fan A et al. Framework for use of toxicity screening tools in context-based decision-making. *Food Chem Toxicol.* 2007, May;45(5):759–796.

134. Low Y, Uehara T, Minowa Y, Yamada H, Ohno Y, Urushidani T et al. Predicting drug-induced hepatotoxicity using QSAR and toxicogenomics approaches. *Chem Res Toxicol.* 2011, August 8;24(8):1251–1262.

135. Zhu H, Rusyn I, Richard A, and Tropsha A. Use of cell viability assay data improves the prediction accuracy of conventional quantitative structure-activity relationship models of animal carcinogenicity. *Environ Health Perspect.* 2008, April;116(4):506–513.

136. Rusyn I, Sedykh A, Low Y, Guyton KZ, and Tropsha A. Predictive modeling of chemical hazard by integrating numerical descriptors of chemical structures and short-term toxicity assay data. *Toxicol Sci.* 2012, May;127(1):1–9.

137. WHO. *WHO Laboratory Manual for the Examination and Processing of Human Semen*, 5th edn. Geneva, Switzerland: WHO Press, 2010.

138. Levitas E, Lunenfeld E, Weiss N, Friger M, Har-Vardi I, Koifman A, and Potashnik G. Relationship between the duration of sexual abstinence and semen quality: Analysis of 9,489 semen samples. *Fertil Steril.* 2005, June;83(6):1680–1686.

139. Keel BA. Within- and between-subject variation in semen parameters in infertile men and normal semen donors. *Fertil Steril.* 2006, January;85(1):128–134.

140. Francavilla F, Barbonetti A, Necozione S, Santucci R, Cordeschi G, Macerola B, and Francavilla S. Within-subject variation of seminal parameters in men with infertile marriages. *Int J Androl.* 2007, June;30(3):174–181.

141. Elzanaty S, Malm J, and Giwercman A. Duration of sexual abstinence: Epididymal and accessory sex gland secretions and their relationship to sperm motility. *Hum Reprod.* 2005, January;20(1):221–225.

142. Carlsen E, Petersen JH, Andersson AM, and Skakkebaek NE. Effects of ejaculatory frequency and season on variations in semen quality. *Fertil Steril.* 2004, August;82(2):358–366.

143. Slama R, Eustache F, Ducot B, Jensen TK, Jørgensen N, Horte A et al. Time to pregnancy and semen parameters: A cross-sectional study among fertile couples from four European cities. *Hum Reprod.* 2002, February;17(2):503–515.

144. Larsen L, Scheike T, Jensen TK, Bonde JP, Ernst E, Hjollund NH et al. Computer-assisted semen analysis parameters as predictors for fertility of men from the general population. The Danish First Pregnancy Planner Study Team. *Hum Reprod.* 2000, July;15(7):1562–1567.

145. Mocarelli P, Gerthoux PM, Patterson DGJ, Milani S, Limonta G, Bertona M et al. Dioxin exposure, from infancy through puberty, produces endocrine disruption and affects human semen quality. *Environ Health Perspect.* 2008, January;116(1):70–77.

3 Exposure Assessment

Everything should be made as simple as possible, but not simpler.

Albert Einstein
The Ultimate Quotable Einstein, 2013

All models are wrong; some are useful.

George E.P. Box
Empirical Model Building and Response Surfaces, 1987

There are two basic aspects to consider in an exposure assessment—the first is what happens to chemicals in the environment and the second is what behaviors of the receptors produce contact with potentially contaminated environmental media. Thus, exposure assessment for both human health risk assessment requires consideration of both physical–chemical aspects of contamination (how chemicals occur in environmental media) and human behavior (how people come in contact with these media). These two aspects will be referred to as the environmental and behavioral realms. Aspects of exposure in ecological risk assessment will be considered in Chapter 6. The need to obtain data that address questions in both these realms and the resulting use of a number of assumptions introduces uncertainty into exposure assessment—evident in numerous examples in this chapter.

In order for a risk to occur, there must be exposure to environmental media, that is, soil, air, water, or sediment, and the medium must contain hazardous materials. The Red Book defines exposure assessment as the "process of measuring or estimating the intensity, frequency, and duration of human exposures to an agent currently present in the environment or of estimating hypothetical exposures that might arise from the release of new chemicals into the environment."[1]

SCENARIOS AND RECEPTORS

Within the behavioral realm of exposure assessment in human health risk assessment, a receptor is a participant in an exposure scenario. All exposure scenarios require that one or more environmental media contain chemicals believed hazardous but also the ways in which populations or individuals might contact these media. These populations and individuals are considered receptors.[2,3]

Generally, risk assessments conducted by EPA involving media other than air use a set of standard exposure scenarios. These scenarios include residential property use, commercial or industrial use, and recreational use. The receptors in these scenarios include adults, children, and workers. EPA produced a supplemental guidance to RAGS in 1991, the *Standard Default Exposure Factors*, that essentially codified four distinct exposure/land use scenarios as

- Residential
- Commercial/industrial
- Agricultural
- Recreational[4]

To comply with the recommendation in the Red Book to produce standard inference guidelines, EPA prescribed a set of default exposure factors for these scenarios (Table 3.1). This document was produced following internal EPA discussions occurring in 1990 and 1991. This guidance is an initial attempt to provide values to use in the standard intake equation:

$$\text{Intake} = \frac{(C \times IR \times EF \times ED)}{(BW \times AT)} \tag{3.1}$$

where

C is the chemical concentration in a medium
IR is the intake/contact rate
EF is the exposure frequency
ED is the exposure duration
BW is the body weight
AT is the averaging time

These variables are known as exposure factors. From 1990 until 1997, EPA released a series of drafts of the *1997 Exposure Factors Handbook*.[5] This final document is extremely comprehensive and provides a very complete description of the data upon which the values of the various exposure factors were based.[5] This report and its updates were the work of Jacqueline Moya of EPA's National Center for Environmental Assessment. The various releases of the *Exposure Factors Handbook* are similarly comprehensive and detailed enough that they have become the de facto standard reference for information on exposure factors. In 2008, also under Moya's direction, EPA released the *Child-Specific Exposure Factors Handbook*[6] and the updated *Exposure Factors Handbook 2011 Edition*.[7]

In a later section of this chapter, a detailed description of the various exposure factors is provided; this description is based on the extensive and very useful information in these EPA publications.

EXTERNAL DOSE, POINT OF CONTACT, AND INTERNAL DOSE

As part of the response to the "Red Book," EPA released a draft version of the *Guidelines for Exposure Assessment* in 1986. This version is no longer available on EPA's website. In 1988, EPA produced the *Superfund Exposure Assessment Manual*.[8] This manual dealt almost exclusively with the environmental realm and models of environmental fate and transport of chemicals. In 1992, EPA updated and finalized the *Guidelines for Exposure Assessment*.[9]

TABLE 3.1
EPA's Standard Default Exposure Factors

	Residential		Commercial/Industrial	Agricultural		Recreational	
	Adult	Child	Adult	Adult	Child	Adult	Child
Water ingestion (L/day)	2		1	2	NA	NA	NA
Soil ingestion (mg/day)	100	200	50	200	100	NA	NA
Total inhalation rate (m³/day)	20	NA	20	20	NA	NA	NA
Indoor inhalation rate (m³/day)	15	NA	NA	15	NA	NA	NA
Exposure frequency (days/year)	350		250	350		350	
Exposure duration (years)	24	6	25	24	6	30	NA
Body weight (kg)	70	15	70	70	15	70	NA
Consumption of home-grown produce (g/day)				42 (fruit) 80 (veg.)		NA	
Consumption of locally caught fish (g/day)				NA		54	

Source: United States Environmental Protection Agency (USEPA), *Risk Assessment Guidance for Superfund Volume 1: Human Health Evaluation Manual. Supplemental Guidance. "Standard Default Exposure Factors,"* PB91-921314. OSWER Directive 9285.6-03 Washington, DC, March 25, 1991.

In contrast to the previous work in epidemiology and industrial hygiene, which focused on external doses only, these guidelines defined two steps occurring in exposure—contact with a chemical followed by entry or absorption into the body to produce an internal dose. Specifically, internal dose was used to refer to the amount of chemical absorbed across an exchange boundary such as the gastrointestinal tract, lungs, or skin. The distinction between internal and external dose was a giant leap forward because it allowed risk assessors to distinguish between systemic effects—produced at locations in the body other than the point at which entry occurs—and portal-of-entry effects that occur at the point of contact. This distinction paved the way for the use of PBPK models as part of the toxicity assessment.

The distinction between various types of doses and exposures is discussed clearly in Chapter 1 of the *Exposure Factors Handbook 2011 Edition*.[7] Exposure does not necessarily produce a dose, but a dose cannot occur without exposure. EPA presents a continuum from exposure to effects; the continuum considers both individuals and populations. This continuum is essentially the same as that described as an AOP.[10] The exposure–effect continuum is shown in Figure 3.1.*

Exposure may occur over a short or long time scale. Toxicity also has a time scale and it is important that these time scales match. For example, exposure to chlorine gas at 1000 ppm is fatal within minutes. However, chronic exposure to other substances at low level for many years may produce chronic effects, such as cancer. Generally, most environmental risk assessments deal with chronic exposures. Previously, EPA has defined three time periods for both exposure and toxicity—chronic, as greater than 7 years; subchronic, between 14 days and 7 years; and acute, less than 14 days.[3] The Agency for Toxic Substances and Disease Registry (ATSDR) in their derivation of minimum risk levels (MRLs) considers a chronic time period to be 1 year or greater, an intermediate time period to be 14–365 days, and an acute time scale to be 1–14 days.[11] In more recent toxicity assessments in EPA's IRIS, four different time periods were specified—acute, meaning less than 24 h; short term, between 1 and 30 days; subchronic, from 30 days to 7 years; and chronic, longer than 7 years.[12]

If contact with a chemical occurs over a chronic time scale, the external dose is usually averaged or integrated over time to obtain an ADD. In most risk assessments, the exposure assessment provides measures of ADD, usually in units of mg of chemical per kilogram of body weight per day or mg/kg/day. In many cases, only a portion of the chemical is absorbed or moves across the exchange boundary. In such a case, the chemical is considered less than 100% bioavailable. While bioavailability may be considered an aspect of the toxicity assessment, EPA's 1992 *Guidelines for Exposure Assessment* made it clear that it could also be considered as part of the exposure assessment—especially when using internal dose as a measure of exposure.

Point-of-contact exposure assessment measures the chemical or stressor concentration at the interface between the body and the environment. The value of this method is that exposure is directly measured rather than estimated, with the

---

* The concept of the exposure–effect continuum, while not incorrect, does not tell the whole story. In the next chapter on the dose–response assessment, the idea of the exposure–effect discontinuum and its relation to dysregulation of homeostasis and thresholds will be discussed.

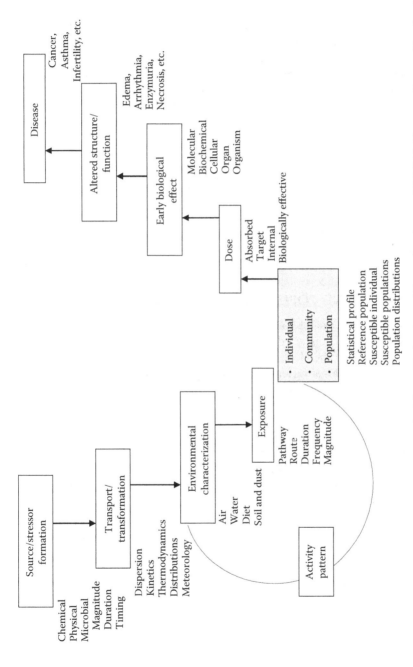

FIGURE 3.1 Exposure effect continuum. (From United States Environmental Protection Agency [USEPA], *Exposure Factors Handbook 2011 Edition (Final)*, EPA/600/R-09/052F. Washington, DC, September, 2011.)

proviso that the measurements are accurate. The most familiar example of such a measurement is a radiation dosimeter, the small film badge worn by x-ray technicians and others exposed routinely to radiation. All point-of-contact measurements suffer from two difficulties—most often, they produce short-term measurements and they are not specific to any source of the stressor or chemical being measured. Hence, a means of source discrimination and extrapolation of the short-term measures to the generally longer time scale of the risk assessment are needed.

Scenario evaluation is the most common means of exposure assessment. Estimates of factors in the behavioral realm that bring a receptor into contact with a contaminated medium are used, along with measurements of the chemical concentration in that medium, to estimate external dose. One of the most controversial and uncertain exposure factors used in scenario evaluation is children's soil and dust ingestion. The assumption underlying this factor is that children will contact outdoor soil and then any soil remaining on their hands after contact will be ingested by incidental hand-to-mouth contact; indoor dust may be partially comprised of outdoor soil and thus provides a similar exposure pathway indoors.

Biomonitoring or measuring internal dose provides a third means of performing an exposure estimate. Biomonitoring data can be used along with knowledge of the toxicokinetics of the chemical to estimate historical doses.

INDIVIDUAL AND POPULATION EXPOSURE

Information may be available about the distribution of exposures within a population. The NHANES conducted by the CDC provides population measures of the levels of a number of chemicals in blood.[13] Individual exposures can be determined for specific individuals at various percentiles within the population by Monte Carlo (MC) simulation or by choosing specific percentiles the values of exposure factors.

For the majority of exposure assessments in environmental risk assessment, exposures will be modeled using equations similar to Equation 3.1. In such cases, these individuals will be hypothetical—assumed to represent specific percentiles of exposure within the target population.

However, when point estimates of the various exposure factors are used in Equation 3.1, the percentiles of the resulting estimates of exposure and risk remain unknown. For example, in Superfund-type risk assessments, the risk estimate used to determine whether a cleanup is warranted is determined for the hypothetical individual at the point of reasonable maximum exposure or RME.[3,14]

The RME concept was also presented in the 1992 *Guidelines for Exposure Assessment*.[2] The guidelines state that the upper end of the distribution of risk should be characterized and high-end estimates of individual risk, such as the hypothetical RME individual, and the high-end estimate should fall at the 90th percentile or above. Additionally, the guidelines provide a detailed and cogent discussion of uncertainty assessment that concludes the following:

> It is fundamental to exposure assessment that assessors have a clear distinction between the variability of exposures received by individuals in a population, and the uncertainty of the data and physical parameters used in calculating exposure.[2]

Within EPA, the RAF, part of the Office of the Science Advisor, produced these guidelines. The RAF does not become involved in regulation. Instead, this group was established to promote consensus on difficult and controversial risk assessment issues within USEPA. As such, the recommendations of the RAF are not necessarily for immediate regulatory application but rather represent long-term policy recommendations. Regarding the statement from the *Guidelines for Exposure Assessment*, the RAF understood that, in many cases, risk assessment practitioners might not be able to estimate variability and uncertainty in a quantitative fashion but wanted to ensure that the goal of doing so was maintained and the qualitative distinction between variability and uncertainty kept in mind.

In cases in which exposure information or percentiles are obtained from biomonitoring data, individual data values may represent actual people—their identities will, of course, not be revealed because of ethical considerations. Assuming the population from which the biomonitoring results were obtained is representative of the target population, the percentiles within the biomonitoring data will thus represent percentiles of exposure within the target population.

In EPA's Food Quality Protection Act (FQPA), the Office of Pesticide Programs (OPP) determined in 1999 to choose the 99.9th percentile of exposure as the regulatory target for single-day acute dietary exposure to pesticides.[15] This was an interim choice because OPP recognized that individuals might be exposed to more than one pesticide with a common mechanism of action. For example, chlorpyrifos and methamidophos are both organophosphate (OP) pesticides that act by inhibition of an enzyme called cholinesterase that has important functions in regulating neuromuscular transmission and brain activity in mammals.[16,17] In 2006, OPP withdrew the guidance that stipulated the use of the 99.9th percentile in favor of a cumulative risk assessment approach.[18,19]

FREQUENCY DISTRIBUTIONS AS A REPRESENTATION OF EXPOSURE FACTORS IN A POPULATION: THE LOGNORMAL DISTRIBUTION AS AN EXAMPLE

To understand the remainder of this chapter, a discussion of statistics is warranted at the point. There exists some fuzziness in terminology about both percentiles of a distribution and the different representations of the central value of the distribution, usually the mean or median. The shape of the distribution is important in understanding these central values. Dr. Dale Hattis, a professor at Clark University in Worcester, Massachusetts, and long-time practitioner of risk assessment, used lognormal distributions to model variability in exposure factors and observed that many of these factors are approximately lognormal.[20] What this means is that most exposure factors have right-skewed distributions with most of the values near the median and a few high values. Income in the United States follows this pattern— most citizens earn near the median and a very few earn over $1M. The high earnings of these few raise the average or arithmetic mean income in the United States to a value quite a bit higher than the median. The other real advantage of the lognormal distribution is that it provides a mathematically tractable and therefore

BOX 3.1   THE LOGNORMAL DISTRIBUTION

This probability distribution is widely used in environmental applications. The lower bound is zero and it has no upper bound. One can think of this distribution as a normal or Gaussian distribution of the logarithms of the data. The normal distribution is the familiar "bell curve."

The two plots in the following show the probability density functions of a sample lognormal distribution on both linear (top) and logarithmic (bottom) x-axes.

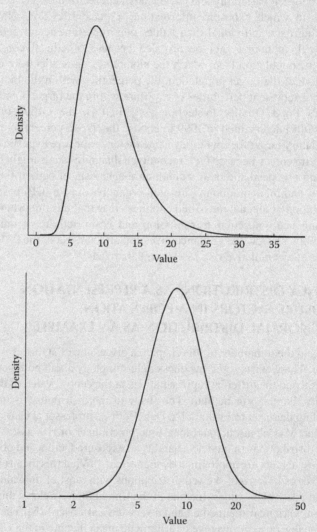

BOX 3.1 (continued)   THE LOGNORMAL DISTRIBUTION

The lognormal distribution can be fully characterized by two parameters—but there are three ways these parameters can be expressed. The most basic parameters are μ and σ, which are the mean and standard deviation of the underlying normal distribution (right plot). The GM and GSD are the exponentiation of μ and σ, respectively. The GM occurs at the median or 50th percentile of the distribution:

$$GM = \exp(\mu); \quad \mu = \ln(GM) \tag{3.2}$$

$$GSD = \exp(\sigma); \quad \sigma = \ln(GSD) \tag{3.3}$$

The formulae for the arithmetic mean and arithmetic standard deviation are slightly more complicated. Recall the example of the distribution of incomes—the average or arithmetic mean income was higher than the median income. These formulae are as follows:

$$\text{Arithmetic mean (AM)} = \exp(\mu + 0.5\ \sigma^2) \tag{3.4}$$

$$\text{Arithmetic standard deviation (ASD)} = \text{SQRT}[\exp(2\mu + \sigma^2) \times (\exp(\sigma^2) - 1)] \tag{3.5}$$

If one knows the AM and ASD, the GM and GSD can be calculated as follows:

$$GM = \frac{AM^2}{\text{SQRT}(AM^2 + ASD^2)} \tag{3.6}$$

$$GSD = \exp\left(\text{SQRT}\left[\ln\left(1 + \frac{ASD^2}{AM^2} \right) \right] \right) \tag{3.7}$$

useful distribution with which to represent most exposure factors. Box 3.1 provides more information and some useful formulae for the lognormal distribution.

COMMON SENSE THINKING ABOUT EXPOSURE

During the 1990s, the journal *Risk Analysis* and the Society of Risk Analysis (SRA) maintained a listserver. One of the posts to this listserver was from Paul S. Price, one of the pioneers of PRA.[21–24] Price posted a list of common sense ideas about exposure assessment presented below.

1. Exposures happen to people:
   a. Model people and how people are exposed.
   b. Don't model exposures and wonder about people.

2. People have properties/characteristics that affect their exposures:
   a. Age, gender, life span, habits, etc.
3. Everything is correlated:
   a. Exposure events depend on many factors. Engaging in a particular activity will likely produce some exposure but will also preclude exposure from another activity.
4. We only know little things:
   a. Short-term measurements are easiest to obtain, for example, activity, food consumption, and body weight on a single day.
5. Little things add up:
   a. For exposure assessment, an extrapolation must be made from short-term measurements to chronic estimates.
6. Exposures are real but unknowable—data are knowable but rarely relevant.
7. Are the data truly random?
   a. True randomness that can be characterized with statistics is easy to account for, but what about randomness characterized by extreme events?
8. How does one account for uncertainty?
   a. Not all uncertainty is the same.
   b. Different types of uncertainty need to be considered in different ways.

COMMON SENSE ABOUT VARIABILITY

The true values of exposure factors change with time—for example, daily soil and dust ingestion rates will be different in a single child at age 2 versus at age 7. Spatial variation in exposure also occurs; for instance, the use of the average concentration of a contaminant in fish in a river might be different in an upstream versus a downstream location. Hence, one would want to know the river location from which consumers obtained fish.

There are, of course, differences in exposure factors between people—interindividual variability. Measurements of a sample of 10 adults selected at random would likely indicate 10 different values for body weight or number of daily servings of fruits and vegetables.

One should also not forget to think about variation within a single individual over time—intraindividual variability. Body weight may change in a single person depending on age and food consumption. Exposure may also change for that individual depending on alterations in activity patterns.

COMMON SENSE ABOUT UNCERTAINTY

Three general types of uncertainty are present in exposure assessment—scenario uncertainty, parameter uncertainty, and model uncertainty.

Scenario uncertainty relates to how closely the qualitative description of exposure represents the true exposure. Does one indeed understand the exposure situation? For example, one might conclude that fish consumption from a small lake could be represented by EPA's default assumption for an individual's daily fish consumption or 54 g/day for 350 days/year, equivalent to a yearly consumption of just under 42 lb.

To see how realistic this value is, EPA's estimates of the total biomass of fish in ponds near the Housatonic River in Massachusetts will be used. These estimates were produced using electrofishing techniques. The measured biomass was 8.3 g/m² of water surface area.[25]

Assuming this value is representative, a 1-acre pond would be about 4100 m² in area and would contain 34 kg or about 75 lb of fish. Two anglers catching and consuming fish at this default consumption rate would decimate the fish population in a year. The fish in the pond are a finite resource. Similarly, a chemical contaminant in groundwater is also present in a limited amount and the exposure concentration may change over time due to degradation of the chemical by microorganisms, groundwater flow, or continued release from an unidentified source. Specific knowledge of how concentrations in environmental media change over time enable one to incorporate this information into a risk assessment.

Parameter uncertainty results from sampling errors and whether the quantitative data used are indeed representative of the target population. A simple example of a sampling error would be the use of average body weight of either National Football League linemen or marathon runners to represent the body weight of the general population. An error of representativeness would result from a mismatch of the sampled population with the target population—such an error could result from assessing soil adherence to skin in children coming from music lessons versus children coming from soccer practice.

Model uncertainty results from the mismatch of the modeled exposure scenario with the real-world exposure situation. Models are simplified representations of reality. In most situations, the dependencies between the various model inputs are unknown. This is especially true in fate-and-transport models in which various estimated chemical properties may have unknown dependencies that are not included in the model. Often, models cannot be tested against empirical data and must be judged on heuristics alone.

The classic example of model uncertainty is that of William Thomson, Lord Kelvin's estimate of the age of the earth. Lord Kelvin imagined the earth to have solidified from an originally gaseous or molten state. In 1844, he used Fourier's equations for heat transfer to estimate that the earth was between 20 and 98 Ma old. Modern radiometric dating indicates the earth is 4.5 Ga old.[26]

Kelvin was dismayed by the doctrine of uniformitarianism, popular among geologists of the late nineteenth century, allowing that the earth was of unlimited age.[27] Kelvin viewed this doctrine as unscientific and inconsistent with thermodynamics. In short, he developed a model based on elegant science and impeccable mathematics that turned out to be spectacularly wrong.[28]

Was it hubris for Lord Kelvin, at that time, to declare as fundamentally flawed any conceptual scheme that allowed for an earth over 1 Ga old? With the same logic, Kelvin dismissed Darwin's theory of natural selection on the grounds that there was insufficient time for natural selection to occur.

This is indeed a cautionary tale that models are wrong, but as the quote from the statistician George E. P. Box at the start of this chapter, some—maybe not Lord Kelvin's—can be downright useful!

COMPOUNDING CONSERVATISM

The concept of RME was codified in the exposure guidelines and RAGS that intake rate (IR), EF, and ED were to be upper percentile values, whereas BW was chosen as a central value (Equation 3.1). The use of upper percentile values in the numerator and central or lower percentile values in the denominator would tend to increase the estimated exposure. Many risk assessors consider this practice to be "compounding conservatism" that produces unrealistic and exaggerated risk estimates.

Discussion of compounding conservatism was prominent in the scientific literature during the 1990s. David Burmaster of Alceon in Cambridge, Massachusetts, was one of the most vocal critics of USEPA's risk assessment policies at that time. Sadly, Burmaster was a man ahead of his time and suffered a great deal of frustration when USEPA risk assessors turned a deaf ear to his requests that they consider variability in exposure and exposure and incorporate this variation using probabilistic risk assessment (PRA).[29–31]

Mathematically, it can be shown that conservatism compounds dramatically for deterministic point estimates of risk constructed from upper percentiles of input parameters (see Box 1.3).[32,33]

Burmaster's influence likely resulted in the publication of *Risk Assessment Guidance for Superfund. Volume III—Process for Conducting Probabilistic Risk Assessment* and a number of other EPA guidance documents. Nonetheless, the practice of probabilistic exposure assessment at EPA occurs relatively infrequently, this plethora of guidance notwithstanding.[34–36]

CONCENTRATION TERM

The concentration term in an exposure assessment is the factor in which aspects from the environmental and behavioral realms interact to the greatest extent. Contamination in an environmental medium is never uniform—similarly, human contact with this medium is not uniform. Hence, developing an appropriate value for concentration depends on knowing the spatial and temporal aspects of the occurrence of contamination as well as the spatial and temporal aspects of behavior by which receptors contact the contaminated medium.

Ideally, a risk assessor would observe the behavior of receptors and then design a sampling plan to determine the concentrations in the environmental media with which these receptors come in contact. There are limits, of course—fitting a group of children with GPS transmitters and following their play might be viewed as intrusive and excessive—although this has been done quite a bit lately, both for environmental exposure assessment and to obtain physical activity data to address childhood obesity.[37,38]

In practice, what generally happens is that environmental sampling plans are designed by environmental scientists or engineers whose main focus is on cleanup.

In the case of contaminated soil, a scientist or engineer whose goal is to understand the nature and extent of contamination will tend to oversample the most contaminated areas to support cleanup efforts. Therefore, for risk assessors, the question is: how closely does that sampling designed to delineate contamination reflect the areas that receptors will contact?

EXPOSURE UNIT CONCEPT

The concept of the exposure unit (EU) is closely associated with the concentration term. If one thinks of the receptor as a sampler of the contaminated medium, the exposure unit represents the extent of that medium contacted by a receptor within a specified period of time.[34]

The EU concept is easiest to think about with regard to soil. If one assumes that the receptor is a child aged 1–6 years in a residential exposure scenario and this hypothetical child can contact no more than 1000 ft^2 in a span of a day, then the EU would be 1000 ft^2. In the time period of a single day, the child may contact only 1000 ft^2 within the 0.5 acre (20,000 ft^2) residential EU. If this child lives in a house on a quarter acre lot (approx. 10,000 ft^2) and the house occupies 2000 ft^2, then there will be eight different exposure units associated with that residence.

Further assuming that the yard has both a sandbox and a swing set for the child's use, the area around each of these may be a "preferred" EU. With the sandbox and swing set, a risk assessor might wish to weight the concentrations in these EUs more heavily when determining the concentration term.

In the absence of preferred EUs, the assumption made is that a child will be exposed to the soil in each of the eight EUs within the yard with equal likelihood and the actual concentration contacted by the child will be the true but unknown mean concentration in the entire yard. In such a case, the long-term average soil concentration contacted by this child will approximate the true but unknown concentration in the yard because of the familiar phenomenon of "regression to the mean."[39]

The time period that has been considered is a single day. If the time period were increased to a month or a year, it might then be appropriate to consider the entire yard as the EU, exactly because of regression to the mean—but in such a case, the child's movements constitute the true averaging mechanism. Please note—it is important to understand that aspects from both the environmental and behavioral realms interact over time and it is also important to take this interaction into account as much as possible.

RANDOM VERSUS NONRANDOM EXPOSURE

Most risk assessments for contaminated land are based on potential future use, and risk assessors are very unlikely to possess information about swing sets or sandboxes. These were mentioned to introduce the concept of random versus nonrandom exposure within a single EU.

In the list of common sense assumptions from Paul Price presented earlier, one of these assumptions was that true randomness is easier to characterize with statistics.

However, the actual behavior of receptors that brings them into contact with contaminated media is very uncertain, and assuming randomness of a receptor's movement within an EU is an oversimplification and a source of uncertainty.

Spatially explicit models of exposure have been developed and refined in the area of ecological risk assessment and have been applied to animal and plant populations.[40–46] Initial attempts have also been made to use spatial techniques in assessing risk to human populations.[47,48]

TEMPORAL VARIATION IN CONCENTRATION

Temporal variability in chemical concentrations needs to be considered. Possible changes in concentration may occur due to wind erosion of soil, leaching of chemicals from soil to groundwater, or bioaccumulation in fish. It is most important that risk assessors consider the time scale of the adverse health effect being considered and make sure that both the derivation of the concentration term and selection of the ED are appropriate to this time scale.

Obviously, there may be considerable variability in short-term exposure within a population of receptors because of spatial and temporal variability within both the environmental and behavioral realms. Such variability may need to be considered for some acute endpoints such as methemoglobinemia in infants caused by nitrates in drinking water.

For risk assessments based on concerns for chronic toxicity with endpoints such as cancer, any short-term variability in the concentration a receptor experiences will average out to the long term chronic exposure concentration within the EU because of regression to the mean.

The underlying and possibly incorrect assumption is that the concentrations do not change over time. If concentrations within an EU do in fact change, such as natural attenuation of solvents in groundwater, as noted, these changes may need to be taken into account when calculating the concentration term.

VARIABILITY OF CONCENTRATION IN VARIOUS ENVIRONMENTAL MEDIA

A number of environmental factors can produce both spatial and temporal variation in the concentration of a chemical within an exposure medium. Most often, the spatial variability occurs by the mechanism by which the contamination occurred.

Variability of Concentrations in Air

With regard to short-term concentration, air is most variable of all exposure media. In addition, single or short-term exposures to chemicals in air may produce very different effects from repeated exposures. Toxicity factors for inhaled chemicals are developed in units of concentration—reference concentrations (RfCs) are generally found in units of mg/m^3 and inhalation unit risk (IUR) concentrations are generally found in units of $(\mu g/m^3)^{-1}$. In addition, single or short-term exposures to chemicals in air may produce very different effects from repeated exposures. Hence, it is important to understand the time scale of the toxic effect when selecting the ED. In many cases, a receptor will spend time in two or more microenvironments with different concentrations in each. EPA defines a microenvironment in terms of the chemical concentration therein but also points out that these microenvironments include homes, schools, vehicles, restaurants, or the outdoors.[49]

When considering variation in air concentrations or the movement of a receptor through multiple microenvironments, the exposure concentration is a time-weighted average (TWA) of the concentrations in the various microenvironments.[50]

For a number of outdoor air pollutants, standards have been established. There exist National Emission Standards for Hazardous Air Pollutants (NESHAPS) for 188 hazardous chemicals. Compliance with these standards is determined for both stationary sources such as smokestacks and mobile sources such as automobiles. Since 1996, EPA has conducted risk assessments based on modeled air concentrations within census blocks that include stationary and mobile sources as well as background from long-range transport of pollutants in the atmosphere.[51]

Obviously, there is one kind of uncertainty in the estimation of concentrations to which a single individual is exposed within a particular microenvironment and another kind of uncertainty in the estimation of modeled concentrations used in the national-scale assessments.

Variability of Concentrations in Groundwater

Receptors experience exposure to groundwater at a fixed point in space—either a well or a spring. Biodegradation by bacteria, volatilization, and other processes can alter chemical concentrations in groundwater. EPA has long been aware of these processes and has developed many guidance documents on natural attenuation of fuels, chlorinated solvents, inorganic chemicals, and radionuclides. These documents are available at http://www.epa.gov/ada/gw/mna.html, and monitored natural attenuation is considered a viable cleanup option for contaminated groundwater.

As part of the development of the concentration term, consideration of temporal changes in groundwater concentrations occurring by natural attenuation may be appropriate in the risk assessment. In such cases, risk assessors may need to obtain the advice of a hydrogeologist.

Variability of Concentrations in Surface Water

The effects of dilution, evaporation, and mixing constantly alter the chemical concentrations in a flowing stream or lake. Therefore, any surface water sample should be considered a "snapshot" in time. Needless to say, there exists considerable uncertainty in determining the long-term average concentration in surface water from environmental measurements.

Variability of Concentrations in Soil

The spatial variability of soil contamination will generally result from the process by which the contamination occurred. For example, spills or pesticide application results in a patchy pattern of contamination, whereas stack emissions will form a more even pattern downwind.

In subsurface soil, concentrations of chemicals in subsurface soil change mostly due to degradation or leaching to groundwater. Clearly wind erosion is not a factor.

On the other hand, surface soil is subject to erosion by wind and surface water runoff. Over time, concentrations in surface soil may change, but generally at a slow rate relative to other media.

Variability of Concentrations in Sediment

Like soil, sediment is subject to being physically moved or transported. Ocean currents, river floods, and ship movements may all contribute to sediment transport. Sediment trend analysis may provide information to determine the concentration term to be applied to sediment.[52] One of the most useful tools used in assessment of the Sangamo–Weston site, discussed in Chapter 1, was a sediment transport model.

Variability of Concentrations in Fish

The availability of food, introduction of a predator species, change in habitat, and intensity of angler harvest all may change the concentration of contaminants within a fish population. The concentrations of chemicals in territorial fish will likely reflect the sediment concentrations in the home range of the individual fish, whereas concentrations in migratory fish will be more unpredictable because of the frequent change in areas with their migrations.

Recreational anglers may harvest fish from different locations within a lake and consume fish of different sizes and species. It is possible to use site-specific information to determine the exposure point concentration over time in fish. An example is provided as one of the exercises at the end of this chapter.

In 2009, a decision was made at the Sangamo–Weston Superfund site to remove the hydroelectric plants on Twelve Mile Creek and thus facilitate covering the contaminated lake sediments with clean sediments transported from upstream locations, as a means of site remediation. Although sediment concentrations of PCBs had declined, measurements of fish tissue concentrations in 2008 indicated little change since the early 1990s. The reason fish concentrations did not change as did those in the upper layer of sediment was that sediment-burrowing *Hexagenia* mayflies still contacted contaminated sediment and were a major part of the diet of smaller forage fish.[53]

ASSESSING EXPOSURE FACTORS AND SUPPORTING DATA

Since 1991, with the publication of *Standard Default Exposure Factors*, EPA has provided default values to use in RME factors in various exposure scenarios.[54] This section will examine the basis for these values and discuss the underlying data.

SOIL AND DUST INGESTION

Estimates of children's soil and dust ingestion have driven the risk at a large number of hazardous waste sites—both in North America and Europe. Because of the deleterious effects of lead on children's neurodevelopment and the accompanying

TABLE 3.2
Comparison of the Most Recent Soil and Dust Ingestion Rates in Children Obtained with Three Different Methods

Methodology	Mean	Median	95th Percentile
Mass balance of fecal tracers[87]	26	33	79
Behavior estimates[105]	68	35.7	224
Lead biokinetics[62]	48	36	128

societal and individual costs, scientists came to understand that lead in soil contributed to children's blood lead levels—likely by incidental ingestion.[55–59]

The first attempt at measuring or estimating children's soil ingestion was based on measurements of soil on the hands of 22 children.[60] The mean amount on the hands was 10 mg, and the children were assumed to put their hands in their mouths 10 times per day for a daily soil intake of 100 mg. Very health-protective/conservative estimates of soil ingestion by children of between 1 and 10 g/day were used in an early risk assessment for dioxin.[61]

The earliest quantitative method used to measure children's soil and dust ingestion was mass balance using fecal tracer studies. Recently, behavioral analysis using video records of children's hand-to-mouth contacts and other microactivity have also produced estimates of soil ingestion. Paired blood lead and soil lead measurements can also be used to provide estimates of soil and dust ingestion in children. Values from all three methods are similar (Table 3.2).[62]

Mass Balance of Fecal Tracers

Historically, the most common method of estimating soil ingestion in children was fecal tracer studies in which a metal present in soil, such as aluminum, silicon, or titanium that was considered to be poorly absorbed by the GI tract, was measured in both feces and soil. With these measurements, soil ingestion among rats could be estimated with assumptions about the amount of time spent indoors and outdoors.

Children's tendencies to play on the floor when indoors, the ground when outside, and to put their hands in their mouths, all contribute to the amount of soil and dust ingested. Obviously, this method cannot distinguish ingestion of outdoor soil from ingestion of indoor dust.[63–83]

The first mass balance studies of soil/dust ingestion in children produced estimates that varied by over 10-fold.[84,85] This variation was due to the tracer metal selected.[63]

There is considerable uncertainty regarding these fecal tracer estimates and statistical analysis has quantified but not reduced this uncertainty. There are likely biological differences in the handling of these metals in the gut. For example, aluminum may be absorbed or may interfere with the absorption of other chemicals.[86,87] Silicon is absorbed and functions to strengthen bone.[88–90]

Titanium is both absorbed and secreted by lymphatic tissue in the gut in a circadian rhythm.[91–93] Examination of the titanium results from fecal tracer studies is consistent with cyclic absorption and release—on some days, fecal titanium is very high and other days very low.[6,84,93]

In addition, the poorly soluble aluminosilicate particles in soil may have a mechanical effect on absorption and may enhance or diminish absorption.[94–96]

A meta-analysis of four large soil/dust ingestion studies in children indicated a mean soil ingestion rate of 26 mg/day with a 95th percentile value of 79 mg/day.[97] Details of the mass balance calculation used in soil ingestion studies is provided in Box 3.2.

BOX 3.2   MASS BALANCE CALCULATION FOR ESTIMATING SOIL INGESTION FROM FECAL TRACERS IN SOIL

When an individual is in equilibrium with regard to a particular tracer, that is, when no net gain or loss of the body burden of tracer is occurring, the intake amount from ingestion and inhalation will equal the amount excreted in urine and feces:

$$I_{air} + I_{food} + I_{soil} + I_{water} = O_{feces} + O_{urine} \qquad (3.8)$$

where

I_{air} = Intake from air = air concentration × amount of air inhaled
$\quad = C_{air} \times A_{air}$
I_{food} = Intake from food = food concentration × amount consumed
$\quad = C_{food} \times A_{food}$
I_{soil} = Intake from soil = soil concentration × amount of soil ingested
$\quad = C_{soil} \times A_{soil}$
I_{water} = Intake from water = water concentration × amount of water drunk
$\quad = C_{water} \times A_{water}$

Generally, the concentrations of tracer metals such as aluminum or silicon in air and water are negligible. However, some tracer metals are present in toothpaste or other personal care items. These can be included in the food term. Hence, the simplified mass balance equation is as follows:

$$A_{soil} = \frac{(O_{feces} - I_{food})}{C_{soil}} \qquad (3.9)$$

As it turned out, when this equation was applied, soil ingestion estimates based on titanium tended to be 10-fold higher than those based on aluminum or silicon—even when toothpaste is taken into account.[63,84,85] The estimates of soil ingestion were very sensitive to the assumed fecal transit time.

BOX 3.2 (continued)   MASS BALANCE CALCULATION FOR
ESTIMATING SOIL INGESTION FROM FECAL TRACERS IN SOIL

In the following are shown the average concentrations of aluminum, silicon, and titanium in food, feces, urine, and soil and the dry weights of consumed food and liquid and fecal dry weights along with the estimates of soil ingestion. The concentrations in drinking water are assumed to be zero. The numerical calculation is also shown.

	Dry Wt. (g/Day)	Concentrations (μg/g)		
		Al	Si	Ti
Food	287.2	27	39.5	9.9
Feces	13.6	650	1100	279
Soil		6.16	26.4	0.6
Calculation		$(650 \times 13.6 - 27 \times 287.2)/6.16$	$(1100 \times 13.6 - 39.5 \times 287.2)/26.4$	$(279 \times 13.6 - 9.9 \times 28.2)/0.6$
Est. soil ingestion (g/day)	176		137	1585

Children's Microactivity Studies

Investigation into the role of children's microactivity, concentrating on hand-to-mouth behavior, became a focus of exposure assessment in response to concern about the use of copper chromated arsenic (CCA) to preserve wood used outdoors. This preservative had been used since the 1930s and had been shown to leach from treated decks. A number of states issued warnings that children playing under or around decks might be exposed to unsafe levels of arsenic.[98]

A variety of hand-loading studies of arsenic from contact with decks were performed. The methodologies in these studies were highly variable.[99–105] In addition to collecting additional data on the transfer efficiency of arsenic from treated wood to skin, a number of studies on children's hand-to-mouth activity were also conducted. These were put together by EPA in an extensive risk assessment for CCA-treated decks.[106,107] The videotape studies were also undertaken to assess children's exposure to pesticides in or on foods.[108–113]

This behavioral information was obtained using videography and can be combined with data on the time children spend in various locations to obtain estimates of soil and dust ingestion.[62,114,115] Young children commonly engage in hand-to-mouth and object-to-mouth behavior and this assumption underlies this methodology. Surveys about children's behavior obtained from caregivers may provide additional data. One advantage of these studies is that soil ingestion can be separated from dust ingestion, whereas mass balance studies can only provide information on combined soil/dust ingestion. These behavioral data were used to determine soil ingestion rates in children by EPA and published in the open scientific literature.[114–124] Unfortunately, these soil ingestion rate estimates were calculated with a version of the Stochastic Human Exposure and Dose Simulation Model (SHEDS) that is no longer available on EPA's website and the publication provides insufficient detail to enable independent checking of the calculations in the paper.

A long-held assumption in risk assessment is that children may receive their entire daily amount of soil ingestion from a single contact with soil. Certainly, a child playing outside will likely have dirty hands and a number of hand-to-mouth contacts may occur before the hands are washed. However, adult volunteers in another soil study noted that as little as 10 mg of soil in the mouth was gritty and unpleasant and even this little soil would likely be spit out.[125] Children exhibited median indoor hand-to-mouth CRs of 3.4–17/h and median outdoor hand-to-mouth CRs of 0–7.1.[124] The hand-to-mouth transfer efficiency of soil was modeled with a beta distribution having a mean of 20%.[114] This value is about the same as observed empirically from palm licking, but higher than observed from either thumb sucking or finger mouthing.[125] Hence, it would take a lot of hand-to-mouth contacts for a child to ingest 200 mg/day of soil.

Lead Biokinetic Estimates of Soil/Dust Ingestion

In the 1970s, a positive correlation between child blood lead levels and lead in indoor dust was observed.[55,58,126–132] Ratios of stable lead isotopes have been used to determine lead sources that contribute most to the body burden of lead in children.[133] Biochemically, lead is similar to calcium and there is a reservoir of lead in bone.[134–136] EPA's *Exposure Factors Handbook 2011 Edition*[7] describes that a study showing the decline in blood lead levels following closure of a smelter and remediation of individual yards as providing soil ingestion rates from these data. Similarly, work done at the Bunker Hill smelter site in Coeur d'Alene, Idaho, was also cited as an example of biokinetic estimation of soil ingestion rate but that work actually calculated lead intake and bioavailability from soil and dust rather than soil ingestion.[137]

EPA routinely uses and recommends the integrated exposure uptake biokinetic (IEUBK) model to assess children's exposure to lead.[138] In contrast to the IEUBK model, which is a polynomial approximation of a PBPK model, the ATSDR uses "slope factors" to estimate blood lead levels in children. In this context, a slope factor is the proportionality constant between lead in a given medium, that is, food, soil, and air, and blood lead. This methodology has been used to calculate a target soil concentration for lead based on an increase in blood lead of 2 µg/dL and has also been used to estimate soil ingestion rates in children and those estimates were in between those obtained from mass balance of fecal tracers and those from behavioral studies.[62] As noted, estimates from all three methods were fairly close.

Adult Soil Ingestion, Human Behavior, and Observational Estimates

For many years, EPA used an estimate of soil ingestion for construction or excavation workers derived from an observational estimate and a few assumptions. The value was 480 mg/day.[4] The assumption was made that all the soil in a 50 µm thick layer on the thumb and fingers of one hand could be transferred to the mouth during activities such as eating or smoking. Indeed, the slightly spongy and uneven surface of a slice of sandwich bread would be ideal for conveying soil to the mouth for ingestion—assuming the excavation worker failed to wash up before lunch. The median surface area of the hands, front and back, is about 0.1 m^2 or 1000 cm^2. The palmar surface of the thumb and fingers would be about one-sixteenth of the total or about 65 cm^2. The volume of a 50 μm layer over this area would be 0.625 cm^3. Assuming a dry bulk density of soil of 1.5 g/cm^3, the mass of soil would be 487 mg, very close to this estimate noted.

Adult Soil Ingestion from Fecal Tracers

Baseline soil ingestion rates were measured in six adult volunteers given a known amount of soil in gelatin capsules.[64] The most valid tracers in this study were Al, Si, Y, and Zr. The same investigators measured soil ingestion rates in 10 other adult volunteers given soil in gelatin capsules for part of the study. In both studies, incidental soil ingestion rate was calculated by subtracting the soil dose in the gelatin capsules. The most valid tracers in this study were identified as Al, Si, Ti, Y, and Zr.

Other workers measured soil ingestion in 19 families consisting of 19 children 7 years or less, 19 adult women, and 19 adult men.[139] The tracers used were Al, Si, and Ti. The mean soil ingestion rates for adult men were 92, 23, and 359 mg/day based on Al, Si, and Ti, respectively. The mean soil ingestion rates for women were 68, 26, and 625 mg/day for the same three tracers.

DRINKING WATER CONSUMPTION

Historically, the drinking water consumption values used for risk assessment were 2 L/day for adults and 1 L/day for children under 10 and infants. These amounts include water in beverages as well as tap water. Those living in warm climates or those with a high level of physical activity may consume more water than others, but there is little quantitative information in this regard.

From 1990 through 2011, water consumption estimates were based on USDA's 1977–1978 Nationwide Food Consumption Survey (NFCS). The mean and 90th percentile results for adults 20 to over 65 were 1.4 and 2.3 L/day. For children 1–3, these values were 0.65 and 1.4 L/day. For children 4–6, these values were 0.74 and 1.5 L/day.[7]

Lognormal distributions were fit to these data and adjusted to represent the 1988 US population.[140] These various distributions are shown in Figure 3.2.

The most recent study of water consumption was based on the US Department of Agriculture's Continuing Survey of Food Intakes by Individuals (CSFII) using data collected between 1994 and 1998.[141] The data were collected using dietary recall from more than 20,000 individuals from all 50 states and the District of Columbia. The water consumed was classified as follows: direct water, consumed directly as a beverage; indirect water, added during food or beverage preparation; intrinsic water, naturally occurring in foods or beverages; and commercial water,

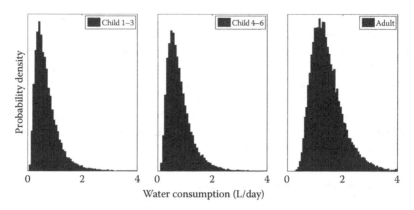

FIGURE 3.2 Lognormal distributions of daily water consumption by adults and two groups of children.

added to processed foods during manufacture. Water source categories were either "community" water obtained from a public water supply, "bottled" water, or "other" water obtained from wells or cisterns.

EPA used similar methodology to analyze water consumption from 2003–2006 NHANES. EPA's *Exposure Factors Handbook 2011* recommends age-specific values for drinking water consumption. EPA's Regional Screening Level Table (RSLT) provided at http://rais.ornl.gov uses default values from Part B of *RAGS*[142] and the *Standard Default Exposure Factors*[4] or 2 L/day for adults and 1 L/day for children in a residential exposure scenario. The default value for workers in a commercial industrial scenario is 1 L/day.

It might seem that drinking water rates expressed in mL/kg/day, in other words, normalized to body weight, might be less useful; however, these have a place in PRA. Additional information on PRA is provided later.

INHALATION EXPOSURE

Defining exposure for inhaled chemicals is more complex than for ingested chemicals. The respiratory system consists of three regions: nasopharyngeal, tracheobronchial, and pulmonary. The upper airway from the nose to the larynx comprises the nasopharyngeal region. The tracheobronchial region consists of the trachea, bronchi, and bronchioles; it forms the conducting airway between the nasopharynx and the deep lung. The pulmonary region consists of respiratory bronchioles, alveolar ducts and sacs, and alveoli. Exchange of oxygen and carbon dioxide occurs within the alveoli.

Materials are removed from inspired air in all regions of the respiratory system. The hairs in the nose filter out large inhaled particles, and the rest of the nasopharyngeal region moderates the temperature and humidity of inhaled air. The surface of the tracheobronchial region is covered with ciliated mucus-secreting cells. The cilia beat and move mucus upward as a means of removing material from the deep lung regions to the mouth—the so-called mucociliary elevator.

A special case exists for fibers. Fibers can deposit along the wall of an airway by a process known as interception. This occurs when a fiber makes contact with an airway wall. The likelihood of interception increases as the airway diameter diminishes. Fiber shape influences deposition too. Long, thin, straight fibers tend to deposit in the deep region of the lung compared to thick or curved fibers.

RAGS Part A[3] indicated that the following equation (identical to Equation 3.1) should be used for inhalation exposure:

$$\text{Inhalation intake}\left(\text{mg/kg/day}\right) = C_{air} \times \left(\frac{IR}{BW}\right) \times \frac{(ET \times EF \times ED)}{AT} \qquad (3.10)$$

where
   C_{air} is the concentration in air (mg/m^3)
   IR is the inhalation rate (m^3/h)
   ET is the exposure time (daily fraction)
   EF is the exposure frequency (days/year)
   ED is the exposure duration (years)
   AT is the averaging time (days)

However, as discussed at length in EPA's 1994 *Methods for Derivation of Inhalation Reference Concentrations and Application of Inhalation Dosimetry*,[143] this method does not account for either species-specific differences in airway anatomy or physicochemical characteristics of the inhaled contaminant that both result in differences in deposited or delivered doses. Estimation of risk from inhalation of chemicals should use toxicity criteria expressed in terms of air concentrations; such toxicity criteria cannot be used with the standard intake equation (Equation 3.1) that expresses dose in mg/kg/day.

Hence, for 15 years, toxicity criteria expressed in terms of concentration were converted to units of intake to be able to be used with Equation 3.1 with the CR being the inhalation rate. This conversion and the method were not correct, and it is unfortunate that EPA's Superfund program took 15 years to remedy this problem. Some state environmental agencies, notably the Georgia Environmental Protection Division, have promulgated the use of RAGS, Part B, equations and thus have legislated incorrect science. At EPA, this situation was explicitly corrected in Part F of RAGS, *Supplemental Guidance for Inhalation Risk Assessment*.[50]

Gases, Vapors, and Particles

Inhaled materials may be gases, vapors, or particulates. Inhaled gases are considered in three categories depending on their water solubility and reactivity:

- Category 1—these gases are highly water-soluble and reactive. Generally, they will affect the surface of the proximal respiratory tract and not be absorbed into the systemic circulation. Corrosive materials such as chlorine are category one gases.

- Category 2—these gases are water-soluble and possibly reactive but nonetheless will penetrate to the blood. There is some overlap between category one and category two gases; however, category two gases may cause portal-of-entry effects in the lung as well as systemic effects following absorption into the circulation.
- Category 3—these gases are sparingly soluble in water and generally not reactive. They are absorbed into the systemic circulation and cause effects at sites other than the lung. Styrene is an example of a category 3 gas.

A vapor is the gaseous phase of a substance below the critical temperature of that substance. Generally, vapors exist in contact with liquid or solid. The concentration of a vapor when nongaseous material is present is determined by the vapor pressure, which, in turn, depends on the ambient temperature. Volatile chemicals dissolved in water may volatilize from the water resulting in measurable vapor concentrations. Volatilization from either soil or water is determined by the Henry's Law constant of the substance.

Particulate matter or aerosol droplets will either penetrate to the deep lung or be deposited in the tracheobronchial region depending on particle size. The site of deposition of particles in the respiratory tract is of profound importance for the potential to produce adverse effects. The size, shape, and density of the particles, either solid or aerosol, all affect which portion of the respiratory system receives the majority of the dose.

Particulate matter and aerosols are characterized by their mass median diameter (MMD). The median value of the diameter of particles or aerosol droplets is the MMD and is generally measured in μm. For homogenous substances with a uniform density, the MMD of an aerosol provides a measure of the median mass of the individual particles. Particulate matter and aerosols may also be characterized by the mass median aerodynamic diameter (MMAD), a measure that accounts for particle shape as well as size and density. If an aerosol particle of unknown shape has an MMAD of 1 μm, the particle behaves in a similar fashion as a spherical particle 1 μm in diameter.

Inhaled particles may be aqueous with dissolved material or solid-insoluble particles. Physical airflow during breathing along with particle size determines where a particle is likely to deposit in the respiratory system. Particles with an MMAD greater than 1 μm and not filtered out in the nasopharynx tend to deposit in the upper respiratory tract at airway branching points. These particles will likely collide with the walls of the airway at branch points in the tracheobronchial tree in a process called impaction. Smaller particles not removed by impaction will likely be deposited in small bronchi and bronchioles by sedimentation, a process in which they settle out due to lower airstream velocity. Particles less than 0.3 μm move through the air randomly, by Brownian motion. They likely will remain suspended to be exhaled but also may deposit on the walls of the alveoli or terminal bronchioles.

Hence, both the size of the particles, the branching pattern and physical dimensions of the airways, and the velocity of the airflow determine the pattern of deposition of airborne particles in the respiratory tract. Once deposited, particles may be engulfed by pulmonary macrophages and removed by the mucociliary elevator.

In general, particles with an MMD or MMAD of 5–30 µm are deposited in the nose and throat. Particles with an MMD or MMAD of 1–5 µm are deposited in the trachea and bronchial regions. Particles 0.3–1 µm tend to be deposited in the alveoli. Very small particles of <0.3 µm are unlikely to be deposited and will be exhaled.[144–146]

Time-Averaging of Exposure Concentrations for Inhalation

Inhalation toxicity criteria are generally adjusted to represent constant exposure. However, because human receptors live and work in different environments or situations, the exposure details of these various situations need to be combined into a TWA that can be compared appropriately with the toxicity criterion.

The US OSHA has used this type of time-weighting since the 1970s. OSHA usually provides several regulatory values that are specific to different time periods. The PEL is generally compared with an 8-h TWA to represent a work day. The short-term exposure limit (STEL) is developed to provide safety for a 15-min exposure.

Similarly, EPA derives RfCs, generally in units of mg/m^3 that can represent chronic, subchronic, or acute time periods. For cancer, the toxicity criterion is the IUR value, generally in units of per $\mu g/m^3$.

The pattern of exposure may be important to consider. There may be different effects from a long-duration low-level exposure than from a series of intermittent higher-level exposures. This may depend on whether the chemical is metabolized, possibly to a toxic byproduct, or whether the chemical accumulates in the body.

Generally, the exposure concentration for acute exposure can be represented by the measured air concentration. However, for longer exposures, these will need to be averaged to represent a time period appropriate to the toxicity factor. In addition, exposure may occur in multiple environments or situations, and these various exposures must be averaged appropriately.

Inhalation Rates

Early studies on inhalation rates used spirometry to measure the actual rate or predicted the rate from measurements of pulse. In the 1990s, a method based on energy expenditures was developed.[147] The general equation used in this method is as follows:

$$V_E = E \times H \times VQ \qquad (3.11)$$

where
   V_E is the ventilation rate (L/min or m^3/h)
   E is the energy expenditure rate (kJ/min or MJ/h)
   H is the volume of oxygen consumed in the production of 1 kJ/min of energy
   VQ is the ventilatory quotient—the ratio of minute volume to oxygen consumption

In the *Exposure Factors Handbook 2011 Edition*, EPA used three approaches to obtain inhalation rates based on this equation.[7] The first approach was to use food intakes from the 1977–1979 USDA-NFCS and adjust these values upward by 20% to account for underreporting. The values for oxygen uptake of 0.05 L O_2/kJ and for the ventilatory quotient were obtained from the original study.[147] The second approach was to obtain basal metabolic rate (BMR) data for various age groups from a range of sources and develop a regression between BMR and body weight. The ratio of the daily BMR to energy expenditure was used to convert this value to daily energy expenditure. The third approach involved developing energy expenditure rates associated with different levels of physical activity and use time–activity data from a survey of how people spend their time in these various activities.

More recently, the disappearance from the body of water doubly labeled with isotopes has been used to measure inhalation rates for periods up to 3 weeks. In this method, water is labeled with the stable isotopes ^2H and ^{18}O (deuterium and heavy oxygen). These isotopes can be measured in urine, saliva, or blood. The disappearance of ^2H is a measure of water output and the disappearance of ^{18}O reflects water output plus CO_2 production. CO_2 production, as a measure of metabolic activity, is then calculated by subtraction. Daily energy expenditures can then be determined from CO_2 production.[7,148–152]

Data from double-labeled water studies have been collected by the IOM of the NAS and the Food and Agriculture Organization of the United Nations. These data are diverse and include subjects with differences in diversity in ethnicity, activity, body type, age, and fitness.[152]

To obtain the actual values for inhalation rates, the reader is referred to Chapter 6 of *Exposure Factors Handbook 2011 Edition* available at http://cfpub.epa.gov/ncea/cfm/recordisplay.cfm?deid=20563.[7] Tables at the back of EPA's chapter provide inhalation rate values from a large number of studies.

DERMAL ABSORPTION OF CHEMICALS

The skin provides a barrier to entry of substances into the body. The outer layer of the skin or epidermis is not vascularized. The stratum corneum is the outer layer of the skin and is about 10–40 μm thick. The stratum corneum is composed of keratinized and dessicated epidermal cells. This outer layer is highly hydrophobic because of its high lipid content, and this hydrophobicity contributes to the barrier function of the skin. Cells in the stratum corneum slough off or desquamate and are replaced by growing keratinocytes in the germinal layer of the epidermis below.

The mature and senescent keratinocytes that comprise the stratum corneum are organized into a "brick and mortar" pattern. These cells form a cornified envelope consisting of cross-linked protein and lipid polymers. The lipid polymers are located on the outside and are comprised of specialized and hydrophobic lipids known as ceramides. The formation of these specialized lipids requires specialized metabolism and many of these lipids display antimicrobial properties.[153] Healthy skin is a barrier to permeation of chemicals—simple contact with the skin is usually insufficient and this is why dermally delivered drugs require adhesive patches or hydrophobic gels and relatively long application times.[154] The stratum corneum has different thicknesses on various parts of the body, for example, the eyelid versus the sole of the foot. These adaptations in the anatomy and physiology of the stratum corneum provide the barrier function of the skin.[155–158]

The inner layer of the skin or dermis is vascular and contains hair follicles and sweat glands. Once a substance has penetrated the epidermis, it will be available to the dermal capillaries for systemic distribution.

Total Skin Surface Area and That of Various Body Parts

In the *Exposure Factors Handbook 2011 Edition*,[7] EPA used data from 1999–2005 and 2005–2006 NHANES surveys to estimate total body surface area as well as that of the various body areas, for example, head, trunk, arms, and hands. Total surface area is estimated using equations with weight and height as dependent variables.[5] A method that used a simple ratio between body weight and total skin surface area yielded approximately similar results and was used to develop distributions of total skin surface area in three age groups.[159] Distributions were also estimated using the height and body weight equations.[160]

To estimate the area of various body parts in adults, EPA developed regression methods with height and weight as the dependent variables. Body weight and height were obtained from the 2005–2006 and 1996–2005 NHANES data.[7] For children and adolescents, measurement data collected for use in ergonomic and product safety design were used as input to a computer model to estimate the surface area of various body parts in children and adolescents from 2 to 18 years.[161]

Possible Fates of Substances Applied to the Skin

Substances applied to the skin may evaporate before penetration or may diffuse through the stratum corneum. Substances that penetrate the stratum corneum may be metabolized in the germinative layer of the epidermis or may continue to the capillaries in the dermis. In addition, some substances may bind irreversibly to lipids or proteins in skin and be sequestered. Sequestered chemicals in the skin may eventually be absorbed or may be lost by desquamation.[162] Hence, studies that measure the amount of material lost from the skin surface will tend to overestimate dermal permeation. Measurements of actual permeation to the blood must also be conducted with care.

Solid Materials Contacting the Skin

Generally, the adherence of soil and sediment to skin is provided in units of mass per unit area, such as mg/cm^2. Instead, what is needed is some measure of skin coverage such as particle layering. These sorts of data are difficult to obtain or estimate because soil particles come in different shapes and sizes. Even a complete monolayer of soil particles on the skin will not cover the entire area of skin because of the shapes of the soil particles. Volatile chemicals in soil, however, may occur as vapor in the spaces between the soil particles and skin and thus be available for absorption into the stratum corneum over much of the contact area.

Unless the soil is wet, smaller particles tend to adhere preferentially to skin.[163–165] For lipophilic chemicals, sorption to soil is governed by the organic carbon content of soils, and the organic carbon content differs in the various particle size fractions. For example, PAHs tend to be correlated with organic carbon in soil, but metals may or may not be, depending on the particular soil chemistry.[166]

"Aging" of a chemical in soil may occur. This means that the chemical in soil may take months or years to reach an equilibrium state. Partitioning of a chemical from soil to skin may depend on the characteristics of aging of the particular chemical. For example, a greater amount of freshly spiked benzo[a]pyrene in soil was absorbed than a sample allowed to "age" for 110 days.[167]

Generally, the movement of a chemical from soil to skin occurs by diffusion and thus is a dynamic process. Hence, the amount of time soil remains on the skin and the physicochemical characteristics of the substances being considered affect the amount of material absorbed. In 1992, EPA's Office of Research and Development published *Dermal Exposure Assessment: Principles and Applications*.[168] This publication included a method for estimating dermal exposure by integrating the flux into the skin over time. However, in 2004, when EPA published *Part E, Supplemental Guidance for Dermal Risk Assessment* of RAGS, the flux had been condensed into a single chemical-specific value for dermal absorption fraction.[169]

Much work has been done on the adherence of soil to skin.[116,163,165,170–177] This factor is highly activity-dependent, as one might imagine. This aspect of exposure is quite amenable to empirical investigation, and the experiments can be as simple as pressing the hands into a pan of soil and then collecting and weighing the adherent material. In *RAGS, Part E*, EPA correctly recommends using the mean or median value of an activity-related adherence factor with an appropriate value for the exposed skin surface area.

Uncertainty in Dermal Exposure to Solid Media

As part of the reregistration of coal tar creosote for compliance with EPA's pesticide program, a dermal exposure assessment was conducted on workers using "whole-body dosimeters." These whole-body dosimeters were nothing more than cotton long underwear worn underneath work clothes and protective gear. Each day the long underwear was sent to the lab for quantitative assessment of PAHs in the adherent particles.

There remains considerable uncertainty regarding the extent of the dermal exposure component for creosote workers. A urinary biomarker of exposure to PAHs is hydroxypyrene.[178] Several studies could not account for the amount of hydroxypyrene in urine from inhalation exposure alone in creosote workers and claimed this was due to dermal exposure without any additional evidence.[179,180]

Although the creosote risk assessment was quite controversial, the use of long underwear in this way represents a novel and relatively innovative way to obtain empirical data on dermal exposure to airborne particulate matter.

Substances Dissolved in Water That Contact the Skin

In RAGS, Part E, EPA presents a model for estimating the dermally absorbed dose per event for a range of chemicals. For organic chemicals, the dose is dependent both on the duration of the event and chemical-specific parameters such as the permeability coefficient K_p and the ratio of the permeability through the stratum corneum to that through the living epidermis beneath.[169]

It is important to point out that the aqueous pathway is highly uncertain—exposure is estimated using an uncertain prediction model based on data from a small number of chemicals. There remains considerable interest in developing better models for skin permeation because of the potential for transdermal delivery of drugs. K_p, the dermal permeability coefficient, is defined as the steady-state flux through the skin normalized to the concentration gradient. For xenobiotic chemicals, the internal concentration will be assumed zero and K_p then is the ratio between flux and concentration on the skin.

Anatomically based physicochemical models describing percutaneous absorption were introduced in the early 1970s.[181] Experimental results for the permeation of organic compounds, large and small, polar and nonpolar, through human skin were assembled. This is the so-called Flynn dataset, named for the scientist who compiled it.[168,169] These data enabled the development of a number of QSAR models for skin permeation. These models generally used both the octanol–water partition coefficient as a measure of lipophilicity and the molecular weight as a measure of size to predict skin permeability.[182,183]

The difficulty with these models for risk assessment is that they were inaccurate for high-profile lipophilic chemicals—PAHs, PCBs, dioxins, and DDT. A comparison of experimentally measured K_p values and those predicted with the QSAR models found that the results often varied by up to two orders of magnitude.[184] As one of the exercises at the end of this chapter, estimates of K_p from dermal application of coal tar in dandruff shampoo and subsequent excretion of PAH metabolites will be compared the a QSAR estimate.

FISH CONSUMPTION

A number of contaminants tend to bioconcentrate in fish through the food chain. Estimating exposure to contaminants in fish becomes complicated due to the need for knowledge about the fish consumption practices of the target population.

The concept of an exposure unit was introduced in the section "Exposure Unit Concept." A common argument among risk assessors dealing with fish consumption is the definition of the appropriate exposure unit—some contend the EU is actually the dinner plate holding the portion of fish to be consumed rather than any water body or part thereof.

More often than not, fish consumption practices are difficult to study. Certainly, avid recreational anglers consume their catch, but how likely is it that individuals support their dietary requirement for protein with self-caught fish—a true subsistence scenario. Fishing and consumption of self-caught fish are long-standing traditions in rural America.[185–187] These traditions are difficult to maintain in the face of population mobility and the continuing concentration of people in urban and suburban areas.[188]

EPA relies on a relatively small number of surveys on fish consumption and presents these estimates as representative of the general population. For consumption of freshwater fish, the key population study was CDC's NHANES data from 2003–2006. EPA's OPP analyzed these data to obtain per capita and consumer-only fish consumption rates (FCRs) for finfish and shellfish. For consumption of saltwater and estuarine fish in coastal regions, USEPA has used data from the National Marine Fisheries Service Marine Recreational Fishery Statistics Survey (NMFS/MRFSS).[7,189] However, in 2006, the NAS strongly criticized the MRFSS and suggested that because of the budgetary and personnel constraints and methodological flaws in data collection and analysis, the data produced by the survey were not reliable.[190]

The "subsistence fish consumer" exposure scenario commonly discussed among risk assessment practitioners is intended to account for those who rely on self-caught fish as a significant proportion of their dietary protein. While this scenario may be valid for Native Americans in remote areas, it is likely not applicable to other populations.[191–193]

In an exercise at the end of the chapter, fish consumption rates recommended by EPA will be compared to the recommended daily allowance (RDA) of protein in order to determine their level of conservatism. In general, fish consumption estimates used in risk assessment are most representative when developed from site-specific data. Although time-consuming and often costly, carefully conducted creel surveys to ascertain actual fish consumption practices are vital to the development to a credible fish consumption risk assessment.

HOW TO ESTIMATE THE CONCENTRATION TERM

In RAGS, Part A, EPA was vague about how to estimate the concentration term in any medium. It is unclear from reading this document exactly what to do to obtain a concentration term that is representative of ongoing exposure. The concentration term is also known as the exposure point concentration.

For soil, it is generally not appropriate to use the highest concentration measured to represent the concentration actually contacted by a receptor that moves. Hence, in 1992, EPA's Superfund program released a document titled *Supplemental Guidance to RAGS: Calculating the Concentration Term.*[194] In this document, the uncertainty between the measured concentrations at a hazardous waste site and the concentrations actually contacted by receptors was recognized and addressed by the suggestion to use the 95% upper confidence limit (UCL) of the arithmetic mean of the measured concentrations. This value would serve as a health-protective estimate of the true but unknown arithmetic mean.

The document also indicated that in the majority of cases, the distribution of concentration, at least in solid media, soil, or sediment, was likely to be lognormal. This claim was based on the theory of successive random dilutions developed by Wayne Ott, an EPA scientist from the agency's inception until the mid-1990s.[195] In this theory, contaminants are diluted by a number of geophysical processes that have a fractional or multiplicative effect on concentrations. Because multiplication is equivalent to addition of logarithms, contamination would occur in a lognormal distribution. In this document, EPA suggested the use of a statistical method for estimating the UCL of the arithmetic mean developed by Charles Land of the NCI.[196]

In many environmental datasets, because the true purpose of data collection was to determine the nature and extent of contamination, there will be a number of nondetect values as well as a number of relatively high values. EPA's policy is to use half the detection limit as a surrogate value for nondetects. When the Land method for calculating the UCL of the mean is used on such a dataset, the calculated UCL is often greater than the highest value in the data. The concentration term guidance suggested that when this occurred, the highest value in the data set should be used in lieu of the calculated UCL. The reason for this policy was to get regulated entities to obtain a greater sample number.

With increasing experience with this method of calculating the UCL of the mean, the realization came that not all environmental datasets are truly lognormal. Dr. Susan Griffin of EPA's Region 8 office in Denver pioneered the use of bootstrap methods for estimating the UCL of the mean when the distribution of the dataset could not be determined.[197] The bootstrap method is often useful and a simple example is provided as an exercise at the end of the chapter.

EPA's Site Characterization and Technical Support Center in Las Vegas produced a number of papers on the calculation of exposure point concentrations as well as EPA's ProUCL software to implement these calculations.[198–201]

ENVIRONMENTAL FATE-AND-TRANSPORT MODELS IN RISK ASSESSMENT

Often, physicochemical properties of a substance can be used to predict how the substance will move in the environment or selectively partition into various environmental media, for example, soil, groundwater, and air. A number of these models are routinely used to predict concentrations in a different medium than that sampled. For example, concentrations of a substance in soil can be used to predict expected concentrations in groundwater due to percolation downward to the water table.

This section is not intended to be a comprehensive survey of all environmental fate-and-transport models but, rather, an introduction to these models and their use. Details of the calculations used in these models can be found elsewhere. For those wishing such detail, all the models discussed here and many others are available on EPA's website.

OUTDOOR AIR MODELING

The simplest air model is a so-called box model. It assumes that air pollutants enter a box from a source within the box or carried on the prevailing winds. Pollutants leave the "box" on the wind or by deposition to the ground. The concentration inside the "box" is the simple average concentration—the amount of pollutant in the "box" divided by the size of the "box." The model is a simple input–output model and is a first step toward obtaining air concentrations for use in exposure assessment.

The Gaussian plume model is also a relatively simple model that is typically applied to point source emitters, such as smoke stacks. The concentration of substances downwind from the source is considered to be spreading outward in three dimensions from the centerline of the plume following a normal or Gaussian statistical distribution. One of the key assumptions of this model is that over short periods of time (such as a few hours), steady-state conditions exist with regard to air emissions and meteorological changes. The model assumes that an idealized plume emanating from the top of the stack is representative of the actual pattern of release. Dispersion then occurs in three dimensions. Downwind dispersion is a function of the mean wind speed blowing across the plume. Crosswind or lateral dispersion is dependent on the relative stability of the surrounding air. Dispersion in the vertical direction will depend on the wind speed and density of the emitted substance relative to that of air.

Generally, the models used today are more complex but are also more uncertain than in the past. For example, the Assessment System for Population Exposure Nationwide (ASPEN) is an air quality dispersion model based on a Gaussian plume simulation and a mapping function that produces a concentration specific to each census tract. Census tracts were developed to contain between 1500 and 8000 residents with an optimum size of 4000. The geographic size of a census tract will vary between urban and rural locations, and the concentration estimates produced by ASPEN are determined by geography and area—not population. Notwithstanding, these estimates are keyed to population and this potential mismatch is a source of uncertainty.

Some air models include nonpoint sources such as automobiles; some models consider the effect of terrain or the built environment; both these factors add to model complexity. EPA's Industrial Source Complex—Plume Rise Model Enhancements (ISC-PRIME) melds a Gaussian plume model with a model of building downwash to include the effects of the built environment.

One of the most challenging areas of air modeling is the modeling of dense gases. Once released, these gases sink to the ground and spread out under their own weight. Both the presence of buildings and the local terrain affect the movement of the ground level plume. Chlorine is a dense gas and is often transported in railroad tankers. Deaths often occur during railroad accidents from chlorine inhalation.[202] The resulting gas cloud remains at ground level and moves downhill, interrupted by larger ground level masses such as buildings. Chlorine tanker trucks have been used as weapons during the Iraq war.[203]

INDOOR AIR MODELING

Substances may occur in indoor air from inside sources or may enter the building from the outside. In Chapter 1, the example of how a journalist affected public perception of the risks from hexavalent chromium was described. In the case of indoor air, a similar situation happened when Mark Obmascik of *The Denver Post* criticized EPA in 2002 for the use of a vapor intrusion model developed by Paul Johnson and Robert Ettinger.[204] This model was used by EPA at the Redfield Rifle Scopes site near Denver as an example of model application. Redfield manufactured rifle scopes and binoculars and used degreaser solvents such as trichloroethylene that leached from the site into groundwater. Denver Water provided water to residents near the Redfield site, but the model was used to attempt to determine if there might be any risk from migration of solvent vapors from the groundwater into indoor air.

On January 6, 2002, Obmascik wrote an article in *The Denver Post* titled "Toxins in Air, Regulations Fail to Protect U.S. Residences from Gases."[205] Obmascik was critical of EPA and Colorado state officials because they did not test air within homes. On January 7, 2002, Obmascik wrote: "With that model, you'd get just as good results flipping a coin. Half the times it's right, and half the time, it's wrong."[206] Obmascik was clearly ignorant of the idea that there could be indoor background sources of vapors. For example, trichloroethylene is found in a number of cleaning products used in arts and crafts and household maintenance.[207]

The real advantage of the model is that it provides a way of distinguishing substances occurring in indoor air and sampling results of indoor air could never distinguish the source. In fact, both EPA and the state of New York have compiled databases on background indoor air concentrations of volatile organic compounds.[208,209] In his enthusiasm for career building, Obmascik clearly lost sight of his responsibility as a journalist to inform himself and the public of the true state of the science.

One happy consequence of the public and private sector scrutiny of vapor intrusion is that the strengths and weaknesses of this model are now well known. Basically, the model develops estimates for concentrations of vapors that might exist underneath a house, potentially trapped by a concrete slab with footers. Next, the model develops estimates for the vapors that could penetrate the house through small cracks in the concrete or elsewhere in the foundation. This is the most uncertain part of the model. In fact, the recommendation of both EPA and the Interstate Technology and Regulatory Council (IRTC) is to use the model as a screening tool and, based on the model results, obtain samples of subslab soil gas or even indoor air.[210,211]

GROUNDWATER PROTECTION MODELS

EPA's *Soil Screening Guidance: User's Guide*, the associated technical background document and supplemental guidance documents provide a methodology for developing estimates of the concentrations of substances in soil that, if transported to the underlying groundwater, would not result in unacceptable levels of these substances in groundwater.[212–214]

The model used is a simple linear equilibrium soil/water partition equation with a number of simplifying assumptions that are biased in a health-protective way. The model is used for screening and as a means of deciding whether to establish monitoring wells and sample groundwater.

ACCOUNTING FOR VARIABILITY AND UNCERTAINTY IN EXPOSURE FACTORS

PRA is viewed as one way to improve risk assessment—both by regulators and those within the regulated community. Probabilistic methods provide tractable means for propagating estimates of uncertainty and variability through an equation or model. There exist analytical solutions for the propagation of variance but these are often difficult to implement. In contrast to the analytical methods,

Monte Carlo analysis (MCA) is a specific probabilistic method that uses computer simulation to combine multiple probability distributions in an equation. MCA is relatively simple to implement with commonly available software. In one of the exercises at the end of the chapter, readers will be able to perform a relatively simple MC exercise implementing the bootstrap method for calculating the 95% UCL on the arithmetic mean as the concentration term.

One area of PRA that continues to provide difficulties is the specification of distributions. This problem has been around ever since the French mathematician Pierre Simon Laplace used the uniform probability distribution for ease of analysis rather than for any metaphysical reason.[215] His lack of preference for a particular value in a range became known as the "principle of indifference."

The use of uniform probability distribution is generally problematic in risk assessment because almost always one can determine additional information to specify a more informed distribution.[216-218]

There are a host of techniques to be used to develop appropriate distributions and the field of probabilistic exposure assessment continues to evolve.[219] One method of performing a PRA deserves mention—microevent exposure analysis.[23] This method was developed by Paul Price, mentioned early in this chapter as a proponent of common sense in risk assessment. In the microevent exposure approach, an individual's chronic dose rate is modeled as the sum of doses received from separate exposure events. Because this approach tracks individuals through time, values of exposure factors appropriate to various life stages can be used. Probabilistic aspects of specific exposure events can be selected from the appropriate distributions, for example, the size of a meal of fish or the number of fish meals per month.

EXERCISES FOR THOUGHT AND DISCUSSION

ESTIMATING K_p FROM EXPERIMENTS WITH COAL TAR SHAMPOO

Urinary 1-hydroxypyrene was measured in the urine of volunteers after they shampooed twice in the evening for 30 s each time using a total of 20 g of shampoo containing 285 mg/kg of pyrene. Urine was collected for 2 days thereafter.

The average total excretion of 1-hydroxypyrene was 30.9 μmol. The molecular weight of pyrene is 202.25 g/mol and the log K_{ow} is 4.88. Now, from EPA's website, download RAGS, Part E, on Dermal Risk Assessment at http://www.epa.gov/oswer/riskassessment/ragse/.

Use Equation 3.2 on page 3–4 of RAGS, Part E. For this exercise, we will assume that the shampoo has the same density as water.

We will calculate K_p for pyrene using the Potts and Guy equation shown on the top of page 3–7 in RAGS, Part E. Other physicochemical parameters were obtained from the Hazardous Substances Databank at http://toxnet.nlm.nih.gov/cgi-bin/sis/htmlgen?HSDB.

The predicted K_p value from the Potts and Guy equation is 0.194 cm/h. The values for B and τ are not provided but those for fluoranthene were used as surrogate values. Fluoranthene has a very similar MW (202.3) and log K_{ow} value (4.95) to those for pyrene. Alternatively, use the equations in Appendix A of RAGS, Part E to calculate values of B, t_{event}, and t* for yourself.

Equation 3.2 in RAGS, Part E, predicts that the dermally absorbed dose is 0.0168 mg/cm²-event for each 30-s shampooing event. Doubling this gives 0.0336 mg/cm²-event. Table 7.2 in EPA's *Exposure Factors Handbook 2011 Edition* gives the area of the head as 0.154 m² for males and 0.121 m² for adult females.[7] Assuming 1/3 of this value represents the scalp, the dermally absorbed dose would be 17 mg for males and 13 mg for females. If all this absorbed dose were metabolized and excreted as hydroxypyrene, then males would excrete 84 µmol and females 65 µmol. Both these values are within a factor of 2–3 of the observed value.

Please work through these calculations yourself. There is great value in becoming familiar with the calculation methods and online resources.

COMPARISON OF EPA'S ESTIMATES OF FISH CONSUMPTION RATES WITH THE RECOMMENDED DAILY ALLOWANCE FOR PROTEIN

To ascertain the degree of conservatism in these estimates, the daily amount of protein consumed from fish was compared to the estimated average daily protein requirement (EAR) and minimal protein requirements for humans. The following narrative and tables provide the data and a description of what to do in order to enable readers to perform this comparison. The estimated average requirement (EAR) on a daily basis and the RDA for protein, both in g/kg/day, can be obtained from *Dietary Reference Intakes for Energy, Carbohydrate, Fiber, Fat, Fatty Acids, Cholesterol, Protein, and Amino Acids* from the NAS.[220]

Age	EAR (Male and Female)	RDA (Male and Female)
1–3	0.87	1.05
4–8	0.76	0.95
9–13	0.76	0.95
14–18	0.73 M, 0.71 F	0.85
19–30	0.66	0.8
31–50	0.66	0.8
51–70	0.66	0.8
>70	0.66	0.8

You should be able to easily find EPA's *Exposure Factors Handbook 2011 Edition*.[7] It is a sufficiently useful document that it is advisable to keep a copy on your own computer. Please consult EPA's Table 10.1 for recommended consumer-only data on total finfish and shellfish consumption. How do these values compare with the EAR and RDA values in the table shown earlier?

BOOTSTRAP SAMPLING

Please go to http://www.crcpress.com/product/isbn/9781466598294 and download the Excel spreadsheet "Simon-Chapter3-bootstrap.xls." This file implements a bootstrap in Excel. The data herein are from an actual hazardous waste site.

Please note that these data are neither normally nor lognormally distributed. If they were, the log-probability plot would be straight.

On the worksheet named "bootstrap," you will note the value "1" in cell K8. Replace this value with any other integer. The reason is to cycle the random number generator. As you do this, the bootstrap sample in columns H and I will change, as will the arithmetic mean calculated using Equation 3.4 in cell K2 and the arithmetic mean of the bootstrap sample in cell K3. Also examine the formulae in the various cells to understand how the bootstrap was implemented.

You may also wish to download EPA's ProUCL software and attempt to work with these data. The raw data are provided in the worksheet named "Raw data." For the bootstrap exercise, only detections were used. The dataset contains nondetects. These can be handled by ProUCL. The bootstrap implemented here is not definitive but rather was done to give you an idea of how MC methods and resampling techniques work.

REFERENCES

 1. National Research Council. *Risk Assessment in the Federal Government: Managing the Process*. Washington, DC: National Academy Press, 1983.
 2. United States Environmental Protection Agency (USEPA). *Final Guidelines for Exposure Assessment*. EPA/600/Z-92/001. Washington, DC, May 29, 1992. http://www.epa.gov/raf/publications/guidelines-for-exposure-assessment.htm
 3. United States Environmental Protection Agency (USEPA). *Risk Assessment Guidance for Superfund: Volume 1: Human Health Evaluation Manual (Part A) (Interim Final) (RAGS)*. EPA/540/1-89/002. Washington, DC, December 1989. http://www.epa.gov/oswer/riskassessment/ragsa/
 4. United States Environmental Protection Agency (USEPA). *Risk Assessment Guidance for Superfund Volume I: Human Health Evaluation Manual. Supplemental Guidance. "Standard Default Exposure Factors."* PB91-921314. OSWER Directive 9285.6-03 Washington, DC, March 25, 1991. http://www.epa.gov/oswer/riskassessment/pdf/oswer_directive_9285_6-03.pdf
 5. United States Environmental Protection Agency (USEPA). *Exposure Factors Handbook (1997 Final Report)*. EPA/600/P-95/002F a-c, Washington, DC, 1997. http://cfpub.epa.gov/ncea/risk/recordisplay.cfm?deid=12464
 6. United States Environmental Protection Agency (USEPA). *Child-Specific Exposure Factors Handbook (Final Report) 2–9*. EPA/600/R-06/096F, Washington, DC, 2008. http://cfpub.epa.gov/ncea/risk/recordisplay.cfm?deid=199243
 7. United States Environmental Protection Agency (USEPA). *Exposure Factors Handbook 2011 Edition (Final)*. EPA/600/R-09/052F. Washington, DC, September, 2011. http://cfpub.epa.gov/ncea/risk/recordisplay.cfm?deid=236252
 8. United States Environmental Protection Agency (USEPA). *Superfund Exposure Assessment Manual*. EPA/540/1-88/001. OSWER Directive 9285.5-1, Washington, DC, April, 1988. http://rais.ornl.gov/documents/Exposure_Assessment_Manual_1988_EPA5401881001.pdf
 9. United States Environmental Protection Agency (USEPA). *Guidelines for Exposure Assessment*. EPA/600/Z-92/001. Washington, DC, May 29, 1992. http://www.epa.gov/raf/publications/guidelines-for-exposure-assessment.htm
10. Ankley GT, Bennett RS, Erickson RJ, Hoff DJ, Hornung MW, Johnson RD et al. Adverse outcome pathways: A conceptual framework to support ecotoxicology research and risk assessment. *Environ Toxicol Chem*. 2010, March;29(3):730–741.

11. Agency for Toxic Substances and Disease Registry. *Minimal Risk Levels.* n.d. http://www.atsdr.cdc.gov/mrls/index.asp. (accessed September 10, 2012).
12. United States Environmental Protection Agency (USEPA). *A Review of the Reference Dose and Reference Concentration Processes.* EPA/630/P-02/002F. Washington, DC, December, 2002. http://www.epa.gov/raf/publications/review-reference-dose.htm
13. Centers for Disease Control and Prevention (CDC). *National Health and Nutrition Examination Survey (NHANES).* n.d. http://www.cdc.gov/nchs/nhanes.htm (accessed September 30, 2012).
14. United States Environmental Protection Agency (USEPA). *National Oil and Hazardous Substances Pollution Contingency Plan (NCP) Proposed Rule.* 53 Federal Register 51394. December 12, 1988. http://www.epa.gov/superfund/policy/remedy/sfremedy/pdfs/ncppropream.pdf (accessed September 30, 2012).
15. United States Environmental Protection Agency (USEPA). *Choosing a Percentile of Acute Dietary Exposure as a Threshold of Regulatory Concern.* Office of Pesticide Programs. Washington, DC, March 16, 2000. http://www.epa.gov/pesticides/trac/science/trac2b054.pdf (accessed September 30, 2012).
16. Timchalk C, Nolan RJ, Mendrala AL, Dittenber DA, Brzak KA, and Mattsson JL. A physiologically based pharmacokinetic and pharmacodynamic (PBPK/PD) model for the organophosphate insecticide chlorpyrifos in rats and humans. *Toxicol Sci.* 2002, March;66(1):34–53.
17. Timchalk C and Poet TS. Development of a physiologically based pharmacokinetic and pharmacodynamic model to determine dosimetry and cholinesterase inhibition for a binary mixture of chlorpyrifos and diazinon in the rat. *Neurotoxicology.* 2008, May;29(3):428–443.
18. United States Environmental Protection Agency (USEPA). *Framework for Cumulative Risk Assessment.* EPA/630/P-02/001F. Risk Assessment Forum. Washington, DC, May, 2003. http://www.epa.gov/raf/publications/pdfs/frmwrk_cum_risk_assmnt.pdf (accessed August 18, 2012).
19. United States Environmental Protection Agency (USEPA). *Opportunities to Improve Data Quality and Children's Health Through the Food Quality Protection Act.* Report No. 2006-P-00009. Office of Inspector General. January 10, 2006. http://www.epa.gov/oig/reports/2006/20060110-2006-P-00009.pdf (accessed September 1, 2012).
20. Hattis D, Banati P, Goble R, and Burmaster DE. Human interindividual variability in parameters related to health risks. *Risk Anal.* 1999, August;19(4):711–726.
21. Wilson ND, Price PS, and Paustenbach DJ. An event-by-event probabilistic methodology for assessing the health risks of persistent chemicals in fish: A case study at the Palos Verdes Shelf. *J Toxicol Environ Health A.* 2001, April 4;62(8):595–642.
22. Sherer RA and Price PS. The effect of cooking processes on PCB levels in edible fish tissue. *Qual Assur.* 1993, December;2(4):396–407.
23. Price PS, Curry CL, Goodrum PE, Gray MN, McCrodden JI, Harrington NW et al. Monte Carlo modeling of time-dependent exposures using a microexposure event approach. *Risk Anal.* 1996;16(3):339–348.
24. Price PS, Conolly RB, Chaisson CF, Gross EA, Young JS, Mathis ET, and Tedder DR. Modeling interindividual variation in physiological factors used in PBPK models of humans. *Crit Rev Toxicol.* 2003;33(5):469–503.
25. United States Environmental Protection Agency (USEPA). *DRAFT FINAL VER II: Fish Biomass Estimate for the Housatonic River Primary Study Area.* DCN: GE-061202-ABBF. Boston, MA, June, 2002. http://www.epa.gov/region1/ge/thesite/restofriver/reports/final_era/SupportingInformationStudies4HousatonicRiverProject/Fish_Biomass.pdf (accessed September 1, 2012).

26. Oreskes N. Evaluation (not validation) of quantitative models. *Environ Health Perspect.* 1998, December;106(Suppl 6):1453–1460.
27. Chamberlain TC. Lord Kelvin's address on the age of the earth as an abode fitted for life. *Science.* 1899, June 6;9(235):889–901.
28. England P, Molnar P, and Richter F. John Perry's neglected critique of Kelvin's age for the Earth: A missed opportunity in geodynamics. *GSA Today.* 2007, January;17(1):4–9.
29. Burmaster DE and Anderson PD. Principles of good practice for the use of Monte Carlo techniques in human health and ecological risk assessments. *Risk Anal.* 1994, August;14(4):477–481.
30. Burmaster DE and von Stackelberg K. Using Monte Carlo simulations in public health risk assessments: Estimating and presenting full distributions of risk. *J Expo Anal Environ Epidemiol.* 1991, October;1(4):491–512.
31. Thompson KM, Burmaster DE, and Crouch EAC. Monte Carlo techniques for quantitative uncertainty analysis in public health risk assessments. *Risk Anal.* 1992;12(1):53–63.
32. Burmaster DE and Harris RH. The magnitude of compounding conservatisms in superfund risk assessments. *Risk Anal.* 1993;13(2):131–134.
33. Cullen AC. Measures of compounding conservatism in probabilistic risk assessment. *Risk Anal.* 1994, August;14(4):389–393.
34. United States Environmental Protection Agency (USEPA). *Risk Assessment Guidance for Superfund (RAGS) Volume III - Part A: Process for Conducting Probabilistic Risk Assessment.* EPA 540-R-02-002. OSWER 9285.7-45. PB2002 963302. Washington, DC, December, 2001. http://www.epa.gov/oswer/riskassessment/rags3adt/ (accessed September 1, 2012).
35. United States Environmental Protection Agency (USEPA). *Policy for the Use of Probabilistic Analysis in Risk Assessment.* Office of the Science Advisor. May 15, 1997. http://www.epa.gov/stpc/2probana.htm. (accessed September 1, 2012).
36. United States Environmental Protection Agency (USEPA). *Guiding Principles for Monte Carlo Analysis.* Office of the Science Advisor. EPA/630/R-97/001. March, 1997. http://www.epa.gov/raf/publications/pdfs/montecar.pdf (accessed September 1, 2012).
37. Oreskovic NM, Blossom J, Field AE, Chiang SR, Winickoff JP, and Kleinman RE. Combining global positioning system and accelerometer data to determine the locations of physical activity in children. *Geospat Health.* 2012, May;6(2):263–272.
38. Buonanno G, Marini S, Morawska L, and Fuoco FC. Individual dose and exposure of Italian children to ultrafine particles. *Sci Total Environ.* 2012, September 9;438C:271–277.
39. Bernstein PL. *Against the Gods: The Remarkable Story of Risk.* New York: John Wiley & Sons, Inc., 1998.
40. Wickwire T, Johnson MS, Hope BK, and Greenberg MS. Spatially explicit ecological exposure models: A rationale for and path toward their increased acceptance and use. *Integr Environ Assess Manag.* 2011, April;7(2):158–168.
41. Schipper AM, Loos M, Ragas AMJ, Lopes JPC, Nolte BT, Wijnhoven S, and Leuven RSEW. Modeling the influence of environmental heterogeneity on heavy metal exposure concentrations for terrestrial vertebrates in river floodplains. *Environ Toxicol Chem.* 2008, April;27(4):919–932.
42. Hope BK. A spatially and bioenergetically explicit terrestrial ecological exposure model. *Toxicol Ind Health.* 2001, June;17(5–10):322–332.
43. Hope BK. A case study comparing static and spatially explicit ecological exposure analysis methods. *Risk Anal.* 2001, December;21(6):1001–1010.
44. Hope BK. Generating probabilistic spatially-explicit individual and population exposure estimates for ecological risk assessments. *Risk Anal.* 2000, October;20(5):573–589.

45. von Stackelberg K, Burmistrov D, Linkov I, Cura J, and Bridges TS. The use of spatial modeling in an aquatic food web to estimate exposure and risk. *Sci Total Environ.* 2002, April 4;288(1–2):97–110.

46. Linkov I, Grebenkov A, and Baitchorov VM. Spatially explicit exposure models: Application to military sites. *Toxicol Ind Health.* 2001, June;17(5–10):230–235.

47. Bell ML. The use of ambient air quality modeling to estimate individual and population exposure for human health research: A case study of ozone in the Northern Georgia Region of the United States. *Environ Int.* 2006, July;32(5):586–593.

48. Cao H, Suzuki N, Sakurai T, Matsuzaki K, Shiraishi H, and Morita M. Probabilistic estimation of dietary exposure of the general Japanese population to dioxins in fish, using region-specific fish monitoring data. *J Expo Sci Environ Epidemiol.* 2008, May;18(3):236–245.

49. United States Environmental Protection Agency (USEPA). *Air Quality Criteria for Particulate Matter: Volume II.* EPA/600/P-95/001bF. Office of Research and Development. Research Triangle Park, NC, October, 2004. http://cfpub.epa.gov/ncea/cfm/recordisplay.cfm?deid=87903#Download (accessed September 8, 2012).

50. United States Environmental Protection Agency (USEPA). *Risk Assessment Guidance for Superfund: Volume 1: Human Health Evaluation Manual (Part F, Supplemental Guidance for Inhalation Risk Assessment) Final.* EPA-540-R-070-002. OSWER 9285.7-82. Office of Superfund Remediation and Technical Innovation. Washington, DC, January, 2009. http://www.epa.gov/oswer/riskassessment/ragsf/index.htm (accessed September 8, 2012).

51. United States Environmental Protection Agency (USEPA). *An Overview of Methods for EPA's National-Scale Air Toxics Assessment.* Office of Air Quality, Planning and Standards. Research Triangle Park, NC, January 31, 2011. http://www.epa.gov/ttn/atw/nata2005/05pdf/nata_tmd.pdf (accessed September 8, 2012).

52. McLaren P and Beveridge RP. Sediment trend analysis of the Hylebos Waterway: Implications for liability allocations. *Integr Environ Assess Manag.* 2006, July;2(3):262–272.

53. Raikow DF, Walters DM, Fritz KM, and Mills MA. The distance that contaminated aquatic subsidies extend into lake riparian zones. *Ecol Appl.* 2011, April;21(3):983–990.

54. United States Environmental Protection Agency (USEPA). *Risk Assessment Guidance for Superfund Volume I: Human Health Evaluation Manual. Supplemental Guidance. "Standard Default Exposure Factors."* PB91-921314. OSWER Directive 9285.6-03 Washington, DC, March 25, 1991. http://www.epa.gov/oswer/riskassessment/pdf/oswer_directive_9285_6-03.pdf

55. Lanphear BP, Matte TD, Rogers J, Clickner RP, Dietz B, Bornschein RL et al. The contribution of lead-contaminated house dust and residential soil to children's blood lead levels. A pooled analysis of 12 epidemiologic studies. *Environ Res.* 1998, October;79(1):51–68.

56. Lanphear BP, Succop P, Roda S, and Henningsen G. The effect of soil abatement on blood lead levels in children living near a former smelting and milling operation. *Public Health Rep.* 2003, March;118(2):83–91.

57. Gasana J, Hlaing WM, Siegel KA, Chamorro A, and Niyonsenga T. Blood lead levels in children and environmental lead contamination in Miami inner city, Florida. *Int J Environ Res Public Health.* 2006, September;3(3):228–234.

58. Mielke HW, Gonzales CR, Powell E, Jartun M, and Mielke PWJ. Nonlinear association between soil lead and blood lead of children in metropolitan New Orleans, Louisiana: 2000–2005. *Sci Total Environ.* 2007, December 12;388(1–3):43–53.

59. Ranft U, Delschen T, Machtolf M, Sugiri D, and Wilhelm M. Lead concentration in the blood of children and its association with lead in soil and ambient air—Trends between 1983 and 2000 in Duisburg. *J Toxicol Environ Health A.* 2008;71(11–12):710–715.

60. Lepow ML, Bruckman L, Rubino RA, Markowtiz S, Gillette M, and Kapish J. Role of airborne lead in increased body burden of lead in Hartford children. *Environ Health Perspect.* 1974, May;7:99–102.

61. Kimbrough RD, Falk H, Stehr P, and Fries G. Health implications of 2,3,7,8-tetrachlorodibenzodioxin (TCDD) contamination of residential soil. *J Toxicol Environ Health.* 1984;14(1):47–93.

62. Simon T. A comparison of three different methods for estimating children's soil and dust ingestion rates #1916. *The Toxicologist CD— An Official Journal of the Society of Toxicology.* 2010, March 3;126:124. http://www.toxicology.org/AI/Pub/Tox/2010Tox.pdf (accessed September 22, 2013).

63. Calabrese EJ, Barnes R, Stanek EJI, Pastides H, Gilbert CE, Veneman P et al. How much soil do young children ingest: An epidemiologic study. *Regul Toxicol Pharmacol.* 1989;10(2):123–137.

64. Calabrese EJ, Stanek EJ, Gilbert CE, and Barnes RM. Preliminary adult soil ingestion estimates: Results of a pilot study. *Regul Toxicol Pharmacol.* 1990;12(1):88–95.

65. Calabrese EJ, Stanek EJ, and Gilbert CE. Evidence of soil-pica behaviour and quantification of soil ingested. *Hum Exp Toxicol.* 1991;10(4):245–249.

66. Calabrese EJ and Stanek EJI. A guide to interpreting soil ingestion studies. II. Qualitative and quantitative evidence of soil ingestion. *Regul Toxicol Pharmacol.* 1991;13(3):278–292.

67. Calabrese EJ and Stanek ES. Distinguishing outdoor soil ingestion from indoor dust ingestion in a soil pica child. *Regul Toxicol Pharmacol.* 1992;15(1):83–85.

68. Calabrese EJ and Stanek EJI. A dog's tale: Soil ingestion by a canine. *Ecotoxicol Environ Saf.* 1995;32(1):93–95.

69. Calabrese EJ and Stanek EJI. Resolving intertracer inconsistencies in soil ingestion estimation. *Environ Health Perspect.* 1995;103(5):454–457.

70. Calabrese EJ, Stanek EJ, Barnes R, Burmaster DE, Callahan BG, Heath JS et al. Methodology to estimate the amount and particle size of soil ingested by children: Implications for exposure assessment at waste sites. *Regul Toxicol Pharmacol.* 1996;24(3):264–268.

71. Calabrese EJ, Stanek EJI, Pekow P, and Barnes RM. Soil ingestion estimates for children residing on a superfund site. *Ecotoxicol Environ Saf.* 1997;36(3):258–268.

72. Calabrese EJ, Stanek EJ, James RC, and Roberts SM. Soil ingestion: A concern for acute toxicity in children. *Environ Health Perspect.* 1997;105(12):1354–1358.

73. Calabrese EJ, Staudenmayer JW, and Stanek EJ. Drug development and hormesis: Changing conceptual understanding of the dose response creates new challenges and opportunities for more effective drugs. *Curr Opin Drug Discov Devel.* 2006;9(1):117–123.

74. Calabrese EJ, Staudenmayer JW, Stanek EJI, and Hoffmann GR. Hormesis outperforms threshold model in National Cancer Institute antitumor drug screening database. *Toxicol Sci.* 2006;94(2):368–378.

75. Calabrese EJ. The road to linearity: Why linearity at low doses became the basis for carcinogen risk assessment. *Arch Toxicol.* 2009, March;83(3):203–225.

76. Stifelman M. Daily soil ingestion estimates for children at a Superfund site. *Risk Anal.* 2006;26(4):863.

77. Stanek EJ, Calabrese EJ, and Barnes R. Soil ingestion estimates for children in anaconda using trace element concentrations if different particle size fractions. *Hum Ecol Risk Assess.* 1999;5(3):547–558.

78. Stanek EJ, Calabrese EJ, and Zorn M. Biasing factors for simple soil ingestion estimates in mass balance studies of soil ingestion. *Hum Ecol Risk Assess.* 2001;7(2):329–355.

79. Stanek EJ, Calabrese EJ, and Zorn M. Soil ingestion distributions for Monte Carlo risk assessment in children. *Hum Ecol Risk Assess.* 2001;7(2):357–368.
80. Stanek EJI and Calabrese EJ. A guide to interpreting soil ingestion studies. I. Development of a model to estimate the soil ingestion detection level of soil ingestion studies. *Regul Toxicol Pharmacol.* 1991;13(3):263–277.
81. Stanek EJI and Calabrese EJ. Daily estimates of soil ingestion in children. *Environ Health Perspect.* 1995;103(3):276–285.
82. Stanek EJI, Calabrese EJ, Barnes R, and Pekow P. Soil ingestion in adults—Results of a second pilot study. *Ecotoxicol Environ Saf.* 1997;36(3):249–257.
83. Stanek EJI and Calabrese EJ. Daily soil ingestion estimates for children at a Superfund site. *Risk Anal.* 2000;20(5):627–635.
84. Clausing P, Brunekreef B, and van Wijnen JH. A method for estimating soil ingestion by children. *Int Arch Occup Environ Health.* 1987;59(1):73–82.
85. Binder S, Sokal D, and Maughan D. Estimating soil ingestion: The use of tracer elements in estimating the amount of soil ingested by young children. *Arch Environ Health.* 1986, November;41(6):341–345.
86. Priest ND. The bioavailability and metabolism of aluminium compounds in man. *Proc Nutr Soc.* 1993, February;52(1):231–240.
87. Priest ND. The biological behaviour and bioavailability of aluminium in man, with special reference to studies employing aluminium-26 as a tracer: Review and study update. *J Environ Monit.* 2004, May;6(5):375–403.
88. Martin KR. The chemistry of silica and its potential health benefits. *J Nutr Health Aging.* 2007, March;11(2):94–97.
89. Jugdaohsingh R, Anderson SH, Tucker KL, Elliott H, Kiel DP, Thompson RP, and Powell JJ. Dietary silicon intake and absorption. *Am J Clin Nutr.* 2002, May;75(5):887–893.
90. Jugdaohsingh R. Silicon and bone health. *J Nutr Health Aging.* 2007, March;11(2):99–110.
91. Powell JJ, Ainley CC, Harvey RS, Mason IM, Kendall MD, Sankey EA et al. Characterisation of inorganic microparticles in pigment cells of human gut associated lymphoid tissue. *Gut.* 1996, March;38(3):390–395.
92. Lomer MCE, Thompson RPH, and Powell JJ. Fine and ultrafine particles of the diet: Influence on the mucosal immune response and association with Crohn's disease. *Proc Nutr Soc.* 2002, February;61(1):123–130.
93. Powell JJ, Jugdaohsingh R, and Thompson RP. The regulation of mineral absorption in the gastrointestinal tract. *Proc Nutr Soc.* 1999, February;58(1):147–153.
94. Reffitt DM, Jugdaohsingh R, Thompson RP, and Powell JJ. Silicic acid: Its gastrointestinal uptake and urinary excretion in man and effects on aluminium excretion. *J Inorg Biochem.* 1999, August 8;76(2):141–147.
95. Dominy NJ, Davoust E, and Minekus M. Adaptive function of soil consumption: An in vitro study modeling the human stomach and small intestine. *J Exp Biol.* 2004, January;207(Pt 2):319–324.
96. Halsted JA. Geophagia in man: Its nature and nutritional effects. *Am J Clin Nutr.* 1968, December;21(12):1384–1393.
97. Stanek Iii EJ, Calabrese EJ, and Xu B. Meta-analysis of mass-balance studies of soil ingestion in children. *Risk Anal.* 2012, March;32(3):433–447.
98. Sibbald B. Arsenic and pressure-treated wood: The argument moves to the playground. *CMAJ.* 2002, January 8;166(1):79. http://www.ncbi.nlm.nih.gov/pmc/articles/PMC99243/ (accessed September 22, 2013).
99. Hemond HF and Solo-Gabriele HM. Children's exposure to arsenic from CCA-treated wooden decks and playground structures. *Risk Anal.* 2004, February;24(1):51–64.

100. Kwon E, Zhang H, Wang Z, Jhangri GS, Lu X, Fok N et al. Arsenic on the hands of children after playing in playgrounds. *Environ Health Perspect.* 2004, October;112(14):1375–1380.

101. Shalat SL, Solo-Gabriele HM, Fleming LE, Buckley BT, Black K, Jimenez M et al. A pilot study of children's exposure to CCA-treated wood from playground equipment. *Sci Total Environ.* 2006, August 8;367(1):80–88.

102. United States Environmental Protection Agency (USEPA). *A Set of Scientific Issues Being Considered By the Environmental Protection Agency Regarding: Preliminary Evaluation of the Non-Dietary Hazard and Exposure to Children from Contact with Chromated Copper Arsenate (CCA)-Treated Wood Playground Structures and CCA-Contaminated Soil.* SAP Report No 2001-12. December 12, 2001. http://www.epa.gov/scipoly/sap/meetings/2001/october/ccawood.pdf (accessed September 22, 2012).

103. Consumer Products Safety Commission (CPSC). *Memorandum from B.C. Lee to E.A. Tyrell: Estimating the Risk of Skin Cancer from Ingested Inorganic Arsenic.* Washington, DC, January 26, 1990. http://www.cpsc.gov/PageFiles/84483/woodpla2.pdf (accessed September 22, 2012).

104. Consumer Products Safety Commission (CPSC). *Memorandum from T.A. Thomas to P.M. Bittner: Determination of Dislodgeable Arsenic Transfer to Human Hands and Surrogates from CCA-Treated Wood.* Washington, DC, January 23, 2003. http://www.cpsc.gov/PageFiles/86756/cca3.pdf (accessed September 22, 2012).

105. American Chemistry Council (ACC). *Assessment of Exposure to Metals in CCA-Preserved Wood.* Arlington, VA, June 20, 2003. http://www.epa.gov/oscpmont/sap/meetings/2003/120303_mtg.htm#materials (accessed September 22, 2012).

106. Xue J, Zartarian VG, Ozkaynak H, Dang W, Glen G, Smith L, and Stallings C. A probabilistic arsenic exposure assessment for children who contact chromated copper arsenate (CCA)-treated playsets and decks, Part 2: Sensitivity and uncertainty analyses. *Risk Anal.* 2006, April;26(2):533–541.

107. Zartarian VG, Xue J, Ozkaynak H, Dang W, Glen G, Smith L, and Stallings C. A probabilistic arsenic exposure assessment for children who contact CCA-treated playsets and decks, Part 1: Model methodology, variability results, and model evaluation. *Risk Anal.* 2006, April;26(2):515–531.

108. Hore P, Zartarian V, Xue J, Ozkaynak H, Wang S-W, Yang Y-C et al. Children's residential exposure to chlorpyrifos: Application of CPPAES field measurements of chlorpyrifos and TCPy within MENTOR/SHEDS-Pesticides model. *Sci Total Environ.* 2006, August 8;366(2–3):525–537.

109. Buck RJ, Ozkaynak H, Xue J, Zartarian VG, and Hammerstrom K. Modeled estimates of chlorpyrifos exposure and dose for the Minnesota and Arizona NHEXAS populations. *J Expo Anal Environ Epidemiol.* 2001, May;11(3):253–268.

110. Zartarian VG, Streicker J, Rivera A, Cornejo CS, Molina S, Valadez OF, and Leckie JO. A pilot study to collect micro-activity data of two- to four-year-old farm labor children in Salinas Valley, California. *J Expo Anal Environ Epidemiol.* 1995;5(1):21–34.

111. Zartarian VG, Ferguson AC, Ong CG, and Leckie JO. Quantifying videotaped activity patterns: Video translation software and training methodologies. *J Expo Anal Environ Epidemiol.* 1997;7(4):535–542.

112. Zartarian VG, Ferguson AC, and Leckie JO. Quantified dermal activity data from a four-child pilot field study. *J Expo Anal Environ Epidemiol.* 1997;7(4):543–552.

113. Zartarian VG, Ozkaynak H, Burke JM, Zufall MJ, Rigas ML, and Furtaw EJ. A modeling framework for estimating children's residential exposure and dose to chlorpyrifos via dermal residue contact and nondietary ingestion. *Environ Health Perspect.* 2000, June;108(6):505–514.

114. Özkaynak H, Xue J, Zartarian VG, Glen G, and Smith L. Modeled estimates of soil and dust ingestion rates for children. *Risk Anal.* 2011, April;31(4):592–608.
115. Reed KJ, Jimenez M, Freeman NC, and Lioy PJ. Quantification of children's hand and mouthing activities through a videotaping methodology. *J Expo Anal Environ Epidemiol.* 1999;9(5):513–520.
116. Ferguson AC, Canales RA, Beamer P, Auyeung W, Key M, Munninghoff A et al. Video methods in the quantification of children's exposures. *J Expo Sci Environ Epidemiol.* 2006, May;16(3):287–298.
117. Black K, Shalat SL, Freeman NC, Jimenez M, Donnelly KC, and Calvin JA. Children's mouthing and food-handling behavior in an agricultural community on the US/Mexico border. *J Expo Anal Environ Epidemiol.* 2005;15(3):244–251.
118. Beamer P, Key ME, Ferguson AC, Canales RA, Auyeung W, and Leckie JO. Quantified activity pattern data from 6 to 27-month-old farmworker children for use in exposure assessment. *Environ Res.* 2008, October;108(2):239–246.
119. Beamer P, Canales RA, and Leckie JO. Developing probability distributions for transfer efficiencies for dermal exposure. *J Expo Sci Environ Epidemiol.* 2009, March;19(3):274–283.
120. Auyeung W, Canales RA, Beamer P, Ferguson AC, and Leckie JO. Young children's hand contact activities: An observational study via videotaping in primarily outdoor residential settings. *J Expo Sci Environ Epidemiol.* 2006, September;16(5):434–446.
121. AuYeung W, Canales RA, and Leckie JO. The fraction of total hand surface area involved in young children's outdoor hand-to-object contacts. *Environ Res.* 2008, November;108(3):294–299.
122. Ko S, Schaefer PD, Vicario CM, and Binns HJ. Relationships of video assessments of touching and mouthing behaviors during outdoor play in urban residential yards to parental perceptions of child behaviors and blood lead levels. *J Expo Sci Environ Epidemiol.* 2007;17(1):47–57.
123. Tulve NS, Suggs JC, McCurdy T, Cohen Hubal EA, and Moya J. Frequency of mouthing behavior in young children. *J Expo Anal Environ Epidemiol.* 2002;12(4):259–264.
124. Xue J, Zartarian V, Moya J, Freeman N, Beamer P, Black K et al. A meta-analysis of children's hand-to-mouth frequency data for estimating nondietary ingestion exposure. *Risk Anal.* 2007;27(2):411–420.
125. Kissel JC, Shirai JH, Richter KY, and Fenske RA. Empirical investigation of hand-to-mouth transfer of soil. *Bull Environ Contam Toxicol.* 1998, March;60(3):379–386.
126. Angle CR, Marcus A, Cheng IH, and McIntire MS. Omaha childhood blood lead and environmental lead: A linear total exposure model. *Environ Res.* 1984, October;35(1):160–170.
127. Langlois P, Smith L, Fleming S, Gould R, Goel V, and Gibson B. Blood lead levels in Toronto children and abatement of lead-contaminated soil and house dust. *Arch Environ Health.* 1996;51(1):59–67.
128. Lanphear BP, Weitzman M, Winter NL, Eberly S, Yakir B, Tanner M et al. Lead-contaminated house dust and urban children's blood lead levels. *Am J Public Health.* 1996, October;86(10):1416–1421.
129. Charney E, Sayre J, and Coulter M. Increased lead absorption in inner city children: Where does the lead come from? *Pediatrics.* 1980, February;65(2):226–231.
130. Thornton I, Davies DJ, Watt JM, and Quinn MJ. Lead exposure in young children from dust and soil in the United Kingdom. *Environ Health Perspect.* 1990;8955–8960.
131. Mielke HW and Reagan PL. Soil is an important pathway of human lead exposure. *Environ Health Perspect.* 1998, February;106(Suppl 1):217–229.

132. Berglund M, Lind B, Sörensen S, and Vahter M. Impact of soil and dust lead on children's blood lead in contaminated areas of Sweden. *Arch Environ Health.* 2000;55(2):93–97.

133. Manton WI, Angle CR, Stanek KL, Reese YR, and Kuehnemann TJ. Acquisition and retention of lead by young children. *Environ Res.* 2000, January;82(1):60–80.

134. O'Flaherty EJ. Physiologically based models for bone-seeking elements. IV. Kinetics of lead disposition in humans. *Toxicol Appl Pharmacol.* 1993, January;118(1):16–29.

135. O'Flaherty EJ. Physiologically based models for bone-seeking elements. V. Lead absorption and disposition in childhood. *Toxicol Appl Pharmacol.* 1995, April;131(2):297–308.

136. Manton WI. Sources of lead in blood. Identification by stable isotopes. *Arch Environ Health.* 1977;32(4):149–159.

137. von Lindern I, Spalinger S, Petroysan V, and von Braun M. Assessing remedial effectiveness through the blood lead:soil/dust lead relationship at the Bunker Hill Superfund Site in the Silver Valley of Idaho. *Sci Total Environ.* 2003, February 2;303(1–2):139–170.

138. United States Environmental Protection Agency (USEPA). *Guidance Manual for the Integrated Exposure Uptake Biokinetic Model for Lead in Children.* PB93.963510. OSWER #9285.7-15-1. Washington, DC, February, 1994. http://www.epa.gov/ superfund/lead/products.htm#user (accessed September 22, 2012).

139. Davis S and Mirick DK. Soil ingestion in children and adults in the same family. *J Expo Sci Environ Epidemiol.* 2006;16(1):63–75.

140. Roseberry AM and Burmaster DE. Lognormal distributions for water intake by children and adults. *Risk Anal.* 1992, March;12(1):99–104.

141. Kahn HD and Stralka K. Estimated daily average per capita water ingestion by child and adult age categories based on USDA's 1994–1996 and 1998 continuing survey of food intakes by individuals. *J Expo Sci Environ Epidemiol.* 2009, May;19(4):396–404.

142. United States Environmental Protection Agency (USEPA). *Risk Assessment Guidance for Superfund: Volume 1: Human Health Evaluation Manual (Part B, Development of Preliminary Remediation Goals) (RAGS, Part B).* EPA/540/R-92/003. Publication 9285.7-01B. Washington, DC, December, 1991. http://www.epa.gov/oswer/ riskassessment/ragsb/ (accessed September 25, 2012).

143. United States Environmental Protection Agency (USEPA). *Methods for Derivation of Inhalation Reference Concentrations and Application of Inhalation Dosimetry.* EPA/600/8-90/066F. Research Triangle Park, NC, October, 1994. http://cfpub.epa. gov/ncea/cfm/recordisplay.cfm?deid=71993 (accessed September 25, 2012).

144. Crowder TM, Rosati JA, Schroeter JD, Hickey AJ, and Martonen TB. Fundamental effects of particle morphology on lung delivery: Predictions of Stokes' law and the particular relevance to dry powder inhaler formulation and development. *Pharm Res.* 2002, March;19(3):239–245.

145. Hickey AJ, Martonen TB, and Yang Y. Theoretical relationship of lung deposition to the fine particle fraction of inhalation aerosols. *Pharm Acta Helv.* 1996, August;71(3):185–190.

146. Martonen TB, Katz I, Fults K, and Hickey AJ. Use of analytically defined estimates of aerosol respirable fraction to predict lung deposition patterns. *Pharm Res.* 1992, December;9(12):1634–1639.

147. Layton DW. Metabolically consistent breathing rates for use in dose assessments. *Health Phys.* 1993, January;64(1):23–36.

148. Brochu P, Ducré-Robitaille J-F, and Brodeur J. Physiological daily inhalation rates for free-living individuals aged 1 month to 96 years, using data from doubly labeled water measurements: A proposal for air quality criteria, standard calculations and health risk assessment. *Hum Ecol Risk Assess.* 2007, January;12(4):675–701.

149. Brochu P, Ducré-Robitaille J-F, and Brodeur J. Physiological daily inhalation rates for free-living pregnant and lactating adolescents and women aged 11 to 55 years, using data from doubly labeled water measurements for use in health risk assessment. *Hum Ecol Risk Assess*. 2007, January;12(4):702–735.
150. Brochu P, Ducré-Robitaille J-F, and Brodeur J. Physiological daily inhalation rates for free-living individuals aged 2.6 months to 96 years based on doubly labeled water measurements: Comparison with time-activity-ventilation and metabolic energy conversion estimates. *Hum Ecol Risk Assess*. 2007, January;12(4):736–761.
151. Brochu P, Brodeur J, and Krishnan K. Derivation of physiological inhalation rates in children, adults, and elderly based on nighttime and daytime respiratory parameters. *Inhal Toxicol*. 2011, February;23(2):74–94.
152. Stifelman M. Using doubly-labeled water measurements of human energy expenditure to estimate inhalation rates. *Sci Total Environ*. 2007, February 2;373(2–3):585–590.
153. Feingold KR. The outer frontier: The importance of lipid metabolism in the skin. *J Lipid Res*. 2009, April;50(Suppl):S417–S422.
154. Paudel KS, Milewski M, Swadley CL, Brogden NK, Ghosh P, and Stinchcomb AL. Challenges and opportunities in dermal/transdermal delivery. *Ther Deliv*. 2010, July;1(1):109–131.
155. Nemes Z and Steinert PM. Bricks and mortar of the epidermal barrier. *Exp Mol Med*. 1999, March 3;31(1):5–19.
156. Lee SH, Jeong SK, and Ahn SK. An update of the defensive barrier function of skin. *Yonsei Med J*. 2006, June 6;47(3):293–306.
157. Madison KC. Barrier function of the skin: "la raison d'être" of the epidermis. *J Invest Dermatol*. 2003, August;121(2):231–241.
158. Marks R. The stratum corneum barrier: The final frontier. *J Nutr*. 2004, August;134 (8 Suppl):2017S–2021S.
159. Phillips LJ, Fares RJ, and Schweer LG. Distributions of total skin surface area to body weight ratios for use in dermal exposure assessments. *J Expo Anal Environ Epidemiol*. 1993;3(3):331–338.
160. Murray DM and Burmaster DE. Estimated distributions for total body surface area of men and women in the United States. *J Expo Anal Environ Epidemiol*. 1992;2(4):451–461.
161. Boniol M, Verriest J-P, Pedeux R, and Doré J-F. Proportion of skin surface area of children and young adults from 2 to 18 years old. *J Invest Dermatol*. 2008, February;128(2):461–464.
162. Spalt EW, Kissel JC, Shirai JH, and Bunge AL. Dermal absorption of environmental contaminants from soil and sediment: A critical review. *J Expo Sci Environ Epidemiol*. 2009, February;19(2):119–148.
163. Choate LM, Ranville JF, Bunge AL, and Macalady DL. Dermally adhered soil: 1. Amount and particle-size distribution. *Integr Environ Assess Manag*. 2006, October;2(4):375–384.
164. Choate LM, Ranville JF, Bunge AL, and Macalady DL. Dermally adhered soil: 2. Reconstruction of dry-sieve particle-size distributions from wet-sieve data. *Integr Environ Assess Manag*. 2006, October;2(4):385–390.
165. Kissel JC, Richter KY, and Fenske RA. Factors affecting soil adherence to skin in hand-press trials. *Bull Environ Contam Toxicol*. 1996, May;56(5):722–728.
166. Li H, Chen J, Wu W, and Piao X. Distribution of polycyclic aromatic hydrocarbons in different size fractions of soil from a coke oven plant and its relationship to organic carbon content. *J Hazard Mater*. 2010, April 4;176(1–3):729–734.
167. Roy TA and Singh R. Effect of soil loading and soil sequestration on dermal bioavailability of polynuclear aromatic hydrocarbons. *Bull Environ Contam Toxicol*. 2001, September;67(3):324–331.

168. United States Environmental Protection Agency (USEPA). *Dermal Exposure Assessment: Principles and Applications.* EPA/600/8-91/011B. Interim Report. Washington, DC, January 1992. http://www.epa.gov/ncea/pdfs/efh/references/ DEREXP.PDF (accessed September 29, 2012).

169. United States Environmental Protection Agency (USEPA). *Risk Assessment Guidance for Superfund: Volume 1: Human Health Evaluation Manual (Part E, Supplemental Guidance for Dermal Risk Assessment) Final.* EPA/540/R/99/005. OSWER 9285.7-02EP. PB99-963312. Washington, DC, July, 2004. http://www.epa. gov/oswer/riskassessment/ragse/ (accessed August 19, 2012).

170. Kissel JC, Richter KY, and Fenske RA. Field measurement of dermal soil loading attributable to various activities: Implications for exposure assessment. *Risk Anal.* 1996, February;16(1):115–125.

171. Holmes KKJ, Shirai JH, Richter KY, and Kissel JC. Field measurement of dermal soil loadings in occupational and recreational activities. *Environ Res.* 1999, February;80(2 Pt 1):148–157.

172. Rodes CE, Newsome JR, Vanderpool RW, Antley JT, and Lewis RG. Experimental methodologies and preliminary transfer factor data for estimation of dermal exposures to particles. *J Expo Anal Environ Epidemiol.* 2001;11(2):123–139.

173. Yamamoto N, Takahashi Y, Yoshinaga J, Tanaka A, and Shibata Y. Size distributions of soil particles adhered to children's hands. *Arch Environ Contam Toxicol.* 2006, August;51(2):157–163.

174. Ferguson AC, Bursac Z, Biddle D, Coleman S, and Johnson W. Soil-skin adherence from carpet: Use of a mechanical chamber to control contact parameters. *J Environ Sci Health A Tox Hazard Subst Environ Eng.* 2008, October;43(12):1451–1458.

175. Ferguson A, Bursac Z, Coleman S, and Johnson W. Comparisons of computer-controlled chamber measurements for soil-skin adherence from aluminum and carpet surfaces. *Environ Res.* 2009, April;109(3):207–214.

176. Ferguson AC, Bursac Z, Coleman S, and Johnson W. Computer controlled chamber measurements for multiple contacts for soil-skin adherence from aluminum and carpet surfaces. *Hum Ecol Risk Assess.* 2009;15(4):811–830.

177. Ferguson A, Bursac Z, Johnson W, and Davis J. Computer controlled chamber measurements for clay adherence relevant for potential dioxin exposure through skin. *J Environ Sci Health A Tox Hazard Subst Environ Eng.* 2012;47(3):382–388.

178. Jongeneelen FJ. Benchmark guideline for urinary 1-hydroxypyrene as biomarker of occupational exposure to polycyclic aromatic hydrocarbons. *Ann Occup Hyg.* 2001, January;45(1):3–13.

179. Elovaara E, Heikkilä P, Pyy L, Mutanen P, and Riihimäki V. Significance of dermal and respiratory uptake in creosote workers: Exposure to polycyclic aromatic hydrocarbons and urinary excretion of 1-hydroxypyrene. *Occup Environ Med.* 1995, March;52(3):196–203.

180. Borak J, Sirianni G, Cohen H, Chemerynski S, and Jongeneelen F. Biological versus ambient exposure monitoring of creosote facility workers. *J Occup Environ Med.* 2002, April;44(4):310–319.

181. Scheuplein RJ and Blank IH. Permeability of the skin. *Physiol Rev.* 1971, October;51(4):702–747.

182. Potts RO and Guy RH. Predicting skin permeability. *Pharm Res.* 1992, May;9(5):663–669.

183. Guy RH and Potts RO. Penetration of industrial chemicals across the skin: A predictive model. *Am J Ind Med.* 1993, May;23(5):711–719.

184. Mitragotri S, Anissimov YG, Bunge AL, Frasch HF, Guy RH, Hadgraft J, et al. Mathematical models of skin permeability: An overview. *Int J Pharm.* 2011, Oct 10;418(1):115–129.

185. Burger J, Stephens WL, Boring CS, Kuklinski M, Gibbons JW, and Gochfeld M. Factors in exposure assessment: Ethnic and socioeconomic differences in fishing and consumption of fish caught along the Savannah River. *Risk Anal.* 1999, June;19(3):427–438.
186. Burger J. Consumption patterns and why people fish. *Environ Res.* 2002, October;90(2):125–135.
187. Burger J. Daily consumption of wild fish and game: Exposures of high end recreationists. *Int J Environ Health Res.* 2002, December;12(4):343–354.
188. Ray R, Craven V, Bingham M, Kinnell J, Hastings E, and Finley B. Human health exposure factor estimates based upon a creel/angler survey of the lower Passaic River (part 3). *J Toxicol Environ Health A.* 2007, March 3;70(6):512–528.
189. Ruffle B, Burmaster DE, Anderson PD, and Gordon HD. Lognormal distributions for fish consumption by the general US population. *Risk Anal.* 1994;14(4):395–404.
190. National Academy of Sciences (NAS). *Review of Recreational Fisheries Survey Methods.* Washington, DC: National Academies Press, 2006. http://www.nap.edu/catalog.php?record_id=11616 (accessed September 29, 2012).
191. Mariën K and Patrick GM. Exposure analysis of five fish-consuming populations for overexposure to methylmercury. *J Expo Anal Environ Epidemiol.* 2001;11(3):193–206.
192. Harris SG and Harper BL. A native American exposure scenario. *Risk Anal.* 1997, December;17(6):789–795.
193. Wolfe RJ and Walker RJ. Subsistence economies in Alaska: Productivity, geography, and development impacts. *Arctic Anthropol.* 1987;24:56–81.
194. United States Environmental Protection Agency (USEPA). *Supplemental Guidance to RAGS: Calculating the Concentration Term.* May 1992.
195. Ott WR. *Environmental Statistics and Data Analysis.* Boca Raton, FL: CRC Press, 1995.
196. Land CE. Confidence intervals for linear functions of the normal mean and variance. *Ann Math Stat.* 1971;42:1187–1205.
197. Schulz TW and Griffin S. Estimating risk assessment exposure point concentrations when the data are not normal or lognormal. *Risk Anal.* 1999, August;19(4):577–584.
198. United States Environmental Protection Agency (USEPA). *On the Computation of a 95% Upper Confidence Limit of the Unknown Population Mean Based Upon Data Sets with Below Detection Limit Observations.* EPA/600/R-06/022. Las Vegas, NV, March, 2006. http://www.epa.gov/osp/hstl/tsc/issue.htm (accessed September 29, 2012).
199. United States Environmental Protection Agency (USEPA). *The Lognormal Distribution in Environmental Applications.* EPA/600/S-97/006. Las Vegas, NV, December, 1997. http://www.epa.gov/osp/hstl/tsc/issue.htm (accessed September 29, 2012).
200. United States Environmental Protection Agency (USEPA). *Estimation of the Exposure Point Concentration Term Using a Gamma Distribution.* EPA/600/R-02/084. October, 2002. http://www.epa.gov/osp/hstl/tsc/issue.htm (accessed September 29, 2012).
201. Singh A and Nocerino J. Robust estimation of mean and variance using environmental data sets with below detection limit observations. *Chemom Intell Lab Syst.* 2002;60(1):69–86.
202. Jones R, Wills B, and Kang C. Chlorine gas: An evolving hazardous material threat and unconventional weapon. *West J Emerg Med.* 2010, May;11(2):151–156.
203. Cave D and Fadam A. Iraq insurgents employ chlorine in bomb attacks. *The New York Times.* February 22, 2007. http://www.nytimes.com/2007/02/22/world/middleeast/22iraq.html (accessed September 30, 2012).
204. Johnson PC and Ettinger RA. Heuristic model for predicting the intrusion rate of contaminant vapors into buildings. *Environ Sci Technol.* 1991, August 8;25(8):1445–1452.
205. Obmascik M. Home deadly home: Toxins in air Regulations fail to protect U.S. residences from gases. *The Denver Post.* 2002, Jan 6;A-01. http://nl.newsbank.com/nl-search/we/Archives?p_action=list&p_topdoc=21 (accessed on LexisNexus, October 22, 2013 at http://www.lexisnexis.com.proxy-remote.galib.uga.edu/hottopics/lnacademic/).

206. Obmascik M. EPA home-toxins test 'crude and limited' Widely used computer model often wrong. *The Denver Post.* 2002, Jan 7;A-01. http://nl.newsbank.com/nl-search/we/Archives?p_action=list&p_topdoc=21 (accessed on LexisNexus, October 22, 2013 at http://www.lexisnexis.com.proxy-remote.galib.uga.edu/hottopics/lnacademic/).

207. United States Department of Health and Human Services (DHHS). *Household Products Database. Trichloroethylene.* n.d. http://hpd.nlm.nih.gov/cgi-bin/household/search?tbl=TblChemicals&queryx=79-01-6 (accessed September 30, 2012).

208. United States Environmental Protection Agency (USEPA). *Background Indoor Air Concentrations of Volatile Organic Compounds in North American Residences (1990–2005): A Compilation of Statistics for Assessing Vapor Intrusion.* EPA 530-R-10-001. June, 2011. http://www.epa.gov/oswer/vaporintrusion/documents/oswer-vapor-intrusion-background-Report-062411.pdf (accessed September 30, 2012).

209. New York State Department of Health (NYSDOH). *Final NYSDOH BEEI Soil Vapor Intrusion Guidance. Appendix C: Volatile Organic Chemicals In Air—Summary of Background Databases.* Troy, NY, October 2006. http://www.health.ny.gov/environmental/investigations/soil_gas/svi_guidance/ (accessed September 30, 2012).

210. Interstate Technology and Regulatory Council (ITRC). *Vapor Intrusion Pathway: A Practical Guideline.* Washington, DC, January 2007. http://www.itrcweb.org/Guidance/ListDocuments?topicID=28&subTopicID=39 (accessed September 30, 2012).

211. United States Environmental Protection Agency (USEPA). *Draft Guidance for Evaluating the Vapor Intrusion to Indoor Air Pathway from Groundwater and Soils (Subsurface Vapor Intrusion Guidance).* EPA530-D-02-004. Washington DC, November, 2002. http://www.epa.gov/epawaste/hazard/correctiveaction/eis/vapor.htm (accessed September 30, 2012).

212. United States Environmental Protection Agency (USEPA). *Supplemental Guidance for Developing Soil Screening Levels for Superfund Sites.* OSWER 9355.4-24. Washington, DC, December, 2002. http://www.epa.gov/superfund/health/conmedia/soil/ (accessed July 4, 2012).

213. United States Environmental Protection Agency (USEPA). *Soil Screening Guidance: Technical Background Document.* EPA/540/R-95/128. Washington, DC, July, 1996. http://www.epa.gov/superfund/health/conmedia/soil/ (accessed September 30, 2012).

214. United States Environmental Protection Agency (USEPA). *Soil Screening Guidance: Users' Guide.* EPA540/R-96/018. Washington, DC, July, 1996. http://www.epa.gov/superfund/health/conmedia/soil/ (accessed July 6, 2012).

215. Stigler SM. *The History of Statistics: The Measurement of Uncertainty Before 1900.* Cambridge, MA: Belknap Press, 1986.

216. Thompson KM. Developing univariate distributions from data for risk analysis. *Hum Ecol Risk Assess.* 1999;5(4):755–783.

217. Seiler FA and Alvarez JL. On the selection of distributions for stochastic variables. *Risk Anal.* 1996;16(1):5–18.

218. Lambert PC, Sutton AJ, Burton PR, Abrams KR, and Jones DR. How vague is vague? A simulation study of the impact of the use of vague prior distributions in MCMC using WinBUGS. *Stat Med.* 2005, August 8;24(15):2401–2428.

219. Bogen KT, Cullen AC, Frey HC, and Price PS. Probabilistic exposure analysis for chemical risk characterization. *Toxicol Sci.* 2009, May;109(1):4–17.

220. National Research Council (NRC). *Dietary Reference Intakes for Energy, Carbohydrate, Fiber, Fat, Fatty Acids, Cholesterol, Protein, and Amino Acids (Macronutrients).* Washington, DC: The National Academies Press. 2005. http://www.nap.edu/catalog/10490.html (accessed October 22, 2013).

4 Dose–Response Assessment

What is there that is not poison? All things are poison and nothing is without poison. Solely, the dose determines that a thing is not a poison.

Paracelsus
History of dose response, J Toxicol Sci., 2010

The dose–response assessment or toxicity assessment provides a means of understanding whether human exposure to environmental contaminants has the potential to produce adverse health effects. The difference between dose–response assessment and HI is that dose–response assessment is the process of quantifying (rather than identifying as merely possible) the relationship between the dose of a particular chemical to which an individual or population is exposed and the likelihood of adverse health effects.[1]

The dose–response assessment most often relies on data from animal studies and extrapolates these data to humans. These studies are most often conducted using very high doses relative to the environmental exposures experienced by humans—hence, two extrapolations are required, one from high doses to low doses and the other from animals to humans. Extrapolation of high doses in an epidemiologic study to lower environmental doses is also needed when the dose–response assessment for a chemical is based on high-exposure epidemiology data, such as that in occupational or observational studies.

Traditionally, the US EPA and other regulatory agencies have developed two types of quantitative toxicity criteria for oral exposure—RfDs and CSFs. RfDs are developed for noncancer effects and embody the concept of a threshold in which a biological factor must be depleted or overcome for disruption of normal homeostatic mechanisms and resulting adverse effects.[1] CSFs are developed based on the idea that the presumption of a threshold may be inappropriate. The nonthreshold approach of a CSF is used because EPA presumes that any level of exposure, even as small as a single molecule, poses a finite probability of generating a carcinogenic response. Right or wrong, this presumption has enjoyed wide acceptance and has had a significant effect on environmental regulation throughout the history of risk assessment. This assumption is known as the linear no-threshold hypothesis (LNT) hypothesis and will be discussed at length later in the chapter.

The first section of this chapter will provide a discussion of the concept of mode of action (MOA) concept. Next, an in-depth description of the calculation of toxicity factors will be provided. Then, a survey of computational methods that have come to be used in dose–response assessment will be presented. MOA will be revisited, and its application to understanding toxicity in the twenty-first century will be explored. Last, the changing nature of how societal concerns, considerations of animal welfare, and our ever increasing knowledge of the biological basis of disease are changing the manner in which toxicity and dose–response are assessed.

MODE OF ACTION

MOA provides the central organizing principle for understanding the biological underpinnings of toxicity. In government guidance documents, MOA was first mentioned in the NRC's 1993 document *Issues in Risk Assessment*.[2] This publication considered three issues: the use of the maximally tolerated dose (MTD) in animal bioassays for cancer, the two-stage initiation/promotion model of carcinogenesis as a regulatory tool, and a paradigm for ecological risk assessment. MOA was mentioned with regard to the use of the MTD in animal bioassays. The report concluded that bioassays employing the MTD would need additional studies to determine "MOA, pharmacokinetics and applicability of results to the human experience."[2]

Chapter 1 discussed at length the considerable uncertainty in determining the human relevance of results from cancer bioassays in animals.[3,4] The NRC report indicated that the use of the MTD was a necessary evil to be able to obtain protective estimates of risk from animal studies. Cancer can be thought of as many diseases—but there are enough similarities that most people still consider it a single disease entity.[5] The history of cancer research bears this out— for example, the "war on cancer" conceived and carried out during the mid-twentieth century.[6] Hence, to understand this dread disease, the causal factors and biological events leading up to cancer need to be known and understood. Doing so is the essence of MOA.

How much detail is needed to specify a MOA for a particular type of cancer—whether in humans or in animals? EPA indicates that data richness is generally a prerequisite for determining MOA and defines it as follows:

> The term "mode of action" is defined as a sequence of key events and processes, starting with interaction of an agent with a cell, proceeding through operational and anatomical changes, and resulting in cancer formation. A "key event" is an empirically observable precursor step that is itself a necessary element of the mode-of-action or is a biologically based marker for such an element. Mode of action is contrasted with "mechanism of action," which implies a more detailed understanding and description of events, often at the molecular level, than is meant by mode-of-action. The toxicokinetic processes that lead to formation or distribution of the active agent to the target tissue are considered in estimating dose but are not part of the mode of action as the term is used here. There are many examples of possible modes of carcinogenic action, such as mutagenicity, mitogenesis, inhibition of cell death, cytotoxicity with reparative cell proliferation, and immune suppression.[7]

EPA's 2005 *Guidelines for Carcinogen Risk Assessment*, from which the earlier passage was taken, indicates that consideration of MOA should be the centerpiece of any cancer risk assessment. While data richness is highly desirable, even sparse data can be considered in a MOA analysis: there are a finite number of mechanisms by which cancer occurs. These mechanisms necessarily limit the number of MOAs that may be operative for a particular tumor type.[8]

Consideration of MOA will generally enable understanding of the human relevance of tumor response seen in animals, the identification of potentially sensitive subgroups or life stages, and, more importantly for regulatory purposes, the determination of whether low-dose extrapolation should be conducted using a linear nonthreshold approach or a nonlinear approach that uses a presumed threshold and application of safety factors.[7]

MODE OF ACTION VERSUS MECHANISM OF ACTION

These two biological concepts—"MOA" and "mechanism of action"—may seem quite similar and it is only the level of detail that distinguishes them. What is important are the biological features that contribute to the adverse outcome—in other words, the key events.

Mechanism of action refers to the specific sequence of events at the molecular, cellular, organ, and organism level leading from the absorption of an effective dose of a chemical to the production of a specific biological response in the target organ.[7,9] To understand the mechanism of action underlying a particular adverse outcome, one would need knowledge on the likely causal and temporal relationships between the events at the various levels of biological organization, including those events that lead to an effective dose of the chemical at the site of biological action. To specify a mechanism of action, data are needed regarding

- Metabolism and distribution of the chemical in the organism affecting the dose delivered to the molecular site of biological action
- Molecular target(s) or sites of biological action
- Biochemical pathways affected by interaction of the chemical with the site of biological action
- Cellular- and organ-level consequences affecting these biochemical pathways
- Target organs/tissues in which the molecular sites of action and biochemical effects occur
- Physiological responses to these biochemical and cellular effects
- Target organ response to the biochemical, cellular, and physiological effects
- The overall effect on the organism
- Likely causal and temporal relationships between these various steps
- Dose–response parameters associated with each step

In contrast, "MOA" is a more general description of the toxic action of a chemical action.[7,10,11] MOA refers to the type of response produced in an exposed organism or to only the key events or necessary critical features of the mechanism required for the particular biological response to occur. Hence, MOA is known if the full mechanism is known, but the reverse is not true. The distinctions between "mode" and "mechanism" are important for understanding and describing the biological effects of chemicals, including both environmental chemicals and drugs. However, it is important to remain aware that many risk assessors may be less than rigorous in the use of these terms.

In April 1996, EPA published the *Proposed Guidelines for Carcinogen Risk Assessment*,[12] the first update to the 1986 guidelines stemming from the recommendations in the "Red Book."[13] These proposed guidelines recommended consideration of the biological events underlying the carcinogenic process and the incorporation of new information, especially given the rapid pace of ongoing cancer research. The 1986 Cancer Guidelines acknowledged that insights gained from this research would be useful and important for MOA considerations and that MOA would play a greater role in cancer risk assessment—unfortunately, the details of how to apply this knowledge to gain an understanding of MOA were left unclear.[14,15]

Specifically, the proposed guidelines expressed a preference for BBDR models based on the MOA rather than the empirical dose–response models that had, hitherto, been used for cancer dose–response. The International Life Sciences Institute (ILSI) convened an expert panel to evaluate the proposed guidelines and apply them in case studies of two specific chemicals—chloroform and dichloroacetate.[15–17] The findings of this expert panel resulted in the eventual acceptance by EPA's NCEA that a nonlinear toxicity criterion for chloroform would be protective of the carcinogenic endpoint—even though achieving this acceptance required legal action.

SIMPLE EXAMPLE OF MODE OF ACTION

Lynn Turner, a 911 operator living in Forsyth County, Georgia, poisoned her husband with ethylene glycol in 1995 and then, in 2001, poisoned her live-in boyfriend the same way. After the boyfriend's death, the husband's body was exhumed and the postmortem discovered that both men died from ethylene glycol poisoning.[18]

Just how does ethylene glycol, the chemical used in the majority of automobile antifreeze/coolant mixtures, produce toxicity and ultimately death? Mammalian kidney cells metabolize ethylene glycol to calcium oxalate monohydrate that forms crystals within renal tubular cells. These cells rupture, the kidneys cease to function, and death quickly ensues.[19]

It is quite easy to understand the MOA of ethylene glycol from this example. Every summer, sadly, dogs may drink spilled antifreeze and also die. The mechanism of action and thus the MOA is the same for both man and beast.

RECENT EXAMPLE OF THE APPLICATION OF MODE OF ACTION

As noted in Chapter 1, in the absence of MOA information, the regulation of carcinogens assumed that the dose–response was linear in the low-dose region. Implicit in the assumption that the dose–response of a chemical is linear all the way down to zero dose is the notion that a single molecule of a substance may produce effects—a health-protective but, as will be seen, biologically incorrect assumption. The US EPA and the state of California adopted an even more stringent but also biologically incorrect assumption—chemicals that cause cancer by a mutagenic MOA, that is, heritable changes in DNA sequence, act rapidly and that exposure during childhood may produce risks that persist throughout life.[20,21]

Unfortunately, EPA never followed the recommendations of the expert panel that reviewed the 1996 revision of the proposed cancer guidelines. The failure to provide detailed guidance regarding the consideration of MOA has led to an extremely simplistic interpretation by many regulators. The most recent example of such a simplistic misinterpretation is the suggestion that hexavalent chromium produces cancer by a mutagenic MOA.[22] While EPA's 2005 *Guidelines for Carcinogen Risk Assessment* provided an unequivocal statement that MOA should be the centerpiece of all risk assessments, details of exactly how to incorporate and use the information were unfortunately lacking.[7]

For years, toxicologists have known that hexavalent chromium is a human carcinogen by the inhalation route. Workers in chromite processing facilities experience slightly higher rates of lung cancer than the general population.[23–25] Correct or not, many regulatory toxicologists believed that hexavalent chromium was likely a human carcinogen by the oral route as well. From the initial supposition that Cr(VI) might be a human carcinogen by the oral route,[26] it took until 2008 for the NTP to publish the results of a 2-year cancer bioassay conducted in mice and rats.[27] Mice developed small intestinal tumors and rats developed tumors of the oral epithelium.

Notwithstanding EPA's unequivocal recommendations to use the consideration of MOA as the centerpiece of all risk assessment,[7] these bioassay results were interpreted in a highly simplistic fashion by a number of regulatory agencies between their publication and the proposal and elucidation of the actual MOA in 2012.[28–31] The details of the MOA were demonstrated in a clever series of experiments and published in 2012 and 2013 along with a risk assessment based on the findings.[32–38]

The lining of the small intestine consists of a myriad of tiny fingerlike projections called villi (singular villus). The villi function to increase the epithelial surface area available for the absorption of nutrients. Between the villi are invaginations called crypts of Lieberkuhn. The enterocytes, epithelial cells of the villi, slough off into the intestinal lumen and are replaced about every 3 days by new cells migrating upward from the crypts. The normal state of crypt cells is to proliferate and replace villous enterocytes. Chemical signaling with specific signaling molecules known as cytokines originating from the villous cells and elsewhere regulates the proliferative activity of the crypt cells.[39–41]

Briefly, the aspects of the MOA are that Cr(VI) is chemically reduced to trivalent chromium in the stomach. Trivalent chromium is poorly absorbed in mammals. Humans have a more acidic stomach than rats and chemically reduce hexavalent chromium to trivalent chromium to a much greater extent than do rats; rats have a more acidic stomach than mice and thus reduce Cr(VI) to a greater extent than do mice. Hence, a greater amount of Cr(VI) is available for absorption by enterocytes in mice than in either rats or humans. The absorbed Cr(VI) produces cytotoxicity of the cells in the intestinal villi, observed histologically as blunted villi of smaller size than normal. The need for replacement of the damaged villous enterocytes produces an increase in proliferation of the intestinal crypt cells. During such a hyperproliferative state, there is less time between cell divisions for DNA repair mechanisms to complete repair of spontaneous DNA damage, and mutations resulting from faulty DNA repair have a greater chance of revealing themselves as tumors.[42-44] In mice, a distinct threshold was observed for crypt cell hyperplasia and an RfD developed for hyperplasia was also protective of cancer.[32-35,37,38,45,46] The risk assessment for Cr(VI) is one of the best examples of the use of MOA information in existence.

HOMEOSTASIS, THRESHOLDS, AND HORMESIS

Continued existence for any organism is a matter of maintaining homeostasis in the face of an unremitting array of stressors. Evolutionarily successful organisms have developed redundant systems and capacities to deal with many different stressors, but these capacities are finite. When one or more of these capacities are exceeded, a departure from homeostasis, usually in the form of disease or death, occurs. The fact of biological thresholds is implicit in Paracelsus' dictum that all things are poison.[47]

In this chapter, the LNT hypothesis will be discussed at some length. As noted, this hypothesis assumes that even an infinitesimal dose of a carcinogen poses a finite, albeit small, risk and has provided the basis for the regulation of carcinogenic chemicals since the 1970s. Much of the animal testing that has been done in the past has had the goal of demonstrating whether a particular chemical could produce cancer in rodents over their lifetime.

HORMESIS

Paracelsus, the "first" toxicologist" whose famous quotation begins this chapter, noted that toxic substances may be beneficial in small amounts.[48] This phenomenon results in a J-shaped dose–response relationship and has been demonstrated in Monera, Protista, and Metazoa.[48,49] As a scientific concept, hormesis has yet to gain general acceptance, and there are likely political and economic agendas for this lack of acceptance.[50]

The phenomenon of hormesis was first noted in 1887 when small applications of disinfectant were observed to stimulate the growth of yeast, whereas large applications killed the yeast. Evidence seems to be mounting for the universality of hormesis as a phenomenon.[51-53] However, the acceptance of hormesis in risk assessment has yet to occur.[54]

HOW TO INTERPRET AND USE MOA INFORMATION

Dose–response analysis and extrapolation to human exposure levels is a central issue in risk assessment. Risk assessment necessarily uses extrapolations—from doses at which effects are observable down to much lower doses and from effects observed in animals to an understanding of the relevance or lack thereof of those same effects in humans.

To accomplish these extrapolations in a scientifically credible way, MOA frameworks have been developed for application of knowledge of MOA as a means of deciding whether effects in animals are relevant to humans and as a way to understand the role of key events within the overall progression to the adverse outcome or apical toxic event.[55-59]

History and Uses of the MOA/Human Relevance Framework

This framework was originally developed over a decade ago by the WHO International Program for Chemical Safety (IPCS) and specifically focused on chemical carcinogenesis.[55] The original framework was expanded to include a human relevance component and information about the susceptibility of various lifestages.[58–60] More recently, the framework has again been enhanced to be able to examine systematically and quantitatively the key events that occur between the initial dose of a bioactive agent and the effect of concern.[61] The goal, of course, is a risk assessment informed by the most appropriate interpretation of the most up-to-date scientific information.

When the MOA by which a chemical produces toxicity is unknown, regulators often believe they must use highly health-protective default assumptions for low-dose and interspecies extrapolations. One of these assumptions is that dose–response for chemical carcinogens is linear in the low-dose range and no dose, however small, exists below which there is no potential for effect. As noted, this is the LNT.

Quantitative MOA information can reduce uncertainty in risk assessments. Where applicable, quantitative data can be used to replace defaults and choose the most appropriate dose–response models. Hence, an understanding of the MOA is becoming a fundamental component of risk assessment, especially when it comes to classifying carcinogens and making judgments about whether a threshold approach is appropriate or whether the default LNT assumption must be used.[7]

Information about toxicokinetics and metabolism can provide information about the active form of the chemical—indeed, if the toxic moiety is a metabolite of an administered parent chemical, then metabolic activation would clearly be a key event in the MOA. On the other hand, metabolism may also function as a mechanism for detoxification. By taking into account metabolic key events as part of the overall MOA, the influence of induction or inhibition of metabolism of the chemical and variations in patterns of toxicity with metabolic profiles across species, strains, and sexes can be factored into the risk assessment in lieu of defaults.

In addition, the tentative identification of metabolic activation as a key event may permit a powerful counterfactual demonstration that strongly supports identification of key events—if metabolic activation is indeed a key event, then blocking metabolism should also block the occurrence of the adverse outcome. This sort of counterfactual information can provide powerful evidence in support of a proposed MOA.

Even in the absence of such a counterfactual demonstration, consideration of MOA also allows for an understanding of potentially susceptible subgroups and different life stages so that the most appropriate adjustments can be factored into quantitative risk assessments. An example is the polymorphism in the folate carrier or glutathione transferase that may predispose certain individuals to colon cancer.[62,63]

MOA Included in Regulatory Guidance

As noted, EPA is well aware of the utility of the MOA, evidenced by the 2005 Cancer Guidelines.[7] In addition, EPA's *Supplemental Guidance for Assessing Susceptibility from Early-Life Exposure to Carcinogens* also relies on assessing the MOA.[64] The *Framework for Determining a Mutagenic Mode of Action for Carcinogenicity* also relies upon MOA—stating that assuming a mutagenic MOA is not a default position but rather requires proof.[64,65] This guidance document is very timely and the fact that it remains in draft form demonstrates the lack of consensus in risk assessment about the use of the LNT, the lack of general understanding of the difference between genotoxicity and mutagenicity, and, more fundamentally, the appropriate degree of conservatism in risk assessment. Many risk assessment practitioners automatically assume that DNA-reactive chemicals are mutagenic as well, even chemicals such as formaldehyde that occur naturally within the body. This assumption is wrong in most cases and is likely a holdover from the 1970s when regulatory scientists adopted the LNT before the fact of DNA repair mechanisms became common knowledge.

The EC has also incorporated using MOA in its risk assessment guidance for industrial chemicals and biocides.[66] The European Food Safety Authority (EFSA) includes a MOA assessment in its guidance on harmonizing cancer and noncancer risk assessment approaches.[67] The consideration of MOA is recommended in the EC REACH guidance for conducting a chemical safety assessment and in the new "classification, packaging, and labeling" (CPL) regulation on chemical substances and mixtures.[68] OECD recommends using MOA to support the building of chemical categories or when using read-across approaches.[69] With the push for using more systematic and WOE approaches in risk assessment, the use of MOA/Human Relevance Framework (MOA/HRF) and Key Events/Dose-Response Framework (KEDRF) will likely increase correspondingly.

MOA Will Be the Foundation of Twenty-First-Century Toxicology

The interpretation of high-dose toxicity studies requires extrapolation to the lower doses relevant to environmental exposures. Considerations of animal welfare and the growing recognition of the lack of useful information regarding low-dose effects in humans from these high-dose animal tests have prompted an increasing number of restrictions on the use of live animals in toxicity testing.

A new vision and strategy was proposed by the NRC in 2007 to incorporate new in vitro and in silico technologies and computational systems biology.[70] The report emphasized the importance of understanding events leading to toxicity in the context of perturbations in biologic functions, some of which may be reversible or capable of adaptive change. Thus, twenty-first-century risk assessment attempts to link perturbations in biological pathways in animals and in vitro models with adverse effects in humans. Biological pathways are a key connection between exposure and risk—but only when they are mechanistically linked to key events in the MOA.

Tools for Understanding MOA

The development of a proposed or hypothesized MOA will necessitate identification of key events and understanding the dose–response and temporal relationships between the various key events and the adverse outcome as well as between the key events themselves. This is the purpose of the dose–time concordance table, which clearly shows the relationships in both dose and time between the hypothesized key events. The dose–time concordance table (Table 4.1) allows one

TABLE 4.1

Dose–Time Concordance Table for the MOA of Hexavalent Chromium and Mouse Small Intestinal Tumors

Dose/Time		8 Days	90 Days	720 Days
Increasing Dose (mg/L in Drinking Water)	Increasing Time →			
	4	Absorption (presumed)	Absorption	No data
	14	Redox changes	Absorption	Absorption Redox changes (presumed) Villous cytotoxicity Crypt proliferation
	60	Absorption (presumed) Redox changes	Absorption Redox changes Villous cytotoxicity	Absorption Redox changes (presumed) Villous cytotoxicity Crypt proliferation
	170	Absorption (presumed) Redox changes Villous cytotoxicity	Absorption Redox changes Villous cytotoxicity Crypt proliferation	Absorption Redox changes (presumed) Villous cytotoxicity Crypt proliferation Tumors
	520	Absorption Redox changes Villous cytotoxicity Crypt proliferation	Absorption Redox changes Villous cytotoxicity Crypt proliferation	Absorption Redox changes (presumed) Villous cytotoxicity Crypt proliferation Tumors

to gain an understanding of dose–response and temporal relationships between the key events and the apical event and also among the various key events.

Table 4.1 shows an example of a dose–time concordance table for the occurrence of small intestinal tumors in mice in response to hexavalent chromium administered in drinking water.[32,33]

The human relevance of a hypothesized MOA may depend on both qualitative and quantitative factors. EPA's Office of Pesticide Programs clearly recognizes this fact and the need for assessing both qualitative and quantitative concordance of key events between animals and humans.[71] For example, in the early 1990s, a technical panel from EPA concluded that male rat renal tubule tumors from chemicals that induced accumulation of $\alpha 2_\mu$-globulin were likely not relevant to humans based on qualitative considerations.[72] Naphthalene produces respiratory tract tumors in rats, but the MOA for these tumors is not relevant to humans for both qualitative and quantitative reasons.[73] Table 4.2 shows

TABLE 4.2

Example of a Dose–Response Species Concordance Table

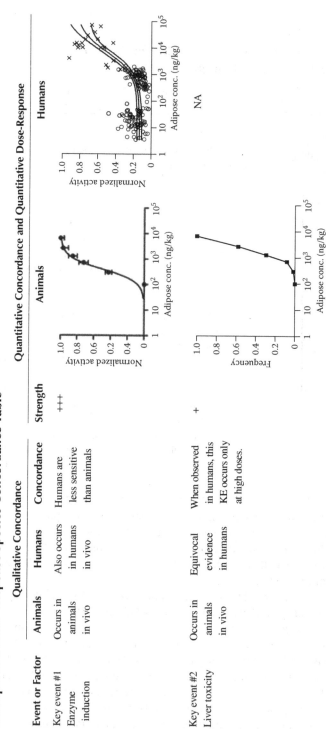

Event or Factor	Qualitative Concordance			Quantitative Concordance and Quantitative Dose-Response	
	Animals	Humans	Concordance	Strength	
Key event #1 Enzyme induction	Occurs in animals in vivo	Also occurs in humans in vivo	Humans are less sensitive than animals	+++	
Key event #2 Liver toxicity	Occurs in animals in vivo	Equivocal evidence in humans	When observed in humans, this KE occurs only at high doses.	+	

Apical event	Occurs in	Has not	Concordance	
Liver tumors	animals in vivo	been observed in humans	cannot be made	—

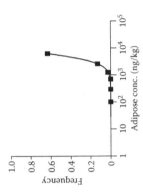

NA

an example (albeit not complete) of one way to set up a dose–response species concordance table. In one place and ideally on a single page, this table can provide information about both qualitative and quantitative concordance of key events between animals and humans and also quantitative dose–response information in both animals and humans.[56,57,61,71,74]

Qualitative Concordance of Key Events between Humans and Animals

Human relevance of the apical endpoint is best determined using a hypothetico-deductive WOE approach.[75] To address human relevance of the MOA, qualitative concordance between humans and animals for each key event needs to be considered. In vitro data may also be available from human or animal cells or tissues; concordance should be considered for these data as well. Ideally, the data will be sufficient to determine which of the key events is relevant to humans, and these data may thus be used to support statements about the relevance to humans of the hypothesized MOA in animals.

Quantitative Concordance of the MOA between Humans and Animals

Quantitative examination of both the dose–response and timing of key events is necessary to determine human relevance. For example, an MOA may be operative in both animals and humans, but extremely unlikely in humans because of quantitative toxicokinetic or toxicodynamic differences.[61] If a key event has the potential to occur in humans, then this quantitative examination can be used to inform animal-to-human extrapolation. Hence, the quantitative concordance should provide information about point-of-departure values such as NOAELs or LOAELs for as many key events as possible in both humans and the animal test species.

Use of Dose–Time and Dose–Response Concordance Information in Understanding the MOA

In general, events that occur at low doses and/or at early stages in the progression toward the apical event may represent (1) the start of a temporal progression, (2) the initial stages of a developing change, or (3) a factor that potentially causes other key events that occur at higher doses or at a later time in the progression. Generally, showing that a particular event is necessary is experimentally difficult; however, this demonstration may be possible in some cases, for example, with transgenic or knockout animals—a powerful counterfactual demonstration supporting the identification of the event as a true key event.[76]

The exact nature of the contribution of a key event to the apical event cannot be necessarily understood from either its dose–response or its timing of occurrence. Some early key events may need to be sustained in order for later key events or the apical event/adverse outcome to occur. Toxicokinetics may affect this timing. For example, lipid-soluble chemicals may be stored in adipose tissue for months or years and produce effects on an ongoing basis; for similar reasons, the dose of a bio-accumulative chemical may be measured as body burden or tissue concentration.

In such a case, the area under the curve (AUC) in units of concentration × time would likely represent the ongoing accumulation in both dose and time better than body burden or tissue concentration at a single time point. Sequestration of a chemical by protein binding may also be represented best by the AUC. A monotonic dose–response relationship between the AUC and a biomarker for a putative key event such as enzyme induction indicates that exploring the quantitative relationship between this biomarker and the apical event/adverse outcome may likely help elucidate details of the MOA.

In other cases, the occurrence of some early key events may trigger a cascade of other events. These early key events then either resolve themselves or become no longer empirically observable. However, the cascade of triggered events continues and leads ultimately to the apical event/adverse outcome.

Weight-of-Evidence Considerations for Understanding the Role of Key Events

A sequence of key events represents a progression over both dose and time. Knowing the relationship between the various key events in both dose and time along with an understanding of biology will contribute to the understanding of the role of a particular key event within the MOA. Often, the counterfactual information discussed earlier is not available; without such information, it may be very difficult to demonstrate the necessity of a particular proposed key event. The understanding of biology can likely contribute, but conclusive support of necessity will be a data gap.

Identifying a key event is based on the confidence one has that this event is necessary for the apical event/adverse outcome and is based on an overall WOE evaluation of qualitative and quantitative aspects of the MOA as well as whether the hypothesized roles of the key events are consistent with the biological basis of the adverse outcome.

The Bradford-Hill considerations have been adopted for use in understanding MOA. Sir Austin Bradford-Hill in 1965 termed these "viewpoints" or "features to consider" rather than criteria.[77] The Bradford-Hill considerations are emphatically not a checklist and necessitate rigorous scientific thinking. They have been quite correctly called "guideposts on the road to common sense."[76] Hence, the consideration of MOA requires a rigorous and reasoned WOE approach to reach an understanding of the overall MOA.[78]

Quantitative Dose–Response Modeling

Examining quantitative dose–response information from as many relevant sources as possible (e.g., human, animal, or in vitro data) is often informative about the progression of events within the MOA. Where possible, the actual dose–response plots should be shown and it is often helpful to show the dose–response of a key event and that of the apical event on the same plot. For clarity, it is helpful to have the same dose range on the x-axis in all the plots. An example of such an analysis based on the NTP bioassay for 2,3,7,8-tetrachlorodibenzodioxin (TCDD) is shown in Figure 4.1a.

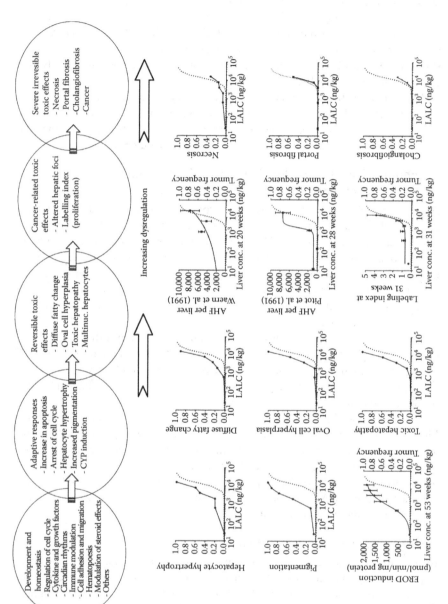

(a)

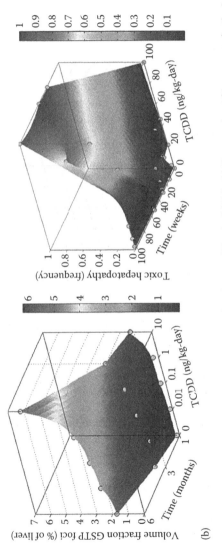

(b)

FIGURE 4.1 Dose–response modeling of key events as a way of understanding the MOA. (a) Progression of key events in the MOA for TCDD-induced liver tumors in rodents. (b) Examples of modeling key events in both dose and time. Left: 3D plot of volume fraction of altered GSTP+ hepatic foci in a 6 month low dose study of TCDD. Right: 3D plot of toxic hepatopathy occurring in rats administered TCDD in a 2-year bioassay. (From Budinsky, R.A., Schrenk, D., Simon, T., Van den Berg, M., Reichard, J.F., Silkworth, J.B., Aylward, L., Brix, A., Gasiewicz, T., Kaminski, N., Perdew, G., Starr, T.B., Walker, N.J., and Rowlands, J.C., Mode of action and dose–response framework analysis for receptor-mediated toxicity: The aryl hydrocarbon receptor as a case study. *Crit. Rev. Toxicol.* (in press). DOI: 10.3109/10408444.2013.835787.)

Modeling key events in dose and time may also be useful. Figure 4.1b shows another example based on TCDD. Again, data in both dose and time were available for an early key event, inhibition of apoptosis within altered hepatic foci, and for another key event, occurring later—toxic hepatopathy. These dose–time relationships helped reach the conclusion that the MOA for hepatocarcinogenesis was nonlinear.

TOXICITY TESTING: PAST AND FUTURE

There can be no doubt that the nature of toxicity testing is undergoing rapid change. The growing realization that high-dose experiments in animals will yield little information with which to predict low-dose effects in humans along with concerns for animal welfare and the increasingly prohibitive cost of animal testing has spurred the development of new methods.

Notwithstanding, there are also significant difficulties with these new methods and traditional animal testing will likely continue to be conducted for high-profile chemicals.

TYPES OF ANIMAL TOXICITY TESTS

The first assumption upon which animal testing is based is that effects observed in laboratory animals are relevant to humans with appropriate qualification. The second assumption that has received considerable scrutiny in recent times is that high-dose experiments in laboratory animals are a valid means of HI and that observed effects will also occur in humans. In Chapter 1, the issues with this assumption were considered in some detail. In this regard, animal toxicity testing cannot demonstrate that a chemical is safe—only what toxic effects could possibly result from exposure to that chemical. In this section, the nature and methods of some of the available animal bioassays will be considered. This selection is far from complete and various toxicology texts will provide additional details of these and other assays. The purpose is to provide an idea of the range of studies that can currently be performed to measure specific aspects of biology.

Chronic Animal Bioassays

These studies are usually performed in rodents and can last from 6 month to 2 years. Chronic bioassays are used to determine whether a chemical is carcinogenic over the lifetime of the animal and generally have a duration of 2 years. Cumulative toxicity can be determined with a shorter-duration study. Dose selection is critical in these studies. Chronic bioassays are most often performed by the NTP. In some studies, many biochemical, histopathological, and other measurements are made in animals sacrificed at interim time points. Doing so requires a large number of animals and significantly increases the cost of the study. However, such information may help to inform the MOA and may be extremely valuable. The NTP bioassay for TCDD was conducted with interim sacrifices and measurements of enzyme induction, cell proliferation, and histopathology and provided a great deal of MOA information.[79] In contrast, the NTP bioassay for hexavalent chromium did not include intermediate sacrifices and thus presented tumor and histopathology data at 2 years only—the unfortunate result was that this bioassay actually raised more questions than it answered.[26,80–84]

Developmental and Reproductive Toxicity Testing

There are several types of endpoints considered in these studies. Teratogenic effects or birth defects may be induced by exposure during development in utero. Reproductive toxicology refers to adverse effects on the male or female reproductive system due to exposure to toxins or other stressors. Developmental toxicology refers to adverse effects that interfere with any developmental process in an organism that results from exposure to either or both parents of the organism or to the organism, during either the pre- or postnatal period.

In fertility and reproductive performance tests, males are given the agent 60 days prior to mating and the females 14 days prior. Administration continues throughout gestation and the period of lactation. The offspring are assessed up to 3 weeks of age for their growth, weight, survival, and general condition. In addition, the fraction of females becoming pregnant and the number of stillborn and live offspring are reported.

Teratogenic effects occur most often when a chemical is administered during organogenesis in the first trimester of pregnancy. Often, fetuses are removed 1 day prior to delivery by Caesarian section. The dams are sacrificed and the uterus examined for resorbed dead and live fetuses. Fetal anomalies, usually of the skeleton, are recorded.

Multigenerational studies are conducted across three generations. At around 30–40 days of age, rodents are administered the chemical, and dosing continues throughout breeding, gestation, and lactation. These parents are the F0 generation. The offspring or F1 generation are also administered the chemical from birth up through breeding, gestation, and lactation. For the F2 generation, the number of live births, litter sizes, and viability counts are recorded. Pup weights are also recorded at intervals up to 21 days of age.

Mutation Assays

The number of types of toxicity studies is growing rapidly. For example, mutations can be assessed using the PIG-A gene. This X-linked gene codes for cell membrane proteins linked to phosphatidylinositol glycans in hematopoietic cells.[85] Somatic mutations in PIG-A can be measured using polymerase chain reaction (PCR) or flow cytometry using antibodies specific for the PIG-A gene products. The PIG-A gene has been used to measure the baseline mutation rate in humans and in laboratory animals administered DNA-reactive chemicals.[86,87]

DOSE–RESPONSE, SPECIES EXTRAPOLATION, AND LOW-DOSE EXTRAPOLATION: LINEAR AND NONLINEAR METHODS

Elucidating and understanding the MOA enables one to determine the type of toxicity factor to develop—linear or nonlinear—depending on the method of low-dose extrapolation. In addition to these characteristics, the route of exposure—oral, inhalation, or dermal—also determines the type of toxicity factor developed. For each critical adverse health effect, MOA is used to determine whether the dose–response relationship in the low-dose region is threshold or nonthreshold. Box 4.1 provides definitions needed for this section.

BOX 4.1   DOSE–RESPONSE CHARACTERISTICS AND DEFINITIONS

Low-dose region: The dose range below the lowest experimental dose. The shape of the dose–response relationship in the low-dose region is unknown and must be presumed.

Point of departure (POD): The POD is that point on the dose–response curve from which low-dose extrapolation in performed. Often, POD for dichotomous effects such as cancer is a 10% response level. For continuous effects, the POD should be a level that reflects a level of response known to be adverse.

Threshold: The dose or exposure below which no adverse or deleterious effect is expected to occur.

Nonlinear dose–response: This type of response shows a pattern of frequency or severity that does not vary directly with dose. If the dose–response relationship for a given chemical is nonlinear, a threshold will likely be observed such that doses below the threshold are not expected to cause adverse effects. Chemicals that produce effects other than cancer are called systemic toxicants and typically exhibit nonlinear dose–response relationships. MOA information indicates that some carcinogens also exhibit a nonlinear dose–response.

Linear dose–response: This type of response shows a pattern of frequency or severity that varies directly with dose. Carcinogens have typically been assumed to exhibit a linear dose–response in the low-dose region. If the dose–response relationship of a chemical is assumed linear, the presumption is that no threshold exists, meaning that any dose, even as small as a single molecule, produces an increase in the probability of a response.

No observed adverse effect level (NOAEL): The highest dose or concentration at which there are no biologically or statistically significant increases in the frequency or severity of an adverse effect over background.

Lowest observed adverse effect level (LOAEL): The lowest dose or concentration at exposure level at which increases in the frequency or severity

BOX 4.1 (continued)   DOSE–RESPONSE CHARACTERISTICS AND DEFINITIONS

of adverse effects occur. These increases should be both biologically and statistically significant.

Benchmark dose (or concentration) (BMD or BMC): A dose or concentration that produces a predetermined level of an adverse response defined by the POD.

BMDL or BMCL: The statistical lower confidence limit on the dose or concentration at the BMD, generally at the 95% level of confidence.

Reference value: An estimate of an exposure, designated by duration and route, to the human population, including susceptible subgroups, that is likely to be without an appreciable risk of adverse health effects over a lifetime.

Reference dose (RfD): An estimate of oral exposure (with uncertainty spanning perhaps an order of magnitude or greater) of a daily oral exposure level for the human population, including sensitive subpopulations, that is likely to be without an appreciable risk of deleterious effects during a lifetime. RfDs may be developed for chronic, subchronic, or acute durations. The units of an RfD are in units of dose, generally mg/kg BW/day.

Reference concentration (RfC): An estimate (with uncertainty spanning perhaps an order of magnitude) of a continuous inhalation exposure to the human population (including sensitive subgroups) that is likely to be without appreciable risk of deleterious effects during a lifetime. The units of an RfC are in units of concentration, generally mg/m^3.

Cancer slope factor (CSF): An upper-bound estimate of the probability of developing cancers per unit intake of a chemical over a lifetime. CSFs are used to estimate the upper-bound probability of an individual developing cancer resulting from a lifetime of exposure to a particular dose of a potential carcinogen. CSFs result from the application of linear low-dose extrapolation and are expressed in units of risk per dose or $(mg/kg/day)^{-1}$.

Inhalation unit risk (IUR): The upper-bound excess lifetime cancer risk estimated to result from continuous exposure to a chemical at a concentration of 1 µg/m^3 in air, expressed in units of $(µg/m^3)^{-1}$.

Nonlinear Toxicity Criteria: Reference Values, Reference Doses, Tolerable Daily Intakes, and Reference Concentrations

The general method of determining nonlinear toxicity factors is to obtain a point of departure(POD) from the dose–response relationship. The POD is that point on the dose–response curve that marks the upper end of the low-dose region and thus the starting point for low-dose extrapolation. The POD should be based on the lowest dose at which an adverse effect is observed and should be within the range of observation. Not all effects are necessarily adverse; for example, exposure to a chemical may result in enzyme induction that may actually be an adaptive response.

Choosing a Point of Departure

Choosing a POD is an absolute requirement for proceeding with dose–response assessment. Yet the choice is dependent on the definition of adversity. EPA defines an adverse effect as "resulting in functional impairment and/or pathological lesions that may affect the performance of the whole organism, or that reduces an organism's ability to respond to an additional challenge."[88]

To understand adversity, the distinction between biological and statistical significance must be considered. For an effect to be biologically significant, it should have a substantial or noteworthy effect on the well-being of the organism. Care is urged with relating a statistical finding to a truly adverse biological effect.[89] The term "adversity" implies some impairment of function or development of pathology that affects the performance of the organism or reduces the organism's capability to withstand additional stressors.[90]

Prior to the explosion of knowledge of the biological basis of health and disease that began around 2000, EPA's Integrated Risk Information System (IRIS) program would, as a matter of science policy, identify a critical effect as that effect occurring at the lowest dose and would select the study using the most sensitive species as the critical study.* From these data, the no-observed-adverse-effect-level (NOAEL) and lowest-observed-adverse-effect-level (LOAEL) would be selected. These two values are experimental bounds on the unknown threshold.

The difficulty in determining adversity can be illustrated by EPA's IRIS database entry for the pesticide oxadiazon (http://www.epa.gov/iris/subst/0253.htm#reforal). Rats were administered diets containing 0, 10, 100, 1000, or 3000 parts per million (ppm) oxadiazon. At a dietary concentration of 100 ppm, increased levels of serum proteins in 18% of females and increased liver weights in 31% of both sexes were observed, and these were chosen as the critical effect. The NOAEL was 10 ppm and the LOAEL was 100 ppm. At 1000 ppm, hepatotoxicity, hemolytic anemia, and kidney effects were observed. Were the serum protein changes and increase in liver weights in the minority of animals tested truly an adverse effect? Perhaps these represented an adaptive or compensatory response within the biological capacity of the animals. Certainly, hepatotoxicity and anemia are adverse effects. Perhaps these higher-dose effects would be more appropriately chosen as adverse rather than adaptive effects. Compare these effects with the critical effect of mortality chosen for dibutyl phthalate (http://www.epa.gov/iris/subst/0038.htm#reforal). Which do you think is truly adverse?

---

* The IRIS program with the National Center for Environmental Assessment (NCEA) at EPA headquarters develops toxicity criteria used by many programs within EPA.

The determination of adversity became such a flashpoint that in 2007, the Agency for Toxic Substances and Disease Registry (ATSDR) of the Centers for Disease Control (CDC) provided a classification of endpoints as nonadverse, less serious, and serious.[91] In 2012, the Texas Commission on Environmental Quality modified this list (as shown in Table 4.3).

TABLE 4.3
Classification of Endpoints by ATSDR and TCEQ

Classification	Endpoint
Nonadverse effects	Weight loss or decrease in body weight gain of less than 10% in adult animals
	Weight loss or decrease in body weight gain in less than 5% in fetuses
	Changes in organ weight of nontarget organ tissues that are not associated with abnormal morphologic or biochemical changes
	Increased mortality over controls that is not statistically significant ($p > 0.05$)
	Some adaptive responses
Less serious effects	Reversible cellular alterations at the ultrastructural level (e.g., dilated endoplasmic reticulum, loss of microvilli, myelin figures) and at the light-microscopy level (e.g., cloudy swelling, hydropic degeneration, fatty change)
	Mild necrosis (dependent upon location, distribution, and magnitude), metaplasia, or atrophy with no apparent decrement of organ function
	Mild to moderate serum chemistry changes (e.g., increased 1–3 and 3–20 times the normal ranges of serum glutamic oxaloacetic transaminase (SGOT) and serum glutamic pyruvic transaminase (SGPT) are considered mild and moderate, respectively)
	Organ weight change in known target organ tissue that is not associated with morphologic or biochemical alterations
	Mild behavioral effects as measured by behavioral function tests
	Weight loss or decrease in body weight gain of 10%–19% (assuming normal food consumption and when weight loss is due to a systemic effect of toxicant)
	Some adaptive responses (e.g., hepatic CYP induction)
Serious effects	Death
	Clinical effects of significant organ impairment (e.g., convulsions, icterus, cyanosis)
	Moderate to severe morphologic changes (such as necrosis, metaplasia, or atrophy) in organ tissues that could result in severe dysfunction (e.g., marked necrosis of hepatocytes or renal tubules)
	Moderate to major behavioral effects as measured by behavioral function tests
	Weight loss or decrease in body weight gain of 20% or greater (assuming normal food consumption)
	Major serum chemistry changes (e.g., increased > 20 times the normal ranges of SGOT and SGPT)
	Major metabolic effects (e.g., ketosis, acidosis, alkalosis)
	Cancer

NOAELs, LOAELs, and Benchmark Doses

The use of the NOAEL or LOAEL value as the POD has been criticized—doing so ignores much of the dose–response relationship. In addition, the values of the NOAEL are dependent on study design and the doses chosen. Hence, some studies can only identify a LOAEL—a so-called "hanging" LOAEL because the value of the unknown NOAEL could be as low as zero. Multiple studies may provide a range of NOAEL and LOAEL values.

Therefore, benchmark dose (BMD) modeling can be used to identify a POD. In BMD modeling, an empirical mathematical model is fit to the entire dose–response data for a chemical. The BMD corresponding to a prescribed level of response (benchmark response or BMR) and the statistical lower limit on the BMD (BMDL) are determined; generally, the BMDL is used as the POD.

Because BMD modeling uses all the data, this method can reduce uncertainty due to small sample sizes. BMDL values are generally comparable to NOAELs.[92–106]

Some dose–response data are not amenable to BMD modeling. The nature of the data collected during the study and the quality of the experimental study both determine whether the data are amenable to modeling. Since the mid-1990s, EPA has provided the Benchmark Dose Software (BMDS), a software used to perform BMD modeling and guidance on dose–response modeling.[107–109] This software uses a range of models and determines parameters by maximum likelihood estimation. The software is available at http://www.epa.gov/ncea/bmds/. A number of the exercises at the end of this chapter will use BMDS. In 2000, EPA released an external review draft of the Benchmark Dose Technical Guidance Document. This draft was never finalized and additional information was provided on BMDS in 2012.[107,110]

Uncertainty/Extrapolation Factors for Low-Dose Extrapolation

Uncertainty factors (UFs) are used to extrapolate from the POD to lower doses to account for a number of uncertainties inherent in the assessment. UFs are intended to account for

- Unknown variation in sensitivity among the members of the human population (intraspecies variability)
- Uncertainty in extrapolating animal data to humans (interspecies variability)
- Uncertainty in extrapolating from data obtained in a shorter-term study to lifetime exposure (subchronic to chronic)
- Uncertainty in extrapolating from an LOAEL rather than from an NOAEL
- Uncertainty associated with extrapolation from animal data when the toxicity database is incomplete

The default value of each of these UFs is 10.* These values are based on an initial estimate from the US Food and Drug Administration, suggesting that acceptable

---

* The reason for this is the same as that for the emergence of mathematics based on a decimal system—humans have 10 fingers! A waggish toxicologist once stood up at a symposium on UFs and held his hands in the air with only three fingers extended. He then claimed he was Zorg from the planet Krypton and that he believed the default UF should be 6!

daily intakes (ADIs) for contaminants in food should be based on a chronic animal NOEL or NOAEL divided by a 100-fold UF. This 100-fold factor was intended to account for both intra- and interspecies variability as well as possible synergistic effects between the many intentional or unintentional food additives or contaminants in the human diet.[111–113]

The majority of work on the conceptual development of UFs was accomplished by Dr. Michael Dourson who began his career at EPA and later founded Toxicology Excellence in Risk Assessment (TERA), an independent nonprofit organization whose mission is to foster scientific collaboration and transparency in support of the science underlying risk assessment.[114] Dourson attempted to refine the scientific basis of UFs.[115,116] During the same period of time, European scientists, also dismayed at the lack of a science basis for UFs, were attempting to provide this basis.[117–131] These scientists recognized that both inter- and intraspecies UFs could be split into toxicokinetic and toxicodynamic components. The interspecies UF was split into factors of 4 and 2.5 to account for TK and TD, respectively; the intraspecies UF was split into two equal factors of 3.16 ($\sqrt{10}$).[132]

This recognition enabled the extensive use of physiologically-based toxicokinetic (PBTK) models in risk assessment for species extrapolation—with such use, only the toxicodynamic portion of the UF would need to be applied.

In 2001, the International Programme on Chemical Safety of the WHO published guidance on chemical-specific adjustment factors, and in 2011, USEPA's RAF published draft guidance on data-derived extrapolation factors. These documents were quite similar and provided guidance on the use of data to derive chemical-specific values for the inter- and intraspecies UFs.[133,134]

Box 4.2 provides an example of the derivation of an RfD from a BMDL representing a NOAEL value.

BOX 4.2   EXAMPLE OF DERIVATION OF A RFD

Nitrobenzene produces methemoglobinemia in which red blood cells have a decreased oxygen carrying capacity due to the formation of metHb from hemoglobin. MetHb is normally maintained at <1% of total Hb by NADH diaphorase. Premature babies and individuals with chronic congenital methemoglobinemia due to an inherited deficiency in this enzyme are especially susceptible to metHb-generating chemicals, such as nitrate or nitrobenzene. In humans, cyanosis or other clinical symptoms become apparent at metHb levels of 6%–10%.

Male rats were exposed to doses of 0, 9.38, 18.75, 37.5, and 75 mg/kg/day. Values (mean ± SD) for % metHb were 1.13 ± 0.58, 2.75 ± 0.58, 4.22 ± 1.15, 5.62 ± 0.85, and 7.31 ± 1.44. These data were fitted to a variety of models for continuous (as opposed to quantal) responses using EPA's BMDS. The results are shown in the following table.

(continued)

BOX 4.2 (continued)   EXAMPLE OF DERIVATION OF A RFD

			GOF Statistics for Models Fitted				
				Scaled Residual for Dose in mg/kg/day			
Model	−2*LL	p-Value	AIC	0	9.38	18.75	37.5	75
Exponential (model 4)	0.8893	0.641	44.68154	−0.0668	−0.1956	0.5592	−0.3704	0.0594
Polynomial	2.02924	0.3625	45.8214	−0.338	0.267	0.846	−0.847	0.116
Power	3.1915	0.2028	46.9837	0.0548	−1.04	0.856	0.675	−0.551
Hill	0.6799	0.4096	52.944	0.0352	−0.248	0.427	−0.299	0.0853
Linear	24.3148	<0.0001	66.107	−2.32	0.84	2.59	1.36	−2.14

BMD and BMDL associated with a 1 standard deviation (SD) increase over the control mean are 2.57 and 2.08 mg/kg/day, respectively. Both the BMD and BMDL are below frank effect levels in humans. Hence, these are interpreted as equivalent to NOAELs. The default UFs for inter- and intraspecies extrapolation are both 10. A subchronic to chronic UF of 3 was applied to account for less than lifetime exposure. The database was judged to be sufficient.

Hence, the UFs used were

UF for interspecies extrapolation = 10
UF for intraspecies extrapolation = 10
UF for LOAEL to NOAEL extrapolation = 1
UF for subchronic to chronic time scale = 3
UF for database deficiencies = 1

The resulting RfD is calculated as 2.08 mg/kg/day ÷ (10 × 10 × 1 × 3 × 1) = 7E-03 mg/kg/day. You may wish to download BMDS from EPA's website and recreate these calculations.

LINEAR TOXICITY CRITERIA: CANCER SLOPE FACTORS AND UNIT RISK VALUES

A linear dose–response is an artificial construct associated with the LNT. This precautionary approach is not based on any science; rather, it represents a policy that attempts to address the fact that the shape of the dose–response relationship in the low-dose region is unknown.

Since 1977, the practice of cancer risk assessment has been to use linear low-dose extrapolation to predict cancer risk at the regulatory target of one in a million based on the observed cancer incidence in high-dose animal experiments. Low-dose extrapolation is necessitated by the limited number of animals in a standard cancer bioassay, typically 50 per dose group, a number sufficient to detect a 10% risk of cancer with a measure of statistical confidence but not nearly enough to meet the regulatory target risk of one in a million. As noted earlier, the primary and biologically incorrect assumption of linear low-dose extrapolation is that even a single molecule—an amount obviously too tiny to measure—will still pose a quantifiable risk of cancer.

The LNT became accepted as the norm for cancer risk assessment during the second half of the twentieth century and persists today. The reason is likely a statement made by Hermann J. Muller in his Nobel Prize acceptance speech on December 12, 1946.

Muller was a pioneer in radiation genetics and in 1926 examined the use of radiation in producing mutations in fruit flies. Although others had tried radiation earlier, Muller was the first to be successful in inducing mutations and rushed his work into publication without presenting his data in 1927. At the same time, Lewis J. Stadler used radiation as a mutagen in corn, but Muller published a few months earlier and established priority. Later in 1927, Muller presented the data in detail at the International Congress of Genetics in Berlin and returned to the United States with an international scientific stature. Even as a doctoral student at Columbia, Muller was a "priority hog" and had difficulty crediting others for having their own ideas and was generally reluctant to share his insights with them.[135]

In his Nobel address, Muller argued that the dose–response for radiation-induced mutations was linear and there was "no escape from the conclusion that there is no threshold." A recent examination of the letters between Muller and his colleagues Ernst Caspari and Curt Stern indicates that Muller was deceptive in his lecture, possibly because of a precautionary philosophy, possibly to protect his scientific reputation.[136] At the time of Muller's speech, correspondence reveals that he knew about data that showed a threshold, inconsistent with the LNT.[136–138]

The story of Hermann Muller is a cautionary tale about the importance of scientific integrity. Muller's deception may have been based on precautionary thinking; his Nobel address was given about a year after the end of World War II and the dropping of the atomic weapons on Hiroshima and Nagasaki. At the time, the destructive effects of these weapons and their known association with radiation were fresh in everyone's mind.

From a moral stance, the precautionary principle was given voice with the best of intentions—to encourage policies that protect human health and the environment in the face of uncertain risks. However, acting in the face of incomplete knowledge also has its perils—that an action taken without sufficient understanding may have unintended consequences that actually make matters worse.[139]

In the early 1930s, at what should have been the pinnacle of Muller's scientific career, his life began to unravel over difficulties in his marriage, professional jealousy of his colleagues at the University of Texas, and his outspoken political stance that caused the FBI to investigate him as a communist. In 1932, he attempted suicide by taking an overdose of barbiturates. Muller was an obviously troubled man.[6] Muller was known as a gadfly and was never hesitant to speak out on science policy issues such as eugenics or nuclear weapons.[135] Although Muller was likely motivated by good and honorable intentions, his views on the hazards of radiation may have blurred his vision of the science–policy interface and inappropriately altered his willingness to alter his scientific judgment when presented with new data.

For chemical risk assessment, the 1977 Safe Drinking Water Committee of the NAS unwittingly accepted Muller's "no escape/no threshold" statement and recommended the adoption of linearity at low dose for risk assessment of chemical carcinogens. The reasons for this adoption are poorly documented in their report and it is likely that this choice was also based on a protectionist–precautionary philosophy.[140]

It is not difficult to understand or sympathize with the fear of cancer that has become ingrained in society. For many years, cancer was poorly understood and the medical treatments in the early twentieth century were horrific, often disfiguring, and largely unsuccessful.[6] The fear of cancer is reflected in the adoption of the Delaney Clause by the US Congress in 1958, which states that no food additive that has been shown to induce cancer in man or experimental animals can be considered safe. This fear was also clear in the anger of Congressman Andy McGuire who represented New Jersey's seventh district from 1975 to 1981 upon learning that nitrosamines were present in pesticide samples that USEPA had failed to withdraw from the marketplace.[139] The adoption of the LNT for cancer dose–response by the Safe Drinking Water Committee is a response to the fear of cancer, but it does not reflect the state of knowledge of cancer biology in the twenty-first century.

Why the Linear No-Threshold Hypothesis Is Wrong

Adoption of the LNT occurred prior to the advent of research on DNA repair. DNA damage is not by itself a mutational event, but in the late 1970s, prior to the advent of research on DNA repair, it seemed reasonable that DNA damage itself was mutagenic.

Even today, many scientists commonly believe that the dose–response of a carcinogen that acts by damaging DNA does not exhibit a threshold and they wrongly equate genotoxicity with mutagenicity. There is, however, ample experimental evidence that carcinogens that act by damaging DNA exhibit dose thresholds and these thresholds are likely due to DNA repair and other compensatory processes.[141]

Evolutionarily successful organisms have developed redundant systems that provide both immediate capacity and fail-safe mechanisms to deal with many different stresses. DNA repair is just such a fail-safe mechanism.[102,142–149] As noted earlier, DNA repair may be incomplete when cells are in a hyperproliferative state. The fact of biological thresholds is implicit in the statement by Paracelsus upon which the science of toxicology is based—the dose makes the poison.

Low-dose linearity has been justified by the notion that normal physiological processes reflect a pathological continuum toward cancer or other diseases and that exposure to a stressor will act in an additive fashion with these ongoing pathological processes.[150] Statistical rather than biological arguments are used to attempt to explain away the occurrence of thresholds, and these arguments ignore the need of all organisms to maintain homeostasis.[151,152]

The reductionist hypotheses inherent in the empirical dose–response models commonly used in risk assessment render them scientifically inadequate given the individual differences in phenotype, exposure history, and defense or repair capacities.[153–155] Indeed, a stronger experimental and regulatory focus on biological mechanisms would enable greater flexibility in the regulation of carcinogens without compromising human health.[156,157] More recently, advances in the biological sciences, including systems biology, high-throughput screening methods, and chemical genomics, suggest that the increased understanding of biological responses from these advances will be consistent with the assumption of thresholds and will also clarify the distinction between adaptive and adverse responses.[51]

Had the Safe Drinking Water Committee been aware of DNA repair, the policy of linear low-dose extrapolation for cancer risk assessment might never have been adopted. In sum, the LNT is incorrect for both radiation and chemical carcinogenesis, and its use has driven risk assessment practice for the past 60 years with the result of unnecessary fear on the part of the general public and needless expenditure of resources to comply with regulations that may do more harm than good.[158,159]

Developing a Toxicity Factor Based on Using Linear Low-Dose Extrapolation

Why even present this material, especially given the foregoing discussion on the lack of a scientific basis for the LNT? The reason is—right or wrong—linear extrapolation is the norm for regulatory agencies throughout the world and those planning to work in the field of risk assessment need to understand linear low-dose extrapolation.

Toxicity factors based on linear low-dose extrapolation are in essence a ratio between risk and dose. Of course, consideration of hazard assessment and MOA will guide the choice of whether to use linear or nonlinear extrapolation. Once these prior steps are complete and the decision to use linear extrapolation is made, the following steps are used to develop an oral CSF or IUR value:

- Dose–response modeling
- Selection of a POD
- Adjustments to the POD to account for animal-to-human differences and differences in ED and timing
- Linear low-dose extrapolation
- Evaluation of potential for a mutagenic MOA

These will each be considered in detail.

Dose–Response Modeling for Dichotomous (Quantal) Data

Dose–response modeling seeks to fit a mathematical model to dose–response data that describes the dataset, especially at the lower end of the observable dose–response range. Empirical dose–response modeling is a curve-fitting exercise and should be clearly distinguished from a BBDR modeling effort.[160]

EPA's BMDS runs under Microsoft Windows that performs dose–response modeling. This software is freely available at http://www.epa.gov/ncea/bmds/. For dichotomous data, such as frequency of cancer from an animal bioassay, the software provides nine models to fit the data. These include the gamma, logistic, log-logistic, multistage, probit, log-probit, quantal linear, Weibull, and dichotomous Hill models.

Box 4.3 provides an example of BMD modeling and CSF derivation for a dichotomous endpoint.

BOX 4.3   EXAMPLE OF CSF DERIVATION

In this example, data from the frequency of hepatoblastoma in female B6C3F1/N mice administered *Gingko biloba* by corn oil gavage at doses of 0, 200, 600, or 2000 mg/kg for 5 days/week for 105 weeks. These data were obtained from the draft report from the NTP. Hepatoblastomas were observed at frequencies of 1/50, 1/50, 8/50, and 11/50.

Figure 4.2 shows a screenshot of the BMDS output session. The three graphs show the output for the gamma, logistic, and log-logistic models. You are encouraged to download the software and run this example yourself. EPA suggests using the Akaike Information Criterion (AIC) to select the model. This criterion combines both goodness of fit and parsimony of the model.

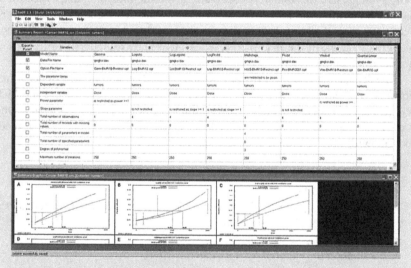

FIGURE 4.2   Screenshot of output from EPA's BMDS.

BOX 4.3 (continued)   EXAMPLE OF CSF DERIVATION

GOF Statistics for Models Fitted

Model	χ^2	p-Value	AIC	Scaled Residual for Dose in mg/kg/day 0	200	600	2000
Log-logistic (restricted)	2.95	0.2286	123.211	0.136	−0.914	1.347	−0.533
Gamma (restricted)	3.36	0.1861	123.5	0.050	−0.871	1.507	−0.575
Multistage (restricted)	3.36	0.1861	123.5	0.050	−0.871	1.507	−0.575
Weibull (restricted)	3.36	0.1861	123.5	0.050	−0.871	1.507	−0.575
Quantal linear	3.36	0.1861	123.5	0.050	−0.871	1.507	−0.575
Log-probit	2.39	0.1219	124.76	0.212	−0.941	1.136	−0.413
Probit	6.37	0.0414	126.169	−0.796	−1.041	2.118	−0.406
Logistic	6.72	0.0348	126.545	−0.875	−1.084	2.157	−0.348

The best fitting model is the log-logistic. Per EPA's 2005 Cancer Guidelines, the lowest POD within the range of observation is chosen. From the following table, the 5% POD would be chosen:

POD	BMD	BMDL
10%	705.8	446.2
5%	334.3	211.3
1%	64.2	40.6

In the absence of other means for species extrapolation, the ratio of body weights to the 1/4 power would be used[161] to obtain an HED at the POD of $211.3 \times 0.144 = 30.43$ mg/kg/day. To perform linear extrapolation and obtain the CSF, the HED is divided into the BMR as follows: $0.05/30.43 = 0.002$ per mg/kg/day. Using this value, the risk-specific dose for *G. biloba* corresponding to a risk of one in a million is $1E-06/0.002 = 5E-04$ mg/kg/day.

The example in Box 4.3 is to illustrate the dose–response modeling and computations involved in development of a CSF. Standard doses of *G. biloba* in humans range from 120 to 800 mg/day.

As an exercise, you may wish to estimate the cancer risk for a 75 kg adult consuming *G. biloba* on a daily basis for, say, 30 years. However, bear in mind that the CSF developed in Box 4.3 was to illustrate dose–response modeling and calculation of a slope factor—as such, performed in the absence of a MOA analysis and without any consideration of the evidence, its strength, or its weight. It is emphatically not provided to suggest that *G. biloba* is a human carcinogen.

Selection of the POD

The POD can be a NOAEL, a LOAEL, or a BMDL, as defined in Boxes 4.2 and 4.3. It is important to ensure that the LOAEL or BMDL is based on truly adverse effects. Generally, the LOAEL will be the lowest exposure at which adverse events occur that are both statistically different in severity or frequency than background and biologically significant.

The approach of using NOAELs or LOAELs as the POD has been criticized because only a single value is used rather than the entire set of dose–response data.[96,97,100,162] Hence, the BMD approach was proposed as an alternative to the NOAEL/LOAEL approach.[94,107] For dichotomous data, the POD is a chosen frequency of the critical effect. Most often, the value of 10% is chosen because it is about the lowest value that can be statistically distinguished from background.[101] For continuous data, EPA suggests the default BMD be one standard deviation above the mean of the controls, but this level should be interpreted in terms of biological significance as well.[108,163,164] The need for biological significance was also recognized in the redefinition of the BMR for continuous data as the "critical effect size."[105,165,166] Alternatively, continuous data can be "dichotomized"— that is, expressed as a percentage or frequency.[101,167]

Because predicting the dose–response relationship in the low-dose region is the focus of the modeling, the lowest POD that is within the range of observation, that is, above the lowest dose, should be used.[7] The lower plot in Figure 4.3 shows an idealized dose–response curve and the details of linear extrapolation from the BMDL. The use of the BMDL is a protective approach that depends on the statistical uncertainty in the BMD and provides an unknown degree of conservatism in the assessment.

ADJUSTMENTS TO THE POD FOR BOTH LINEAR AND NONLINEAR EXTRAPOLATION

ADJUSTING FOR TIME DIFFERENCES

The POD may need to be adjusted for differences in exposure between that used in the study and that in exposed humans. For example, in an inhalation bioassay, animals may be exposed for 6 h a day and 5 days a week. This exposure would be adjusted to represent continuous exposure by multiplying by 6/24 and 5/7. In animal bioassays employing oral gavage for 5 days/week, the administered daily dose would be similarly multiplied by 5/7.

These adjustments reflect the application of Haber's rule, developed early in the twentieth century to address issues of the toxicity of poison gas used as a weapon of war. At that time, scientists observed that both concentration and time interacted to produce toxicity, and the rule indicates that concentration and exposure time are both factors in producing an effect—briefly, $c \times t = k$, where k is proportional to the effect. In essence, k represents the AUC for exposure. Hence, an inhalation exposure to 10 ppm for 1 h will be equivalent to an exposure to 1 ppm for 10 h. Haber's rule has been shown to be applicable for both the inhalation and oral routes of exposure.[168]

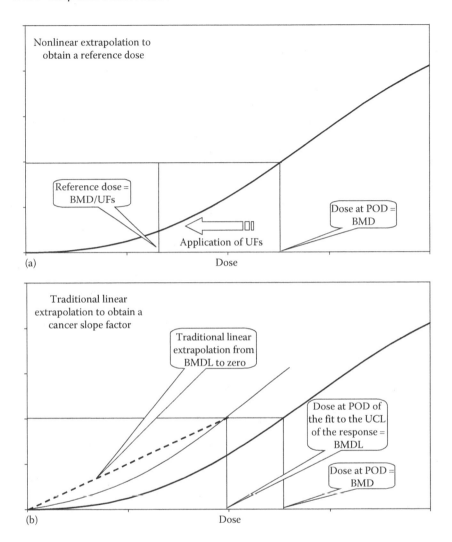

FIGURE 4.3 Nonlinear versus linear low-dose extrapolation: (a) Shows the procedure for nonlinear extrapolation in which UFs are applied to a POD. The POD may be a NOAEL or a LOAEL. If the dose–response modeling is used such as that in EPA's BMDS, the POD may be the dose corresponding to a specific response benchmark, that is, the BMD. (b) Shows the procedure for linear extrapolation in which the slope of the line from the lower confidence limit on the BMD and the corresponding POD to the origin at zero dose and zero response is used as an estimate of the risk of exposure in the low-dose region.

In fact, the application of Haber's rule is valid only when repeated exposures used in an inhalation study result in steady-state internal concentrations. Rapid clearance may render the duration of exposure negligible.[169] Notwithstanding, Haber's rule is used by most regulatory agencies as a default procedure when chemical-specific information is not available.

ADJUSTING FOR SPECIES DIFFERENCES

For inhalation exposure, mathematical dosimetry models are used to account for differences in pulmonary anatomy, disposition of chemical within the airway, and target tissue interactions.[88,170–172]

Adjustments to the POD may also constitute species extrapolation. If the POD represented a dose metric (see next section) from a PBPK model, a model with parameter sets specific to both humans and animals could be used for the toxicokinetic portion of species extrapolation.

DEVELOPING TOXICITY CRITERIA FROM EPIDEMIOLOGIC STUDIES

Generally, epidemiologists speak (or write) of exposure–response rather than dose–response. Many epidemiology studies are qualitative and descriptive; they consider what factors may be risk factors for a disease based on the occurrence of the disease among individuals in a population. More quantitative epidemiology studies attempt to determine the exposure–response relationship in terms of duration and dose/intensity. Co-exposures and possible confounding factors are also considered. Population-level responses are quantitated as standardized mortality or incidence rates (SMRs/SIRs), relative risks (RRs), and odds ratios (ORs). A full discussion of this topic is beyond the scope of this book and those interested are directed to recent guidance from the Texas Commission on Environmental Quality available at http://www.tceq.texas.gov/publications/rg/rg-442.html.

COMPUTATIONAL METHODS IN TOXICITY ASSESSMENT

Computational methods have been applied in toxicology for many years with growing sophistication. In the 1980s, PBPK models were used to predict the distribution of volatile organic chemicals in the body from inhalation exposure. A number of these models were used in risk assessments.[173–176]

Pharmacokinetics is the study of the quantitative relationships between the absorption, distribution, metabolism and excretion (ADME) of chemicals in biological systems. This study is also called toxicokinetics when applied to toxic chemicals. Pharmacodynamics or toxicodynamics is the study of events at the cellular, biochemical, and molecular levels that occur in response to perturbation by a chemical agent or other stressor.

Mathematical models of ADME of a large number of chemicals have been developed for use in environmental risk assessment and in the pharmaceutical industry. Early models concentrated on the estimation of vapors from volatile organic chemicals.[173,174,177–179] These models range in complexity and are referred to as either PBPK models or PBTK models. Recently, BBDR models have incorporated both toxicokinetics and toxicodynamics. As biological measurements become more sophisticated, the field of systems biology has emerged. Systems biology is the use of mathematical and computational models that may incorporate not only toxicokinetics and toxicodynamics but also data from bioinformatics, genomics, proteomics, and newer technologies such as chromatin immunoprecipitation (ChIP) or fluorescence resonance energy transfer (FRET) from which the direct interaction of molecules inside cells may be obtained. Organisms, tissues, cells, and molecules are all biological components with dynamic and complex behavior. The description and prediction of this behavior is the field of systems biology.

An Early PBPK Model Is Still Used for Risk Assessment of Lead

Early in its history, the USEPA recognized that children were exposed to lead through a variety of environmental media and exposure routes. To account for multimedia exposures (e.g., air, water, food, and soil), a model of lead pharmaco-kinetics would be needed. Hence, Alan Marcus of EPA created a set of lead PBPK models during the 1980s.[180–183] Other PBPK models of lead in children were also developed around this time.[184–187]

At the same time, evidence was mounting that exposure to lead affected mental and social development of children, and there was no evidence of a threshold for these effects.[188,189] Until recently, the US Centers of Disease Control provided an action level for blood lead in children of 10 µg/dL for continued monitoring. However, in 2012, this action level was reduced to 5 µg/dL because research since the 1980s has not revealed a blood lead level in children without effect.[190–192]

The MOA for the neurodevelopmental effects of lead likely involves disruption of synaptic transmission and changes in calcium homeostasis that, via a number of mechanisms, lead to alterations in the development of neural networks in the developing brain.[193–197] Both the complexity of these mechanisms and their individual variability likely contribute to the inability to observe a threshold.

In the 1980s, EPA developed the IEUBK model to assess multimedia lead exposure in children. This model is still used and represents one of the successful pharmacokinetic models in risk assessment.

Toxicokinetics and PBPK Models

PBPK models are mathematical representations of biological tissues, organs, and physiological processes occurring in the body and affecting the ADME of chemicals. For use in risk assessment, PBPK models are used to estimate tissue concentrations, movements of the chemical through various tissues (called fluxes), and other quantities that may be related to the effective dose of the toxic moiety within the target tissue. Finding an appropriate surrogate dose metric may present one of the main challenges in adapting a model for use in risk assessment. Choosing a dose metric is important—this quantity must be related in a quantitative fashion to one or more key events and the adverse outcome being assessed.[198]

For example, PBPK models help greatly to ascertain the dose metric in the target tissue most appropriate for the toxicity being considered. For example, if a particular chemical produces liver toxicity, then oral exposure will likely produce effects at lower administered doses than either dermal or inhalation exposure. Chemicals absorbed from the gastrointestinal tract enter the hepatic portal system and move first to the liver. Also, highly lipid-soluble chemicals may tend to bio-accumulate in adipose tissue and possibly produce effects on an ongoing basis. Possibly the AUC for adipose tissue concentration might be the most appropriate dose metric for such bioaccumulative chemicals. Another chemical might be absorbed by the intestine and thus produce cellular damage in the enterocytes before entering the bloodstream; in this case, flux of the chemical from the intestinal lumen into the tissue would be an appropriate dose metric.[31]

TOXICODYNAMIC CONSIDERATIONS

Toxicodynamics is the consideration of how a xenobiotic chemical interacts with tissues, cells, or biomolecules as part of the toxic response. Once a chemical distributes to the target tissue via ADME processes, it interacts with cells of that tissue to produce effects. For example, the binding of a DNA-reactive chemical to nucleic acid would be a toxicodynamic process. These initial biochemical events have been referred to as the molecular initiating event (MIE).[199] This toxicodynamic event will certainly be a key event in the MOA.

The difference between toxicodynamics and toxicokinetics is evident in a very recent risk assessment of the organophosphate (OP) pesticide chlorpyrifos. The MOA for OPs is well known—inhibition of cholinesterases with toxicity manifested as central and peripheral cholinergic effects.[200] Thionophosphorus OPs such as chlorpyrifos do not directly inhibit acetylcholinesterase (AChE) but must first be metabolized to the oxygen analog or oxon by cytochrome p450 mixed function oxidases, mainly occurring in the liver. Paraoxonase 1 (PON1) is an arylesterase that metabolizes OP compounds. Chlorpyrifos oxon is inactivated by PON1 in the liver and other tissues.[201,202] Genetic polymorphisms exist in the PON1 gene and lifestyle factors such as the use of cholesterol-lowering medications and alcohol consumption may increase PON1 activity.[203–206]

The chlorpyrifos risk assessment used a BBDR model that incorporated both the toxicokinetics of chlorpyrifos and its toxicodynamics in terms of cholinesterase inhibition. However, the daily intake of chlorpyrifos has been estimated at less than 11 or 3.4 ng/kg/day in children and adults, respectively. In 3-year-old children, the greatest percent reduction in cholinesterase activity for typical dietary intake was 0.001%. In addition, the intakes were too low for genetic or lifestyle variations in sensitivity to have an effect.[207,208]

The value of this example is that it clearly shows the difference between toxicokinetics and toxicodynamics. The interaction of chlorpyrifos oxon with cholinesterase would be represented by the boxes on the right of Figure 4.4, whereas the ADME considerations would be on the left of the figure.

Species and Route-to-Route Extrapolation

Two methods are currently used for species extrapolation. The first is the application of UFs as discussed earlier in this chapter. The most scientifically sound approach for animal-to-human dosimetric adjustment involves use of a validated PBPK model or chemical-specific adjustment factors to estimate the human

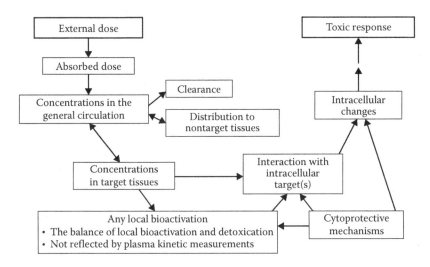

FIGURE 4.4 Schematic of processes underlying a toxic response. "Concentrations" refers to the relevant active form delivered by the general circulation and may be the parent compound or an active metabolite produced in another tissue and delivered to the target tissue or organ.

external dose (mg/kg-day) corresponding to an appropriate dose metric identified from consideration of the MOA.[7,198]

For toxicity factors derived by linear extrapolation, most regulatory agencies use body weight scaling to the 3/4 power. This adjustment may be performed on the doses in an animal experiment, on the POD value, or on the slope factor; however, interspecies scaling should only be performed once. Body weight to a fractional power is generally representative of surface area and is generally predictive for scaling of toxicity data.[161,209–212] Recently, EPA's RAF recommended the combined use to BW[3/4] as a scaling factor along with a reduced value of UFs to derive human equivalent doses (HEDs) for RfDs.[213]

The following equation shows the method for scaling slope factors derived from unadjusted doses in animals:

$$SF_{oral/human} = SF_{oral/animal} \times \left(\frac{BW_{human}}{BW_{animal}} \right)^{1/4} \tag{4.1}$$

SAR/QSAR, READ-ACROSS, AND OTHER COMPUTATIONAL METHODS

Structure-Activity Relationships/Quantitative Structure Activity Relationships (SAR/QSAR) are computational methods that use chemical properties to predict toxicity. These provide a means of estimating toxicity without animal or in vitro testing. The OECD has embraced QSAR as one means of addressing the large number of untested chemicals in commerce.

The OECD Validation Principles for SAR/QSAR both provide a useful framework for interpreting SAR/QSAR information in the context of regulatory purposes and offer a comparable and additional perspective of how to validate and interpret assays and data from emerging technologies.[214]

The five SAR/QSAR validation principles are as follows:

1. A defined endpoint
2. An unambiguous algorithm
3. A defined domain of applicability
4. Appropriate measures of goodness of fit, robustness, and predictivity
5. A mechanistic interpretation, if possible

SAR/QSAR methods provide a means to understand the link between chemical structure and biological activity.[215] SAR/QSAR methods provide a means for preliminary screening of chemicals.[216] The field has grown considerably with increasing reliance on data mining, statistics, and artificial intelligence. As part of an overall strategy to address chemical hazard, prediction models need to be validated whether these models are based on chemical properties or in vitro testing results.[217]

Read-across assumes that chemical structure determines toxicity and, in the analog approach, the method uses data from one or more tested chemicals to predict the toxicity of an untested chemical. In the category approach, read-across uses chemical categories having a number of tested chemicals and trends in the chemical properties to increase the confidence in the toxicity predictions for the untested chemicals.[218,219] A necessary part of read-across is the identification of analogs or categories.[220,221]

As experience with systems biology and computational modeling increases, the number of computational methods for predicting toxicity will likely increase as well. The QSAR validation principles at present provide a common platform for assessing the confidence in the prediction of these models.[214]

MIXTURES

For a number of years, the response provided by risk assessors to the following question was less than satisfactory: Is exposure to chemical mixtures more or less toxic than exposure to single chemicals? Usually, the answer would be something that stated that the science underlying risk assessment was not sufficiently advanced to meet the needs and challenges of modern-day problems. There are reasons for this nonanswer—the study of mixtures has been plagued with generalizations from too few data and with ambiguous use of terminology and consequent imprecise interpretation.[222]

There are several ways of thinking about how chemicals in a mixture might interact to produce toxicity. Additivity suggests that the effect of the mixture can be estimated directly from the sum of the doses (assuming equal potency) or the sum of the responses of the individual chemicals in the mixture—the former is called dose addition and the latter response addition.

Dose addition assumes a common MOA, and if the MOA is not known, EPA recommends separating chemicals by the target organs.[1] This situation is not always true. For example, hydrogen sulfide and cyanide both form methemoglobin (metHb) adducts and both prevent both oxygen transport by the erythrocyte and inhibit electron transport in the mitochondria. Hence, these two chemicals possess a common MOA and dose addition would be expected to predict their combined effects. However, cyanide is detoxified by rhodanese, an enzyme that uses thiosulfate to convert cyanide to thiocyanate, which is much less toxic. Hydrogen sulfide is detoxified by sulfide oxidase, an enzyme that produces thiosulfate from hydrogen sulfide. The increased levels of thiosulfate accelerate detoxification of cyanide. Cyanide antidote kits used by poison control centers contain injectable thiosulfate, and here is an instance where two chemicals with a common MOA fail to be dose additive.[9]

Response addition assumes that the chemicals in the mixture have a dissimilar MOA. An example of response addition is the common practice of summing cancer risks from individual chemicals estimated from CSFs. Since these risks are probabilities, this practice is mathematically and conceptually correct—what is missing is any information about MOA. Dose addition is the basis of the toxic equivalence factors for dioxin-like chemicals that have been used for many years.[223]

Interactions between chemicals may also occur. Mixtures producing risks greater than expected by additivity are called synergistic, and those producing risks less than expected from additivity are called antagonistic. Hence, depending on whether one is thinking about dose addition or response addition, the appearance of synergism or antagonism may be quite different. Interactions are quantitative relationships, and demonstration of interactions requires a quantitative determination based on the quantitative dose–response of single agents in the mixture. Further, interactions are defined quantitatively and their presence or absence can be determined by experiment—whether or not the MOA is known. Recently, five criteria were stated for determining whether interaction between two substances actually occurred. These criteria were quite stringent and meeting all of them would require both large amounts of data and rigorous thinking.[224]

TOXICITY FACTORS FOR REGULATION

What is vital for regulation is the availability of high-quality peer-reviewed toxicity factors that enjoy wide support among regulators, the regulated community, and the public. A number of governments, international groups, and nongovernmental organizations have developed sets of toxicity values.

Toxicity Databases in the United States and around the World

In 1985, EPA's IRIS was created to develop consensus opinions about the health effects that may result from chronic exposure to various substances in the environment and to provide these opinions and accompanying quantitative toxicity factors in an accessible database. The goal was to reduce inconsistency in toxicity assessments. For the next 10 years, IRIS consensus opinions were documented in

IRIS summaries, and EPA, state regulatory agencies, and others came to rely on IRIS information in decision making. In 1997, EPA began publishing comprehensive Toxicological Review support documents and incorporating additional peer review into the development of IRIS toxicity values.

However, recently, there were several highly controversial assessments that prompted requests for review of several chemical-specific assessments and the entire IRIS process by the NAS. The highly controversial assessments include those for TCDD, the solvent trichloroethylene, and formaldehyde. In fact, in the last chapter of the NAS review of EPA's formaldehyde toxicity assessment, the panel recommended a comprehensive revamping of the IRIS process.[225] In 2012, EPA requested that the NAS begin a comprehensive review of the entire IRIS process that is currently ongoing.

During the 1990s, EPA considered IRIS the "gold standard" for toxicity criteria. However, data of sufficient quality were not available for all chemicals, and toxicity values based on this poorer-quality data were assembled in another database called the Health Effects Assessment Summary Tables (HEAST). HEAST can still be found on EPA's website.

A small number of states in the United States had toxicologists on staff and issued their own toxicity factors. The most active state in this regard was California with a database of toxicity factors that was developed by the Office of Environmental Health Hazard Assessment (OEHHA). This database can be found at http://oehha.ca.gov/tcdb/index.asp. The state of Texas also has a vigorous program for developing toxicity factors with extensive guidance that provides procedures for toxicity factor development based on both animal and epidemiologic data. This guidance is available at http://www.tceq.texas.gov/toxicology/esl/guidelines/about.html.

What is interesting is that California and Texas regulators both have disagreements with some of the toxicity factors in IRIS. One of the exercises at the end of this chapter is to compare some of the toxicity values among these three sources.

Health Canada provides toxicity reference values at http://www.hc-sc.gc.ca/contact/order-pub-commande-eng.php, but this document provides no details of how the values were derived. The EU derives its own toxicity criteria as well.

In the United States, TERA maintains a database named International Toxicity Estimates for Risk Assessment (ITER). ITER is available at http://iter.ctcnet.net/publicurl/pub_search_list.cfm and also through the National Library of Medicine at http://toxnet.nlm.nih.gov/. ITER provides chronic human health risk assessment data from a variety of organizations worldwide in a side-by-side format, explains differences in risk values derived by different organizations, and links directly to each organization's website for more detailed information. Furthermore, it is the only database that includes risk information from independent parties whose risk values have undergone independent peer review.[226]

The EU and the OECD provide links to toxicity databases around the world—some from USEPA, some from universities, and others from private consultants. These can be seen at http://www.eea.europa.eu/publications/GH-07-97-595-EN-C2/iss2c1h.html. The International Agency for Research on Carcinogens (IARC) of the World Health Organization (WHO) does not develop quantitative toxicity criteria, but their monographs provide much information about both cancer and noncancer effects of various chemicals. A good way to find these IARC documents is through the ITER database mentioned earlier.

With its growing economy, China is becoming the largest importer of chemicals in the world and has realized the need for a governmental role in environmental risk assessment.[227,228] However, China realizes the societal cost of the LNT; whether this hypothesis becomes entrenched in Chinese government environmental policies remains to be seen.[229]

CHANGING LANDSCAPE OF TOXICITY TESTING

For many years, regulators considered animal toxicity testing to be the so-called gold standard. In response to concerns about animal welfare, the need to assess the growing number of chemicals in commerce, and the increasing realization that high-dose experiments in animals could not address potential effects in humans exposed to much lower doses, the prospects for the use of in vitro testing for risk assessment have also grown.[70,230]

EPA has initiated a number of activities in an effort to incorporate the use of high throughput/high content (HT/HC) assays into risk assessment. The most visible of these is ToxCast™, consisting of a battery of both commercial and publically developed HT/HC assays. Initially, the ToxCast™ approach has been designed to utilize the vast array of commercially available HT/HC assays to screen substances of interest to EPA.

However, as noted, there are also significant disadvantages to ToxCast™. One of the most problematic and challenging aspects of this approach is that many of these commercial methods are proprietary, so details about development, replicability, sensitivity, and specificity of the individual assays are not necessarily available for independent evaluation and scientific peer review. Hence, these proprietary assays are "black boxes" in many ways. From a scientific point of view, the choice of assays based on convenience does an "end run" around the concept of MOA. Ideally, the selection of assays would begin with the identification of TPs and assays would then be chosen because their results reflect the occurrence of key events in those pathways.

What is lacking here is the knowledge of those pathways. For example, how many TPs exist? As the number of enzymes and cellular targets for toxicity are finite, the number of pathways is also likely finite.[70,231] Although there are evolutionary and energetic constraints on the complexity of human biology,[232] a question that remains unanswered is whether the ToxCast assays cover the entire domain of TPs—or indeed, just what fraction of that domain is represented.

The prediction models used for screening, HI, or hazard characterization of in vitro data are not yet mature. Dosimetry and ADME need to be considered.[233,234] For example, when the predictive performance of more than 600 in vitro assays was examined across 60 in vivo endpoints using 84 different statistical classification methods and compared to the predictions based solely on chemical descriptors, the predictive power of the in vitro assays was no better than that of the chemical descriptors.[235]

Twenty-first century toxicology presents an exciting era for toxicologists, risk assessors, and researchers. Programs such as the USEPA's ToxCast™ are laudable in that they demonstrate just how new technologies can be exploited to address the challenges of risk assessment. However, the new challenges are credible validation/evaluation methods and understanding of the strengths and limitations for specific uses. In order to interpret and make use of in vitro data for risk assessment and regulation, what will be needed includes the following:

- An assessment of how well the assay results represent changes in a biomarker and the key events with which the biomarker is associated
- A validated prediction model that incorporates details of dosimetry, bioavailability, ADME, and other relevant factors
- Scientific consensus among regulators, regulated entities, and stakeholders in order to provide confidence in the results for decision-making
- Peer review, communication, and outreach to all stakeholders

One way to approach the development of a plethora of new methods in toxicology is to apply evidence-based methodology to both the data and prediction models. This approach is loosely based on the work of the Cochrane Collaboration in medicine.[236–238] What is clear is that implementing such an evidence-based approach will not be an easy task.[239]

EXERCISES FOR THOUGHT AND DISCUSSION

BENCHMARK DOSE MODELING FOR CANCER

Please download and install EPA's BMDS, available at http://www.epa.gov/ncea/bmds/index.html. Then find the animal bioassay reports from the NTP at http://ntp-server.niehs.nih.gov/. Navigate to the section on reports and publications and find the long-term study reports. You can obtain data from any of the reports, but the recommended ones are TR-521 on TCDD administered by corn oil gavage and TR-546 on hexavalent chromium in drinking water. Both of these have plenty of data to model. BMDS is pretty self-explanatory and you should be able to fit a number of models to these data. This exercise will help you become familiar with both the NTP reports and BMDS.

BENCHMARK DOSE MODELING FOR NONCANCER EFFECTS

Obtain the NTP report on Wy-14,643 (TOX-62) from http://ntp-server.niehs.nih.gov/index.cfm?objectid=8668C3FE-F1F6-975E-77A5E53978C3C502. Use the continuous models in BMDS to model the dose–response data in Table 7 on page 38 of the NTP report showing effects on blood lipids, estradiol and follicle stimulating hormone. Think about your experience modeling these continuous data versus that modeling the quantal (frequency) data in the first exercise.

Comparison of Toxicity Criteria

Pick your favorite five chemicals. Use the Internet to try to find at least three toxicity criteria (e.g., http://iter.ctc.com/publicURL/pub_search_list.cfm). Compare the basis of these criteria. What are the strengths and weaknesses of each?

REFERENCES

1. United States Environmental Protection Agency (USEPA). *Risk Assessment Guidance for Superfund: Volume 1: Human Health Evaluation Manual (Part A) (Interim Final) (RAGS)*. EPA/540/1-89/002. Washington, DC, December, 1989. http://www.epa.gov/oswer/riskassessment/ragsa/
2. National Research Council (NRC). *Issues in Risk Assessment*. Washington, DC: The National Academy Press, 1993. http://www.nap.edu/catalog.php?record_id=2078
3. Mantel N and Bryan WR. "Safety" testing of carcinogenic agents. *J Natl Cancer Inst*. 1961, August;27:455–470.
4. American Cancer Society (ACS). Cancer And The Environment. ACS Pressroom Blog. Atlanta, GA, May 6, 2010. http://acspressroom.wordpress.com/2010/05/06/cancer-and-the-environment/
5. Hanahan D and Weinberg RA. Hallmarks of cancer: The next generation. *Cell*. 2011, March 3;144(5):646–674.
6. Mukherjee S. *The Emperor of All Maladies: A Biography of Cancer*. New York: Scribner, 2011.
7. United States Environmental Protection Agency (USEPA). Risk Assessment Forum. *Guidelines for Carcinogen Risk Assessment*. EPA/630/P-03/001F. March, 2005. http://www.epa.gov/raf/publications/pdfs/CANCER_GUIDELINES_FINAL_3-25-05.PDF
8. Vineis P, Schatzkin A, and Potter JD. Models of carcinogenesis: An overview. *Carcinogenesis*. 2010, October;31(10):1703–1709.
9. Borgert CJ, Quill TF, McCarty LS, and Mason AM. Can mode of action predict mixture toxicity for risk assessment? *Toxicol Appl Pharmacol*. 2004, December 12;201(2):85–96.
10. Dellarco VL and Wiltse JA. US Environmental Protection Agency's revised guidelines for Carcinogen Risk Assessment: Incorporating mode of action data. *Mutat Res*. 1998, September 9;405(2):273–277.
11. Schlosser PM and Bogdanffy MS. Determining modes of action for biologically based risk assessments. *Regul Toxicol Pharmacol*. 1999, August;30(1):75–79.
12. United States Environmental Protection Agency (USEPA). *Guidelines for Carcinogen Risk Assessment*. EPA/630/R-00/004. Washington, DC, September 24, 1986. http://www.epa.gov/cancerguidelines/guidelines-carcinogen-risk-assessment-1986.htm
13. National Research Council (NRC). *Risk Assessment in the Federal Government: Managing the Process*. Washington, DC: The National Academies Press, 1983. http://www.nap.edu/catalog.php?record_id=366
14. United States Environmental Protection Agency (USEPA). Risk Assessment Forum. *Proposed Guidelines for Carcinogen Risk Assessment*. EPA/600/P-92/003C. Washington, DC, April, 1996. http://www.epa.gov/raf/publications/pdfs/propcra_1996.pdf
15. Andersen ME, Meek ME, Boorman GA, Brusick DJ, Cohen SM, Dragan YP et al. Lessons learned in applying the U.S. EPA proposed cancer guidelines to specific compounds. *Toxicol Sci*. 2000, February;53(2):159–172.
16. Boorman GA. Drinking water disinfection byproducts: Review and approach to toxicity evaluation. *Environ Health Perspect*. 1999, February;107(Suppl 1):207–217.

17. Schmidt CW. Chloroform: An EPA test case chloroform: An EPA test case. *Environ Health Perspect.* 1999;107(7):7.

18. Morgan BW, Geller RJ, and Kazzi ZN. Intentional ethylene glycol poisoning increase after media coverage of antifreeze murders. *West J Emerg Med.* 2011, July;12(3):296–299.

19. McMartin K. Are calcium oxalate crystals involved in the mechanism of acute renal failure in ethylene glycol poisoning? *Clin Toxicol (Phila).* 2009, November;47(9):859–869.

20. United States Environmental Protection Agency (USEPA). *DRAFT Toxicological Review of Hexavalent Chromium in Support of Summary Information on the Integrated Risk Information System (IRIS).* EPA/635/R-10/004A. Washington, DC, September, 2010. http://cfpub.epa.gov/ncea/iris_drafts/recordisplay.cfm?deid=221433

21. Office of Environmental Health Hazard Assessment (OEHHA) California Environmental Protection Agency (Cal-EPA). *Public Health Goal for Hexavalent Chromium in Drinking Water.* 2011, July. http://www.oehha.ca.gov/water/phg/072911Cr6PHG.html

22. McCarroll N, Keshava N, Chen J, Akerman G, Kligerman A, and Rinde E. An evaluation of the mode of action framework for mutagenic carcinogens case study II: Chromium (VI). *Environ Mol Mutagen.* 2010, March;51(2):89–111.

23. Cole P and Rodu B. Epidemiologic studies of chrome and cancer mortality: A series of meta-analyses. *Regul Toxicol Pharmacol.* 2005, December;43(3):225–231.

24. Birk T, Mundt KA, Dell LD, Luippold RS, Miksche L, Steinmann-Steiner-Haldenstaett W, and Mundt DJ. Lung cancer mortality in the German chromate industry, 1958 to 1998. *J Occup Environ Med.* 2006, April;48(4):426–433.

25. Park RM, Bena JF, Stayner LT, Smith RJ, Gibb HJ, and Lees PSJ. Hexavalent chromium and lung cancer in the chromate industry: A quantitative risk assessment. *Risk Anal.* 2004, October;24(5):1099–1108.

26. Alexeeff GV, Satin K, Painter P, Zeise L, Popejoy C, and Murchison G. Chromium carcinogenicity: California strategies. *Sci Total Environ.* 1989, October 10;86(1–2):159–168.

27. National Toxicology P. NTP toxicology and carcinogenesis studies of sodium dichromate dihydrate (CAS No. 7789–12–0) in F344/N Rats and B6C3F1 mice (drinking water studies). *Natl Toxicol Program Tech Rep Ser.* 2008, July;(546):1–192.

28. New Jersey Department of Environmental Protection NJDEP. *Derivation of Ingestion-Based Soil Remediation Criteria for Cr+6 based on the NTP Chronic Bioassay Data for Sodium Dichromate Dihydrate.* Report by Alan Stern, Dr. P.H., DABT, Division of Science, Research and Technology. April 8, 2009. http://www.state.nj.us/dep/dsr/chromium/soil-cleanup-derivation.pdf

29. Office of Environmental Health Hazard Assessment CEPAO. *Public Health Goal for Hexavalent Chromium (Cr VI) in Drinking Water.* July, 2011. http://www.oehha.ca.gov/water/phg/072911Cr6PHG.html

30. United States Environmental Protection Agency (USEPA). *DRAFT ToxicologicalReview of Hexavalent Chromium in Support of Summary Information on The IntegratedRisk Information System (IRIS).* EPA/635/R-10/004A. Washington, DC, September, 2010. http://cfpub.epa.gov/ncea/iris_drafts/recordisplay.cfm?deid=221433

31. Thompson CM, Proctor DM, Suh M, Haws LC, and Harris MA. Assessment of the mode of action underlying development of rodent small intestinal tumors following oral exposure to hexavalent chromium and relevance to humans. *Crit Rev Toxicol.* 2013. 43(3):244–274. DOI: 10.3109/10408444.2013.768596. http://informahealthcare.com/doi/pdf/10.3109/10408444.2013.768596

32. Thompson CM, Haws LC, Harris MA, Gatto NM, and Proctor DM. Application of the U.S. EPA mode of action framework for purposes of guiding future research: A case study involving the oral carcinogenicity of hexavalent chromium. *Toxicol Sci.* 2011, January;119(1):20–40.

33. Thompson CM, Proctor DM, Haws LC, Hebert CD, Grimes SD, Shertzer HG et al. Investigation of the mode of action underlying the tumorigenic response induced in B6C3F1 mice exposed orally to hexavalent chromium. *Toxicol Sci.* 2011, June 6;123(1):58–70.

34. Thompson CM, Proctor DM, Suh M, Haws LC, Hébert CD, Mann JF et al. Comparison of the effects of hexavalent chromium in the alimentary canal of F344 rats and B6C3F1 mice following exposure in drinking water: Implications for carcinogenic modes of action. *Toxicol Sci.* 2012;125(1):79–90.

35. Proctor DM, Suh M, Aylward LL, Kirman CR, Harris MA, Thompson CM et al. Hexavalent chromium reduction kinetics in rodent stomach contents. *Chemosphere.* 2012, June 6;89(5):487–493.

36. Kopec AK, Kim S, Forgacs AL, Zacharewski TR, Proctor DM, Harris MA et al. Genome-wide gene expression effects in B6C3F1 mouse intestinal epithelia following 7 and 90days of exposure to hexavalent chromium in drinking water. *Toxicol Appl Pharmacol.* 2011, December 12;259(1):13–26.

37. Kopec AK, Thompson CM, Kim S, Forgacs AL, and Zacharewski TR. Comparative toxicogenomic analysis of oral Cr(VI) exposure effects in rat and mouse small intestinal epithelia. *Toxicol Appl Pharmacol.* 2012, July 7;262(2):124–138.

38. Kirman CR, Hays SM, Aylward LL, Suh M, Harris MA, Thompson CM et al. Physiologically based pharmacokinetic model for rats and mice orally exposed to chromium. *Chem Biol Interact.* 2012, October 10;200(1):45–64.

39. Valentin-Vega YA, Okano H, and Lozano G. The intestinal epithelium compensates for p53-mediated cell death and guarantees organismal survival. *Cell Death Differ.* 2008, November;15(11):1772–1781.

40. Gagné D, Groulx J-F, Benoit YD, Basora N, Herring E, Vachon PH, and Beaulieu J-F. Integrin-linked kinase regulates migration and proliferation of human intestinal cells under a fibronectin-dependent mechanism. *J Cell Physiol.* 2010, February;222(2):387–400.

41. Benoit YD, Paré F, Francoeur C, Jean D, Tremblay E, Boudreau F et al. Cooperation between HNF-1alpha, Cdx2, and GATA-4 in initiating an enterocytic differentiation program in a normal human intestinal epithelial progenitor cell line. *Am J Physiol Gastrointest Liver Physiol.* 2010, April;298(4):G504–G517.

42. Cohen SM and Arnold LL. Chemical carcinogenesis. *Toxicol Sci.* 2011, March;120(Suppl 1):S76–S92.

43. Trosko JE and Upham BL. The emperor wears no clothes in the field of carcinogen risk assessment: Ignored concepts in cancer risk assessment. *Mutagenesis.* 2005, March;20(2):81–92.

44. Ellinger-Ziegelbauer H, Aubrecht J, Kleinjans JC, and Ahr H-J. Application of toxicogenomics to study mechanisms of genotoxicity and carcinogenicity. *Toxicol Lett.* 2009, April 4;186(1):36–44.

45. Thompson CM, Gregory Hixon J, Proctor DM, Haws LC, Suh M, Urban JD, and Harris MA. Assessment of genotoxic potential of Cr(VI) in the mouse duodenum: An in silico comparison with mutagenic and nonmutagenic carcinogens across tissues. *Regul Toxicol Pharmacol.* 2012, October;64(1):68–76.

46. Thompson CM, Fedorov Y, Brown DD, Suh M, Proctor DM, Kuriakose L et al. Assessment of Cr(VI)-induced cytotoxicity and genotoxicity using high content analysis. *PLoS One.* 2012;7(8):e42720.

47. Amdur MOM, Doull JJ, and Klassen CDC. *Cassarett & Doull's Toxicology the Basic Science of Poisons*, 4th edn. New York: McGraw-Hill, 1993.

48. Stebbing AR. Hormesis—The stimulation of growth by low levels of inhibitors. *Sci Total Environ.* 1982, February;22(3):213–234.

49. Calabrese EJ and Baldwin LA. Chemical hormesis: Its historical foundations as a biological hypothesis. *Toxicol Pathol.* 1999;27(2):195–216.

50. Calabrese EJ and Baldwin LA. The marginalization of hormesis. *Toxicol Pathol.* 1999, March 3;27(2):187–194.

51. Holsapple MP and Wallace KB. Dose response considerations in risk assessment—An overview of recent ILSI activities. *Toxicol Lett.* 2008, August 8;180(2):85–92.

52. Stebbing ARD. Interpreting 'dose-response' curves using homeodynamic data: With an improved explanation for hormesis. *Dose Response.* 2009;7(3):221–233.

53. Fukushima S. Carcinogenic risk assessment: Are there dose thresholds for carcinogens? *Asian Pac J Cancer Prev.* 2010;11(1):19–21.

54. Gray G. Hormesis in regulatory risk assessment—Science and Science Policy. *Dose Response.* 2011;9(2):158–164.

55. Sonich-Mullin C, Fielder R, Wiltse J, Baetcke K, Dempsey J, Fenner-Crisp P et al. IPCS conceptual framework for evaluating a mode of action for chemical carcinogenesis. *Regul Toxicol Pharmacol.* 2001, October;34(2):146–152.

56. Meek ME, Bucher JR, Cohen SM, Dellarco V, Hill RN, Lehman-McKeeman LD et al. A framework for human relevance analysis of information on carcinogenic modes of action. *Crit Rev Toxicol.* 2003;33(6):591–653.

57. Meek ME. Recent developments in frameworks to consider human relevance of hypothesized modes of action for tumours in animals. *Environ Mol Mutagen.* 2008, March;49(2):110–116.

58. Seed J, Carney EW, Corley RA, Crofton KM, DeSesso JM, Foster PM et al. Overview: Using mode of action and life stage information to evaluate the human relevance of animal toxicity data. *Crit Rev Toxicol.* 2005;35(8–9):664–672.

59. Boobis AR, Cohen SM, Dellarco V, McGregor D, Meek ME, Vickers C et al. IPCS framework for analyzing the relevance of a cancer mode of action for humans. *Crit Rev Toxicol.* 2006, November;36(10):781–792.

60. Boobis AR, Doe JE, Heinrich-Hirsch B, Meek ME, Munn S, Ruchirawat M et al. IPCS framework for analyzing the relevance of a noncancer mode of action for humans. *Crit Rev Toxicol.* 2008;38(2):87–96.

61. Julien E, Boobis AR, Olin SS, and Ilsi Research Foundation Threshold Working Group. The Key Events Dose-Response Framework: A cross-disciplinary mode-of-action based approach to examining dose-response and thresholds. *Crit Rev Food Sci Nutr.* 2009, September;49(8):682–689.

62. Pool-Zobel B, Veeriah S, and Böhmer F-D. Modulation of xenobiotic metabolising enzymes by anticarcinogens—Focus on glutathione S-transferases and their role as targets of dietary chemoprevention in colorectal carcinogenesis. *Mutat Res.* 2005, December 12;591(1–2):74–92.

63. Ulrich CM, Curtin K, Potter JD, Bigler J, Caan B, and Slattery ML. Polymorphisms in the reduced folate carrier, thymidylate synthase, or methionine synthase and risk of colon cancer. *Cancer Epidemiol Biomarkers Prev.* 2005, November;14(11 Pt 1):2509–2516.

64. United States Environmental Protection Agency (USEPA). *Supplemental Guidance for Assessing Susceptibility from Early-Life Exposure to Carcinogens.* EPA/630/R-03/003F. Washington, DC, March, 2005. http://www.epa.gov/cancerguidelines/guidelines-carcinogen-supplement.htm

65. United States Environmental Protection Agency (USEPA). *Framework for Determining a Mutagenic Mode of Action for Carcinogenicity: Using EPA's 2005 Cancer Guidelines and Supplemental Guidance for Assessing Susceptibility to Early-Life Exposure to Carcinogens.* EPA/120/R-07/002-A. External Peer Review Draft. Washington, DC, September, 2007. http://www.epa.gov/osa/mmoaframework/

66. European Commission Joint Research Centre (ECJRC). *Technical Guidance Document on Risk Assessment, Part II. Environmental Risk Assessment. in support of Commission Directive 93/67/EEC on Risk Assessment for new notified substances Commission Regulation (EC) No 1488/94.* 2003. http://ihcp.jrc.ec.europa.eu/our_activities/public-health/risk_assessment_of_Biocides/doc/tgd

67. European Food Safety Authority EFSA. Opinion of a scientific committee on a request from EFSA related to a harmonised approach for risk assessment of substances which are both genotoxic and carcinogenic. *EFSA J.* 2005;2821–2831. doi: 10.2903/j.efsa.2005.28. http://www.efsa.europa.eu/en/efsajournal/pub/282.htm

68. European Commission EC. Regulation (EC) no. 1272/2008 of the European Parliament and of the Council of 16 December 2008 on classification, labelling and packaging of substances and mixtures, amending and repealing Directives 67/548/EEC and 1999/45/EC, and amending Regulation (EC) no. 1907/2006. *Official J European Union.* 2008;3531–1355. http://eur-lex.europa.eu/LexUriServ/LexUriServ.do?uri=OJ:L:2008:353:0001:1355:en:PDF

69. Organisation for Economic Cooperation and Development (OECD). *Manual for the Assessment of Chemicals.* http://www.oecd.org/env/ehs/risk-assessment/manualfortheassessmentofchemicals.htm.

70. National Research Council (NRC). *Toxicity Testing in the 21st Century: A Vision and a Strategy.* Washington, DC: The National Academies Press, 2007. http://www.nap.edu/catalog.php?record_id=11970

71. Dellarco V and Fenner-Crisp PA. Mode of action: Moving toward a more relevant and efficient assessment paradigm. *J Nutr.* 2012, December;142(12):2192S–2198S. doi: 10.3945/jn.111.157396

72. Rodgers IS and Baetcke KP. Interpretation of male rat renal tubule tumors. *Environ Health Perspect.* 1993, December;101(Suppl 6)45–52.

73. Piccirillo VJ, Bird MG, Lewis RJ, and Bover WJ. Preliminary evaluation of the human relevance of respiratory tumors observed in rodents exposed to naphthalene. *Regul Toxicol Pharmacol.* 2012, April;62(3):433–440.

74. Backhaus T and Faust M. Predictive environmental risk assessment of chemical mixtures: A conceptual framework. *Environ Sci Technol.* 2012, March 6;46(5):2564–2573.

75. Rhomberg LR, Bailey LA, and Goodman JE. Hypothesis-based weight of evidence: A tool for evaluating and communicating uncertainties and inconsistencies in the large body of evidence in proposing a carcinogenic mode of action—Naphthalene as an example. *Crit Rev Toxicol.* 2010, September;40(8):671–696.

76. Phillips CV and Goodman KJ. Causal criteria and counterfactuals; nothing more (or less) than scientific common sense. *Emerg Themes Epidemiol.* 2006;35.

77. Hill AB. The environment and disease: Association or causation? *Proc R Soc Med.* 1965, May;58:295–300.

78. Phillips CV and Goodman KJ. The missed lessons of Sir Austin Bradford Hill. *Epidemiol Perspect Innov.* 2004;1(1):3.

79. National Toxicology Programs (NTP). NTP toxicology and carcinogenesis studies of 2,3,7,8-tetrachlorodibenzo-p-dioxin (TCDD) (CAS No. 1746–01–6) in Female Harlan Sprague-Dawley rats (Gavage Studies). *Natl Toxicol Program.Tech.Rep. Ser.* 2006. NTP TR521. NIH Publication No. 06-4468:1. http://ntp-server.niehs.nih.gov/?objectid=070B69A9-BC89-4234-E4AFAE94C636CC5D

80. Steinpress MG and Ward AC. The scientific process and Hollywood: The case of hexavalent chromium. *Ground Water*. 2001, May;39(3):321–322.
81. Phillips ML. Chromium paper retracted unfairly, author says [Internet]. *Scientist*. 2006, December 22; http://www.the-scientist.com/news/home/38457/
82. Phillips ML. Journal retracts chromium study [Internet]. *Scientist*. 2006, June 7; http://www.the-scientist.com/news/display/23590/
83. Brandt-Rauf P. Editorial retraction. Cancer mortality in a Chinese population exposed to hexavalent chromium in water. *J Occup Environ Med*. 2006, July;48(7):749.
84. National Toxicology Program. Toxicology and carcinogenesis studies of sodium dichromate dihydrate (Cas No. 7789–12–0) in F344/N rats and B6C3F1 mice (drinking water studies). *Natl Toxicol Program Tech Rep Ser*. 2008, July;(546):1–192.
85. Araten DJ, Golde DW, Zhang RH, Thaler HT, Gargiulo L, Notaro R, and Luzzatto L. A quantitative measurement of the human somatic mutation rate. *Cancer Res*. 2005, September 9;65(18):8111–8117.
86. Araten DJ and Luzzatto L. The mutation rate in PIG-A is normal in patients with paroxysmal nocturnal hemoglobinuria (PNH). *Blood*. 2006, July 7;108(2):734–736.
87. Phonethepswath S, Bryce SM, Bemis JC, and Dertinger SD. Erythrocyte-based Pig-a gene mutation assay: Demonstration of cross-species potential. *Mutat Res*. 2008, December 12;657(2):122–126.
88. United States Environmental Protection Agency (USEPA). *Methods for Derivation of Inhalation Reference Concentrations and Application of Inhalation Dosimetry*. EPA/600/8-90/066F. Research Triangle Park, NC, October,1994. http://cfpub.epa. gov/ncea/cfm/recordisplay.cfm?deid=71993 (accessed September 25, 2012).
89. Lewis RW, Billington R, Debryune E, Gamer A, Lang B, and Carpanini F. Recognition of adverse and nonadverse effects in toxicity studies. *Toxicol Pathol*. 2002;30(1):66–74.
90. National Research Council (NRC). *Scientific Review of the Proposed Risk Assessment Bulletin from the Office of Management and Budget*. 2007.
91. Karbe E, Williams GM, Lewis RW, Kimber I, and Foster PMD. Distinguishing between adverse and non-adverse effects. *Exp Toxicol Pathol*. 2002, July;54(1):51–55.
92. Barnes DG, Daston GP, Evans JS, Jarabek AM, Kavlock RJ, Kimmel CA et al. Benchmark Dose Workshop: Criteria for use of a benchmark dose to estimate a reference dose. *Regul Toxicol Pharmacol*. 1995, April;21(2):296–306.
93. Dakeishi M, Murata K, Tamura A, and Iwata T. Relation between benchmark dose and no-observed-adverse-effect level in clinical research: Effects of daily alcohol intake on blood pressure in Japanese salesmen. *Risk Anal*. 2006, February; 26(1):115–123.
94. Davis JA, Gift JS, and Zhao QJ. Introduction to benchmark dose methods and U.S. EPA's benchmark dose software (BMDS) version 2.1.1. *Toxicol Appl Pharmacol*. 2011, July 7;254(2):181–191.
95. Dorato MA and Engelhardt JA. The no-observed-adverse-effect-level in drug safety evaluations: Use, issues, and definition(s). *Regul Toxicol Pharmacol*. 2005, August;42(3):265–274.
96. Filipsson AF, Sand S, Nilsson J, and Victorin K. The benchmark dose method—Review of available models, and recommendations for application in health risk assessment. *Crit Rev Toxicol*. 2003;33(5):505–542.
97. Gaylor DW and Chen JJ. Precision of benchmark dose estimates for continuous (nonquantal) measurements of toxic effects. *Regul Toxicol Pharmacol*. 1996, August;24(1 Pt 1):19–23.
98. Izadi H, Grundy JE, and Bose R. Evaluation of the benchmark dose for point of departure determination for a variety of chemical classes in applied regulatory settings. *Risk Anal*. 2012, May;32(5):830–835.

99. Kuljus K, von Rosen D, Sand S, and Victorin K. Comparing experimental designs for benchmark dose calculations for continuous endpoints. *Risk Anal.* 2006, August;26(4):1031–1043.
100. Leisenring W and Ryan L. Statistical properties of the NOAEL. *Regul Toxicol Pharmacol.* 1992, April;15(2 Pt 1):161–171.
101. Murrell JA, Portier CJ, and Morris RW. Characterizing dose-response: I: Critical assessment of the benchmark dose concept. *Risk Anal.* 1998, February;18(1):13–26.
102. Pottenger LH, Becker RA, Moran EJ, and Swenberg JA. Workshop report: Identifying key issues underpinning the selection of linear or non-linear dose-response extrapolation for human health risk assessment of systemic toxicants. *Regul Toxicol Pharmacol.* 2011, April;59(3):503–510.
103. Sand S, Filipsson AF, and Victorin K. Evaluation of the benchmark dose method for dichotomous data: Model dependence and model selection. *Regul Toxicol Pharmacol.* 2002, October;36(2):184–197.
104. Sand S, Victorin K, and Filipsson AF. The current state of knowledge on the use of the benchmark dose concept in risk assessment. *J Appl Toxicol.* 2008, May;28(4):405–421.
105. Slob W, Moerbeek M, Rauniomaa E, and Piersma AH. A statistical evaluation of toxicity study designs for the estimation of the benchmark dose in continuous endpoints. *Toxicol Sci.* 2005, March;84(1):167–185.
106. Yanagawa T, Kikuchi Y, and Brown KG. Statistical issues on the no-observed-adverse-effect level in categorical response. *Environ Health Perspect.* 1994, January;102(Suppl 1):95–101.
107. United States Environmental Protection Agency (USEPA) Risk Assessment Forum. *Benchmark Dose Technical Guidance Document (External Review Draft).* EPA/630/R-00/001. Washington, DC, October, 2000. http://www.epa.gov/raf/publications/pdfs/BMD-EXTERNAL_10_13_2000.PDF
108. United States Environmental Protection Agency (USEPA). Risk Assessment Forum. *Benchmark Dose Technical Guidance.* EPA/100/R-12/001. Washington, DC, June, 2012. http://www.epa.gov/raf/publications/pdfs/benchmark_dose_guidance.pdf
109. United States Environmental Protection Agency (USEPA). Risk Assessment Forum. *Report on the Benchmark Dose Peer Consultation Workshop.* EPA/630/R96/001. November, 1996. http://www.epa.gov/raf/publications/rpt-benchmark-dose-workshop.htm
110. United States Environmental Protection Agency (USEPA). Risk Assessment Forum. *Benchmark Dose Technical Guidance.* EPA/100/R-12/001. Washington, DC, June, 2012. http://www.epa.gov/raf/publications/pdfs/benchmark_dose_guidance.pdf
111. Barnes DG and Dourson M. Reference dose (RfD): Description and use in health risk assessments. *Regul Toxicol Pharmacol.* 1988, December;8(4):471–486.
112. Dourson ML and Stara JF. Regulatory history and experimental support of uncertainty (safety) factors. *Regul Toxicol Pharmacol.* 1983, September;3(3):224–238.
113. Lehman AJ and Fitzhugh AG. 100-Fold margin of safety. *Assoc. Food Drug Officials U.S. Quart. Bull.* 1954;1833–1835.
114. Toxicology Excellence in Risk Assessment (TERA). Home Page. n.d. http://tera.org/ (accessed September 21, 2012).
115. Dourson ML, Knauf LA, and Swartout JC. On reference dose (RfD) and its underlying toxicity data base. *Toxicol Ind Health.* 1992;8(3):171–189.
116. Dourson ML, Felter SP, and Robinson D. Evolution of science-based uncertainty factors in noncancer risk assessment. *Regul Toxicol Pharmacol.* 1996, October;24(2 Pt 1):108–120.

117. Kroes R, Renwick AG, Cheeseman M, Kleiner J, Mangelsdorf I, Piersma A et al. Structure-based thresholds of toxicological concern (TTC): Guidance for application to substances present at low levels in the diet. *Food Chem Toxicol*. 2004, January;42(1):65–83.

118. Walton K, Dorne JLCM, and Renwick AG. Species-specific uncertainty factors for compounds eliminated principally by renal excretion in humans. *Food Chem Toxicol*. 2004, February;42(2):261–274.

119. Dorne JLCM, Walton K, and Renwick AG. Human variability in the renal elimination of foreign compounds and renal excretion-related uncertainty factors for risk assessment. *Food Chem Toxicol*. 2004, February;42(2):275–298.

120. Renwick AG. Risk characterisation of chemicals in food. *Toxicol Lett*. 2004, April 4;149(1–3):163–176.

121. Renwick AG, Dorne JL, and Walton K. An analysis of the need for an additional uncertainty factor for infants and children. *Regul Toxicol Pharmacol*. 2000, June;31(3):286–296.

122. Renwick AG. The use of safety or uncertainty factors in the setting of acute reference doses. *Food Addit Contam*. 2000, July;17(7):627–635.

123. Dybing E, Doe J, Groten J, Kleiner J, O'Brien J, Renwick AG et al. Hazard characterisation of chemicals in food and diet. Dose response, mechanisms and extrapolation issues. *Food Chem Toxicol*. 2002;40(2–3):237–282.

124. Renwick AG, Barlow SM, Hertz-Picciotto I, Boobis AR, Dybing E, Edler L et al. Risk characterisation of chemicals in food and diet. *Food Chem Toxicol*. 2003, September;41(9):1211–1271.

125. Renwick AG and Lazarus NR. Human variability and noncancer risk assessment—An analysis of the default uncertainty factor. *Regul Toxicol Pharmacol*. 1998, Februry;27(1 Pt 2):3–20.

126. Renwick AG. Duration of intake above the ADI/TDI in relation to toxicodynamics and toxicokinetics. *Regul Toxicol Pharmacol*. 1999, October;30(2 Pt 2):S69–S78.

127. Renwick AG. Data-derived safety factors for the evaluation of food additives and environmental contaminants. *Food Addit Contam*. 1993;10(3):275–305.

128. Renwick AG and Walker R. An analysis of the risk of exceeding the acceptable or tolerable daily intake. *Regul Toxicol Pharmacol*. 1993, December;18(3):463–480.

129. Renwick AG. The use of an additional safety or uncertainty factor for nature of toxicity in the estimation of acceptable daily intake and tolerable daily intake values. *Regul Toxicol Pharmacol*. 1995, December;22(3):250–261.

130. Renwick AG. Safety factors and establishment of acceptable daily intakes. *Food Addit Contam*. 1991;8(2):135–149.

131. Vermeire T, Stevenson H, Peiters MN, Rennen M, Slob W, and Hakkert BC. Assessment factors for human health risk assessment: A discussion paper. *Crit Rev Toxicol*. 1999, September;29(5):439–490.

132. Dorne JLCM and Renwick AG. The refinement of uncertainty/safety factors in risk assessment by the incorporation of data on toxicokinetic variability in humans. *Toxicol Sci*. 2005, July;86(1):20–26.

133. International Programme on Chemical Safety (IPCS). World Health Organization. *Chemical-Specific Adjustment Factors (CSAFs) for Interspecies Differences and Human Variability: Guidance Document for the Use of Data in Development of In Dose/Concentration Response Assessment*. Geneva, 2005. http://whqlibdoc.who.int/publications/2005/9241546786_eng.pdf

134. United States Environmental Protection Agency (USEPA). Risk Assessment Forum. *Guidance for Applying Quantitative Data to Develop Data-Derived Extrapolation Factors for Interspecies and Intraspecies Extrapolation*. External Review Draft. EPA/100/J-11/001. Washington, DC, May, 2011. http://www.epa.gov/raf/files/ddef-external-review-draft05-11-11.pdf

135. Carlson EA. *Hermann Joseph Muller 1890–1967: A Biographical Memoir.* National Academy of Sciences. Washington, DC, 2009. http://www.nasonline.org/publications/biographical-memoirs/memoir-pdfs/muller-hermann.pdf
136. Calabrese EJ. Muller's Nobel Prize Lecture: When ideology prevailed over science. *Toxicol Sci.* 2012, March;126(1):1–4.
137. Calabrese EJ. Key studies used to support cancer risk assessment questioned. *Environ Mol Mutagen.* 2011, October;52(8):595–606.
138. Calabrese EJ. Muller's Nobel lecture on dose-response for ionizing radiation: Ideology or science? *Arch Toxicol.* 2011, December;85(12):1495–1498.
139. Simon T. Just who is at risk? The ethics of environmental regulation. *Hum Exp Toxicol.* 2011, August;30(8):795–819.
140. Calabrese EJ. The road to linearity: Why linearity at low doses became the basis for carcinogen risk assessment. *Arch Toxicol.* 2009, March;83(3):203–225.
141. Waddell WJ. Critique of dose response in carcinogenesis. *Hum Exp Toxicol.* 2006, July;25(7):413–436.
142. Swenberg JA, Barrow CS, Boreiko CJ, Heck HD, Levine RJ, Morgan KT, and Starr TB. Non-linear biological responses to formaldehyde and their implications for carcinogenic risk assessment. *Carcinogenesis.* 1983, August;4(8):945–952.
143. Swenberg JA, Richardson FC, Boucheron JA, and Dyroff MC. Relationships between DNA adduct formation and carcinogenesis. *Environ Health Perspect.* 1985, October;62:177–183.
144. Swenberg JA and Fennell TR. DNA damage and repair in mouse liver. *Arch Toxicol Suppl.* 1987;10:162–171.
145. Swenberg JA, Richardson FC, Boucheron JA, Deal FH, Belinsky SA, Charbonneau M, and Short BG. High- to low-dose extrapolation: Critical determinants involved in the dose response of carcinogenic substances. *Environ Health Perspect.* 1987, December;76:57–63.
146. Swenberg JA, La DK, Scheller NA, and Wu KY. Dose-response relationships for carcinogens. *Toxicol Lett.* 1995, December;82–83:751–756.
147. Swenberg JA, Fryar-Tita E, Jeong Y-C, Boysen G, Starr T, Walker VE, and Albertini RJ. Biomarkers in toxicology and risk assessment: Informing critical dose-response relationships. *Chem Res Toxicol.* 2008, Jan;21(1):253–265.
148. Jarabek AM, Pottenger LH, Andrews LS, Casciano D, Embry MR, Kim JH et al. Creating context for the use of DNA adduct data in cancer risk assessment: 1. Data organization. *Crit Rev Toxicol.* 2009;39(8):659–678.
149. Swenberg JA, Lu K, Moeller BC, Gao L, Upton PB, Nakamura J, and Starr TB. Endogenous versus exogenous DNA adducts: Their role in carcinogenesis, epidemiology, and risk assessment. *Toxicol Sci.* 2011, March;120(Suppl 1):S130–S145.
150. Crump KS, Hoel DG, Langley CH, and Peto R. Fundamental carcinogenic processes and their implications for low dose risk assessment. *Cancer Res.* 1976, September;36(9 Pt 1):2973–2979.
151. Crump KS. Use of threshold and mode of action in risk assessment. *Crit Rev Toxicol.* 2011, September;41(8):637–650.
152. Crump KS, Chiu WA, and Subramaniam RP. Issues in using human variability distributions to estimate low-dose risk. *Environ Health Perspect.* 2010, March;118(3):387–393.
153. Gori GB. Cancer risk assessment: The science that is not. *Regul Toxicol Pharmacol.* 1992, August;16(1):10–20.
154. Purchase IF and Auton TR. Thresholds in chemical carcinogenesis. *Regul Toxicol Pharmacol.* 1995, December;22(3):199–205.

155. Clayson DB and Iverson F. Cancer risk assessment at the crossroads: The need to turn to a biological approach. *Regul Toxicol Pharmacol.* 1996, August;24(1 Pt 1):45–59.
156. Conolly RB, Gaylor DW, and Lutz WK. Population variability in biological adaptive responses to DNA damage and the shapes of carcinogen dose-response curves. *Toxicol Appl Pharmacol.* 2005, September 9;207(2 Suppl):570–575.
157. Lutz WK, Gaylor DW, Conolly RB, and Lutz RW. Nonlinearity and thresholds in dose-response relationships for carcinogenicity due to sampling variation, logarithmic dose scaling, or small differences in individual susceptibility. *Toxicol Appl Pharmacol.* 2005, September 9;207(2 Suppl):565–569.
158. Calabrese EJ. NEPA, EPA and risk assessment: Has EPA lost its way? *Regul Toxicol Pharmacol.* 2012, November;64(2):267–268.
159. Calabrese EJ, Cook RR, and Hanekamp JC. Linear no threshold (LNT)—The new homeopathy. *Environ Toxicol Chem.* 2012, December;31(12):2723.
160. Lovell DP. Dose-response and threshold-mediated mechanisms in mutagenesis: Statistical models and study design. *Mutat Res.* 2000, January 1;464(1):87–95.
161. Clewell HJ, Andersen ME, and Barton HA. A consistent approach for the application of pharmacokinetic modeling in cancer and noncancer risk assessment. *Environ Health Perspect.* 2002, January;110(1):85–93.
162. Brown KG and Erdreich LS. Statistical uncertainty in the no-observed-adverse-effect level. *Fundam Appl Toxicol.* 1989, August;13(2):235–244.
163. Sand SJ, von Rosen D, and Filipsson AF. Benchmark calculations in risk assessment using continuous dose-response information: The influence of variance and the determination of a cut-off value. *Risk Anal.* 2003, October;23(5):1059–1068.
164. Sand S, von Rosen D, Victorin K, and Filipsson AF. Identification of a critical dose level for risk assessment: Developments in benchmark dose analysis of continuous endpoints. *Toxicol Sci.* 2006, March;90(1):241–251.
165. Slob W and Pieters MN. A probabilistic approach for deriving acceptable human intake limits and human health risks from toxicological studies: General framework. *Risk Anal.* 1998, December;18(6):787–798.
166. Slob W. Dose-response modeling of continuous endpoints. *Toxicol Sci.* 2002, April;66(2):298–312.
167. Kalliomaa K, Haag-Grönlund M, and Victorin K. A new model function for continuous data sets in health risk assessment of chemicals using the benchmark dose concept. *Regul Toxicol Pharmacol.* 1998, April;27(2):98–107.
168. Rozman KK and Doull J. The role of time as a quantifiable variable of toxicity and the experimental conditions when Haber's c × t product can be observed: Implications for therapeutics. *J Pharmacol Exp Ther.* 2001, March;296(3):663–668.
169. Bogdanffy MS and Jarabek AM. Understanding mechanisms of inhaled toxicants: Implications for replacing default factors with chemical-specific data. *Toxicol Lett.* 1995, December;82–83:919–932.
170. Jarabek AM. The application of dosimetry models to identify key processes and parameters for default dose-response assessment approaches. *Toxicol Lett.* 1995, September;79(1–3):171–184.
171. Jarabek AM, Asgharian B, and Miller FJ. Dosimetric adjustments for interspecies extrapolation of inhaled poorly soluble particles (PSP). *Inhal Toxicol.* 2005;17(7–8):317–334.
172. Hanna LM, Lou SR, Su S, and Jarabek AM. Mass transport analysis: Inhalation rfc methods framework for interspecies dosimetric adjustment. *Inhal Toxicol.* 2001, May;13(5):437–463.

173. Fiserova-Bergerova V, Vlach J, and Cassady JC. Predictable "individual differences" in uptake and excretion of gases and lipid soluble vapours simulation study. *Br J Ind Med.* 1980, February;37(1):42–49.
174. Reitz RH, Quast JF, Schumann AM, Watanabe PG, and Gehring PJ. Non-linear pharmacokinetic parameters need to be considered in high dose/low dose extrapolation. *Arch Toxicol Suppl.* 1980;3:79–94.
175. Reitz RH, Fox TR, and Quast JF. Mechanistic considerations for carcinogenic risk estimation: Chloroform. *Environ Health Perspect.* 1982, December;46:163–168.
176. Reitz RH, Mendrala AL, Park CN, Andersen ME, and Guengerich FP. Incorporation of in vitro enzyme data into the physiologically-based pharmacokinetic (PB-PK) model for methylene chloride: Implications for risk assessment. *Toxicol Lett.* 1988, October;43(1–3):97–116.
177. Fiserova-Bergerova V. Application of toxicokinetic models to establish biological exposure indicators. *Ann Occup Hyg.* 1990, December;34(6):639–651.
178. Ramsey JC and Andersen ME. A physiologically based description of the inhalation pharmacokinetics of styrene in rats and humans. *Toxicol Appl Pharmacol.* 1984, March 3;73(1):159–75.
179. Bus JS and Reitz RH. Dose-dependent metabolism and dose setting in chronic studies. *Toxicol Lett.* 1992, December;64–65:Spec No. 669–76.
180. Marcus AH. The body burden of lead: Comparison of mathematical models for accumulation. *Environ Res.* 1979, June;19(1):79–90.
181. Marcus AH. Multicompartment kinetic models for lead. I. Bone diffusion models for long-term retention. *Environ Res.* 1985, April;36(2):441–458.
182. Marcus AH. Multicompartment kinetic models for lead. II. Linear kinetics and variable absorption in humans without excessive lead exposures. *Environ Res.* 1985, April;36(2):459–472.
183. Marcus AH. Multicompartment kinetic model for lead. III. Lead in blood plasma and erythrocytes. *Environ Res.* 1985, April;36(2):473–489.
184. O'Flaherty EJ. A physiologically based kinetic model for lead in children and adults. *Environ Health Perspect.* 1998, December;106(Suppl 6):1495–1503.
185. O'Flaherty EJ, Inskip MJ, Franklin CA, Durbin PW, Manton WI, and Baccanale CL. Evaluation and modification of a physiologically based model of lead kinetics using data from a sequential isotope study in cynomolgus monkeys. *Toxicol Appl Pharmacol.* 1998, March;149(1):1–16.
186. Pounds JG and Leggett RW. The ICRP age-specific biokinetic model for lead: Validations, empirical comparisons, and explorations. *Environ Health Perspect.* 1998, December;106(Suppl 6):1505–1511.
187. Bert JL, van Dusen LJ, and Grace JR. A generalized model for the prediction of lead body burdens. *Environ Res.* 1989, February;48(1):117–127.
188. Fulton M, Raab G, Thomson G, Laxen D, Hunter R, and Hepburn W. Influence of blood lead on the ability and attainment of children in Edinburgh. *Lancet.* 1987, May 5;1(8544):1221–1226.
189. Thomson GO, Raab GM, Hepburn WS, Hunter R, Fulton M, and Laxen DP. Blood-lead levels and children's behaviour—Results from the Edinburgh Lead Study. *J Child Psychol Psychiatry.* 1989, July;30(4):515–528.
190. Gilbert SG and Weiss B. A rationale for lowering the blood lead action level from 10 to 2 microg/dL. *Neurotoxicology.* 2006, September;27(5):693–701.
191. Surkan PJ, Zhang A, Trachtenberg F, Daniel DB, McKinlay S, and Bellinger DC. Neuropsychological function in children with blood lead levels <10 microg/dL. *Neurotoxicology.* 2007, November;28(6):1170–1177.
192. CDC. *Preventing Lead Poisoning in Young Children.* 2005.

193. Peterson SM, Zhang J, Weber G, and Freeman JL. Global gene expression analysis reveals dynamic and developmental stage-dependent enrichment of lead-induced neurological gene alterations. *Environ Health Perspect.* 2011, May; 119(5):615–621.

194. Gilbert ME, Mack CM, and Lasley SM. Chronic developmental lead exposure and hippocampal long-term potentiation: Biphasic dose-response relationship. *Neurotoxicology.* 1999, February;20(1):71–82.

195. Liu MC, Liu XQ, Wang W, Shen XF, Che HL, Guo YY et al. Involvement of microglia activation in the lead induced long-term potentiation impairment. *PLoS One.* 2012;7(8):e43924.

196. Perkins GA, Scott R, Perez A, Ellisman MH, Johnson JE, and Fox DA. Bcl-xL-mediated remodeling of rod and cone synaptic mitochondria after postnatal lead exposure: Electron microscopy, tomography and oxygen consumption. *Mol Vis.* 2012;18:3029–3048.

197. Galzigna L, Ferraro MV, Manani G, and Viola A. Biochemical basis for the toxic effects of triethyl lead. *Br J Ind Med.* 1973, April;30(2):129–133.

198. United States Environmental Protection Agency (USEPA). *Approaches for the Application of Physiologically Based Pharmacokinetic Models and Supporting Data in Risk Assessment.* EPA/600/R-05/043F. Washington, DC, August, 2006. http://cfpub.epa.gov/ncea/cfm/recordisplay.cfm?deid=157668

199. Ankley GT, Bennett RS, Erickson RJ, Hoff DJ, Hornung MW, Johnson RD et al. Adverse outcome pathways: A conceptual framework to support ecotoxicology research and risk assessment. *Environ Toxicol Chem.* 2010, March;29(3):730–741.

200. Mileson BE, Chambers JE, Chen WL, Dettbarn W, Ehrich M, Eldefrawi AT et al. Common mechanism of toxicity: A case study of organophosphorus pesticides. *Toxicol Sci.* 1998, January;41(1):8–20.

201. Smith JN, Timchalk C, Bartels MJ, and Poet TS. In vitro age-dependent enzymatic metabolism of chlorpyrifos and chlorpyrifos-oxon in human hepatic microsomes and chlorpyrifos-oxon in plasma. *Drug Metab Dispos.* 2011, August;39(8):1353–1362.

202. Timchalk C, Nolan RJ, Mendrala AL, Dittenber DA, Brzak KA, and Mattsson JL. A physiologically based pharmacokinetic and pharmacodynamic (PBPK/PD) model for the organophosphate insecticide chlorpyrifos in rats and humans. *Toxicol Sci.* 2002, March;66(1):34–53.

203. Adkins S, Gan KN, Mody M, and La Du BN. Molecular basis for the polymorphic forms of human serum paraoxonase/arylesterase: Glutamine or arginine at position 191, for the respective A or B allozymes. *Am J Hum Genet.* 1993, March;52(3):598–608.

204. Costa LG, Giordano G, and Furlong CE. Pharmacological and dietary modulators of paraoxonase 1 (PON1) activity and expression: The hunt goes on. *Biochem Pharmacol.* 2011, February 2;81(3):337–344.

205. Richter RJ, Jarvik GP, and Furlong CE. Paraoxonase 1 (PON1) status and substrate hydrolysis. *Toxicol Appl Pharmacol.* 2009, February 2;235(1):1–9.

206. Sierksma A, van der Gaag MS, van Tol A, James RW, and Hendriks HFJ. Kinetics of HDL cholesterol and paraoxonase activity in moderate alcohol consumers. *Alcohol Clin Exp Res.* 2002, September;26(9):1430–1435.

207. Hinderliter PM, Price PS, Bartels MJ, Timchalk C, and Poet TS. Development of a source-to-outcome model for dietary exposures to insecticide residues: An example using chlorpyrifos. *Regul Toxicol Pharmacol.* 2011, October;61(1):82–92.

208. Price PS, Schnelle KD, Cleveland CB, Bartels MJ, Hinderliter PM, Timchalk C, and Poet TS. Application of a source-to-outcome model for the assessment of health impacts from dietary exposures to insecticide residues. *Regul Toxicol Pharmacol.* 2011, October;61(1):23–31.

209. Krasovskii GN. Extrapolation of experimental data from animals to man. *Environ Health Perspect.* 1976, February;13:51–58.
210. Travis CC and White RK. Interspecific scaling of toxicity data. *Risk Anal.* 1988, March;8(1):119–125.
211. Schneider K, Oltmanns J, and Hassauer M. Allometric principles for interspecies extrapolation in toxicological risk assessment—Empirical investigations. *Regul Toxicol Pharmacol.* 2004, June;39(3):334–347.
212. Travis CC and Morris JM. On the use of 0.75 as an interspecies scaling factor. *Risk Anal.* 1992, June;12(2):311–313.
213. United States Environmental Protection Agency (USEPA). Risk Assessment Forum. *Recommended Use of Body Weight*3/4 as the Default Method in Derivation of the Oral Reference Dose. Final.* EPA/100/R11/0001. Washington, DC, February, 2011. http://www.epa.gov/raf/publications/interspecies-extrapolation.htm
214. Organisation for Economic Co-operation and Development (OECD). *Assessment of Chemicals—Validation of (Q)SAR Models.* n.d. http://www.oecd.org/env/ehs/risk-assessment/validationofqsarmodels.htm (accessed January 22, 2013).
215. Doull J, Borzelleca JF, Becker R, Daston G, DeSesso J, Fan A et al. Framework for use of toxicity screening tools in context-based decision-making. *Food Chem Toxicol.* 2007, May;45(5):759–796.
216. Sedykh A, Zhu H, Tang H, Zhang L, Richard A, Rusyn I, and Tropsha A. Use of in vitro HTS-derived concentration-response data as biological descriptors improves the accuracy of QSAR models of in vivo toxicity. *Environ Health Perspect.* 2011, March;119(3):364–370.
217. Eriksson L, Jaworska J, Worth AP, Cronin MTD, McDowell RM, and Gramatica P. Methods for reliability and uncertainty assessment and for applicability evaluations of classification- and regression-based QSARs. *Environ Health Perspect.* 2003, August;111(10):1361–1375.
218. Vink SR, Mikkers J, Bouwman T, Marquart H, and Kroese ED. Use of read-across and tiered exposure assessment in risk assessment under REACH—A case study on a phase-in substance. *Regul Toxicol Pharmacol.* 2010, October;58(1):64–71.
219. Patlewicz G, Roberts DW, Aptula A, Blackburn K, and Hubesch B. Workshop: Use of "read-across" for chemical safety assessment under REACH. *Regul Toxicol Pharmacol.* 2012, December 12;65(2):226–228.
220. Patlewicz G, Chen MW, and Bellin CA. Non-testing approaches under REACH—Help or hindrance? Perspectives from a practitioner within industry. *SAR QSAR Environ Res.* 2011, March;22(1–2):67–88.
221. Gallegos-Saliner A, Poater A, Jeliazkova N, Patlewicz G, and Worth AP. Toxmatch—A chemical classification and activity prediction tool based on similarity measures. *Regul Toxicol Pharmacol.* 2008, November;52(2):77–84.
222. McInnes GT and Brodie MJ. Drug interactions that matter. A critical reappraisal. *Drugs.* 1988, July;36(1):83–110.
223. Van den Berg M, Birnbaum LS, Denison M, De Vito M, Farland W, Feeley M et al. The 2005 World Health Organization reevaluation of human and Mammalian toxic equivalency factors for dioxins and dioxin-like compounds. *Toxicol Sci.* 2006, October;93(2):223–241.
224. Borgert CJ, Borgert SA, and Findley KC. Synergism, antagonism, or additivity of dietary supplements: Application of theory to case studies. *Thromb Res.* 2005;117(1–2):123–132; discussion 145–151.
225. National Research Council (NRC). *Review of the Environmental Protection Agency's Draft IRIS Assessment of Formaldehyde.* Washington, DC: The National Academies Press, 2011. http://www.nap.edu/catalog.php?record_id=13142

226. Wullenweber A, Kroner O, Kohrman M, Maier A, Dourson M, Rak A et al. Resources for global risk assessment: The International Toxicity Estimates for Risk (ITER) and Risk Information Exchange (RiskIE) databases. *Toxicol Appl Pharmacol.* 2008, November 11;233(1):45–53.

227. Price OR, Jones KC, Li H, Liu Z, Lu Y, Wang H et al. Chemicals management and environmental assessment of chemicals in China. *Environ Pollut.* 2012, June;165,169.

228. Wang H, Yan Z-G, Li H, Yang N-Y, Leung KMY, Wang Y-Z et al. Progress of environmental management and risk assessment of industrial chemicals in China. *Environ Pollut.* 2012, June;165:174–181.

229. Meng X, Zhang Y, Zhao Y, Lou IC, and Gao J. Review of Chinese environmental risk assessment regulations and case studies. *Dose Response.* 2012;10(2):274–296.

230. Patlewicz G, Simon T, Goyak K, Phillips RD, Craig Rowlands J, Seidel S, and Becker RA. Use and validation of HT/HC assays to support 21(st) century toxicity evaluations. *Regul Toxicol Pharmacol.* 2013, March;65(2):259–268.

231. Hartung T and McBride M. Food for Thought… on mapping the human toxome. *ALTEX.* 2011;28(2):83–93.

232. Mayr E. *The Growth of Biological Thought: Diversity, Evolution and Inheritance.* Cambridge, MA: Harvard University Press, 1982.

233. Aylward LL and Hays SM. Consideration of dosimetry in evaluation of ToxCast™ data. *J Appl Toxicol.* 2011, November;31(8):741–751.

234. Aylward LL, Becker RA, Kirman CR, and Hays SM. Assessment of margin of exposure based on biomarkers in blood: An exploratory analysis. *Regul Toxicol Pharmacol.* 2011, October;61(1):44–52.

235. Thomas RS, Black M, Li L, Healy E, Chu T-M, Bao W et al. A comprehensive statistical analysis of predicting in vivo hazard using high-throughput in vitro screening. *Toxicol Sci.* 2012, August;128(2):398–417.

236. Bigby M. Evidence-based medicine in a nutshell. A guide to finding and using the best evidence in caring for patients. *Arch Dermatol.* 1998, December;134(12):1609–1618.

237. Shah HM and Chung KC. Archie Cochrane and his vision for evidence-based medicine. *Plast Reconstr Surg.* 2009, September;124(3):982–988.

238. Guzelian PS, Victoroff MS, Halmes NC, James RC, and Guzelian CP. Evidence-based toxicology: A comprehensive framework for causation. *Hum Exp Toxicol.* 2005, April;24(4):161–201.

239. Stephens ML, Andersen M, Becker RA, Betts K, Boekelheide K, Carney E et al. Evidence-based toxicology for the 21st century: Opportunities and challenges. *ALTEX.* 2013;30(1/13):74–103.

240. Budinsky RA, Schrenk D, Simon T, Van den Berg M, Reichard JF, Silkworth JB, Aylward L, Brix A, Gasiewicz T, Kaminski N, Perdew G, Starr TB, Walker NJ, and Rowlands JC. Mode of action and dose–response framework analysis for receptor-mediated toxicity: The aryl hydrocarbon receptor as a case study. *Crit. Rev. Toxicol.* (in press) DOI: 10.3109/10408444.2013.835787.

5 Risk Characterization

Basil Exposition: What's the other thing that scares you?

Austin Powers: Carnies.

Basil Exposition: What?

Austin Powers: Carnies. Circus folk. Nomads, you know. Smell like cabbage.

Michael York and Mike Myers
Austin Powers: International Man of Mystery, directed by Jay Roach,
(1997, Burbank, CA; New Line Home Entertainment)

The 1983 Red Book specifies risk characterization as the final component in a risk assessment and indicates that this activity is "the description of the nature and often the magnitude of human risk, including attendant uncertainty." However, the document does almost nothing to define what a risk characterization should look like[1], except in the vaguest of terms, much like the film quotation above. However, the 1994 Blue Book provides much helpful guidance on risk characterization.[2] Four elements make up a risk characterization:

- Quantitative estimates of risk
- Qualitative and, if available, quantitative descriptions of uncertainty
- Presentation of the risk estimates in their appropriate context
- Communication of the results of the risk analysis

These four elements will be discussed in detail in the succeeding text.

QUANTITATIVE ESTIMATES OF RISK

Two different methods of developing quantitative risk estimates exist—one for chemicals producing adverse effects considered to have a threshold and the other for chemicals for which no risk-free level of exposure is believed to exist based on the LNT hypothesis. Generally, EPA uses the threshold method and RfDs for noncarcinogens and slope factors or unit risk levels for carcinogens.

The general equation for estimating risk for noncarcinogens is

$$HQ = \left(\frac{1}{RfD} \right) \times \frac{(C \times CR \times EF \times ED)}{(BW \times ED \times 365)} \tag{5.1}$$

where
    HQ is the hazard quotient (unitless)
    RfD is the reference dose (toxicity criterion) (mg/kg-day)
    C is the concentration (mg/kg or mg/L)
    CR is the contact rate or ingestion rate (amount per day)
    EF is the exposure frequency (days/year)
    ED is the exposure duration (years)
    BW is the body weight (kg)

The general equation for estimating risk for carcinogens is

$$\text{Risk} = \text{CSF} \times \frac{(C \times CR \times EF \times ED)}{(BW \times AT)} \tag{5.2}$$

where
    CSF is the cancer slope factor $(\text{mg/kg-day})^{-1}$
    AT is the averaging time, usually 25,550 days or 70 years

Equation 5.1 yields a unitless value for the HQ after dividing the ADD in units of mg/kg-day by the RfD, also in units of mg/kg-day. Values greater than unity indicate the potential for systemic toxicity leading to adverse effects. Equation 5.2 yields a unitless value for the probability of cancer, that is, risk, after multiplying the lifetime ADD in units of mg/kg-day by the CSF in units of $(\text{mg/kg-day})^{-1}$. You may wish to conduct unit analysis on these equations to satisfy yourself that all units cancel to yield a unitless value for hazard or risk.

ESTIMATING RISK FOR NONCARCINOGENS

For each chemical producing systemic toxicity, the dose estimate is divided by the RfD to obtain an HQ (Equation 5.1). If the HI is less than 1, the risk for that chemical is considered unlikely to lead to adverse health effects. If the HI is greater than 1, adverse health effects are more likely and suggest that risk management should be considered. For multiple chemicals, HQs can be summed to estimate an overall HI. For exposure by the dermal route, the RfD is adjusted to reflect an absorbed dose as detailed in Appendix A of EPA's *RAGS, Volume I, Part A*.[3,4] For exposure by the inhalation route, the RfC is used in lieu of the RfD.[5]

    EPA's *RAGS, Volume I, Part A*,[3] indicates that HQs should be summed by either target organ or mechanism of action. Further, this document points out that the HI is thus not an actual measure of risk and that summing the HQ values over chemicals that act by different mechanisms would likely overestimate the potential for adverse effects.

    The real difficulty with the RfD concept is that while HI values appear to be quantitative measures of risk, they are more accurately regulatory "bright lines" appropriate for the determination of highly protective cleanup values but inappropriate as accurate or "best" estimates of human health risk. The use of uncertainty or safety factors in RfD derivation results in toxicity values that are protective estimates of a human threshold with an unknown degree of protection. The ATSDR of the CDC develops minimum risk levels (MRLs) by a process almost identical to that used to develop RfDs. MRLs are used in public health assessments to determine if people are likely to experience adverse effects. Toxicity criteria based on safety factors are inappropriate for such a public health assessment—indeed, the use of safety factors yields toxicity criteria that will result in protective cleanup levels, appropriate for engineers and environmental scientists engaged in remediation or standard setting but inappropriate as a tool for predictive toxicology or public health.

ESTIMATING RISK FOR CARCINOGENIC CHEMICALS

For each carcinogenic chemical, the intake estimated as a dose, usually in units of mg/kg body weight/day is multiplied by the slope factor also in units of (mg/kg/day)$^{-1}$ (Equation 5.2). The resulting value will be a unitless probability value of the incremental risk of an individual developing cancer over a lifetime of exposure.

When exposure occurs by inhalation, unit risk values are used instead of slope factors. Usually, the most appropriate exposure value is a lifetime weighted average of air concentration, usually in units of μg/m^3. Multiplying by a unit risk value in units of (μg/m^3)$^{-1}$ yields a unitless probability value of the incremental lifetime risk.

Because cancer risks are expressed as unitless probabilities, their summation across multiple chemicals and exposure routes is mathematically appropriate. Nonetheless, EPA cautions that summing risks across multiple exposure pathways should be carefully considered, and the authors of *RAGS, Part A*, were clearly aware of the problem of compounding conservatism discussed in Chapter 1.[3]

UNCERTAINTY IN RISK ESTIMATES

Knowledge will continue to be imperfect. Risk assessment is predictive—it tries to make statements about the future, but knowledge is based on the past and the interpretation of past events and uncertainty is inescapable.

Risk estimates calculated using the Red Book paradigm of combining exposure and toxicity are conditional estimates based on many assumptions about exposure and toxicity. Hence, characterization of uncertainty is also a necessary part of the overall risk characterization. Transparency in communicating the uncertainties and assumptions provides appropriate perspective and may also identify data gaps for which additional research or data collection might be advantageous.

In some cases, quantitative statistical uncertainty analysis can provide some insights. A full treatment of quantitative uncertainty analysis is beyond the scope of this book. For those wishing to know more, Appendix A on MC simulation is a place to begin. In that appendix, methods for quantitative uncertainty analysis are presented.

However, in most cases, the results of such an analysis may be challenging to communicate to a general audience. In many cases, quantitative uncertainty analysis may not add much insight to the risk characterization. Value of information (VOI) is a type of uncertainty analysis that attempts to assess "bang for the buck" regarding the question of whether to conduct additional data collection or analysis.

Nature and Classification of Uncertainty

There are a number of ways to classify the types of uncertainties in a risk assessment. A number of authors have provided typologies of uncertainty over the past three decades and no one scheme has come into general usage.[6–8]

Aleatory uncertainty or variability refers to the variation inherent in a population and just how well or how poorly the risk assessment represents this target population. For example, how well does the value of 70 kg for the body weight of an adult represent a target population in Memphis, Tennessee, reported to have the highest obesity rate of US cities versus Portland, Oregon, reported to be the most fit city in the United States. Aleatory uncertainty can be quantified but not reduced with additional data collection.

Epistemic uncertainty or incertitude refers to lack of knowledge or ignorance and is most often called simply, uncertainty.[9] How sure can one be that the parameter values or model structure used to evaluate risks are correct? This type of uncertainty can be reduced by additional data collection. For example, one might wish to know just how much time children spend outside in their yards (front or back). The default generally used by EPA in Superfund-type risk assessments is 350 days/year. How accurate is this value today when children have many indoor activities, such as video games and online activities, on which to spend time? Recent data obtained from children wearing GPS transmitters and video recording or sampled by dedicated cell phones with a method called ecological momentary assessment suggest that the true value of time spent outside by children in the twenty-first century is much less.[10–15] The epidemic of childhood obesity in the developed world is ample testimony to this fact.

Another classification of uncertainty exists as well. Parameter uncertainty arises from measurement error or whether the parameter values used in a risk assessment represent the target population accurately. Model uncertainty arises from lack of an adequate scientific basis for the theory underlying some aspect of the risk assessment. For example, page 165 of the Blue Book highlights the validity or lack thereof of the LNT as an example of model uncertainty.[2]

Some uncertainty can be addressed on a purely statistical basis—conducting an Monte Carlo (MC) assessment or PRA is one way to address the issue of aleatory uncertainty. Some data are inherently uncertain and have been developed at great cost and effort. For example, the data on children's soil and dust ingestion discussed in Chapter 3 are uncertain. Additional data collection may support these data but this quantity and a number of other issues in risk assessment represent "deep" uncertainties, as recently characterized by the National Academy's Institute of Medicine (IOM).[16]

Deep uncertainties are those that are unlikely to be resolved in the time frame in which a risk management decision is needed. This type of uncertainty occurs when disagreements exist among scientists about either the fundamental nature of biological or environmental processes or the methods to characterize these processes. Although expert elicitation (EE) may be considered as a means of addressing these deep uncertainties, that process (discussed in the succeeding text) does not always work.

When the stakeholders in a decision process cannot reach agreement, a situation arises in which the decision may be subject to undue political influence. In such cases, risk assessors may be pressured to alter or revise the results of their analyses to support a desired outcome.

Hence, a caution is provided for the readers of this textbook—if you find yourself in such a situation and have provided an honest, transparent, and good faith risk characterization, resist any political pressure to change your results. Stick to your analysis. If you change your mind without a sufficient science-based reason, you will be perceived as waffling or indecisive—or worse, dishonest. New or additional information may alter your conclusions; in such a case, you must be able to present both the reasoning behind your original conclusion and your reason for the change of mind in a forthright, easy-to-understand, and transparent fashion.

IDENTIFICATION AND QUANTIFICATION OF UNCERTAINTY

In all environmental risk assessments, considerable uncertainty exists regarding the numerical values of inputs and the quantitative estimates of risk—a range of an order of magnitude or more. Generally, identification of the key factors and assumptions that contribute most to the overall uncertainty will provide as much information as any attempt to quantify the overall uncertainty.

Uncertainty exists in both the exposure assessment and the toxicity assessment. Uncertainty regarding exposure exists in chemical monitoring data, in the understanding of the environmental fate of chemicals, and in the nature and extent of human contact with chemicals in various environmental media. Uncertainty regarding toxicity exists, largely because of the lack of information on the MOA for most chemicals. Future changes in the science base of risk assessment also introduce uncertainty—no one can predict the future, and to a large degree, predicting the future is the primary task of risk assessment.

PRESENTATION OF RISK ESTIMATES WITH INCLUSION OF UNCERTAINTY

The results of risk assessments are often boiled down to single numbers. The Red Book, the Blue Book, and the 1992 risk characterization memo from F. Henry Habicht of EPA all opined on this inappropriate overcondensation of information.[1,2,17] William Ruckelshaus, EPA administrator during the Reagan administration, also lamented the use of what he derogatorily called "magic numbers." Ruckelshaus indicated that risk managers must "insist on risk calculations being expressed as distributions" and ranges of probabilities. By "magic numbers," he likely meant easily manipulated and poorly documented risk estimates for which the scientific basis remained unclear. Uncertainty should also be included and he also called for "new tools for quantifying and ordering sources of uncertainty and for putting them into perspective."[2]

Chapter 6 of RAGS Volume 3, the *Guidance for Probabilistic Risk Assessment* provides a discussion of how to present quantitative estimates of risk and uncertainty to a variety of audiences. This source provides a good starting point.

However, there is an overarching difficulty—the general innumeracy of many audiences; these audiences would include the general public, upper-level managers, many attorneys, legislators, and others. According to the National Adult Literacy Survey, almost half of the general population are challenged by relatively simple numeric tasks.[18] Numeracy is an essential skill for understanding risk–benefit information and making appropriate judgments, and unfortunately, communicating quantitative risk information, especially information about uncertainty, to most audiences will require considerable "dumbing down."[19] The field of risk communication evolved to develop ways to communicate complex information in ways that a variety of audiences can understand. In general, successful risk communication requires empathy, compassion, humility, and insight.

QUANTITATIVE ASSESSMENT OF VARIABILITY AND UNCERTAINTY

Later in this chapter, examples are provided of full-risk assessments. Appendix A also provides details on the methodology for MC analysis including quantitative assessment of both uncertainty and variability. In the past, a piecemeal approach to uncertainty analysis has been adopted. In RAGS Volume 3, EPA indicated that application of the probabilistic methods to the toxicity assessment was not justified.[20] Box 1.3 provides a relatively simple way to address quantitative uncertainty in both exposure and toxicity.

RISK ASSESSMENT AS A DESIGN PROBLEM

Uncertainties in the risk assessment need to be described and made fully transparent to all stakeholders. The communication of these uncertainties to the risk manager is most often a difficult task. Risk assessors should be aware that a risk manager wants to arrive at the end of the process with a clear path forward— either a rationale for no action, a strategy for cleanup, or an explicit plan for additional data collection to address one or more of the areas of uncertainty. VOI may also be challenging to communicate, but VOI may be the best means of deciding whether to address uncertainties by additional data gathering.

VALUE OF INFORMATION ANALYSIS

A practical and robust approach to environmental decision making requires that the risk manager understand in which areas uncertainty is irreducible and in which uncertainty can be lessened by more data. The value of this new information is required by the manager—only with a VOI analysis can the manager choose whether to delay a decision in order to collect more data or whether to proceed in order to obtain immediate but uncertain benefits in terms of public health protection.[21]

This conflict between delaying a decision to wait for more data or to continue with a plan to address an environmental hazard with the current state of knowledge is inherent in any science-based decision process. There is no end to the scientific process and uncertainty, as noted in an earlier chapter, is its ever-constant handmaiden.

In a popular science book titled *Doubt Is their Product: How Industry's Assault on Science Threatens your Health*, David Michaels characterizes any science produced by industry as flawed.[22] In 2012 and 2013, scientists at the Environmental Working Group (EWG) and the National Resources Defense Council (NRDC) suggested to the NAS in a public forum that any research program funded by the chemical industry should be considered fatally flawed and should not be used in any IRIS assessments conducted by EPA. These activists also suggested that EPA should move ahead as quickly as possible with IRIS assessments with whatever information was available. In general, the chemical industry is concerned that their products are safe. The industry is in a unique position—in contrast to regulatory agency or nongovernmental organizations—they are able to afford the cost of research. This conflict puts regulatory agencies such as EPA in the role of arbiter of this most basic of quandaries when applying scientific information for decision making—how much is enough?

There are no established "stopping criteria" in pure science. The difficulty in knowing how much is enough adds to this inherent societal conflict in which those who would attempt to limit the extent of scientific inquiry and debate to meet a regulatory deadline and those who believe that there may be significant value in the information expected from ongoing and incomplete research.[21] The Scylla and Charybdis of this dilemma represent the choice between blithely forging ahead with a decision that has unknown consequences versus "paralysis by analysis."

The decision-theoretic process begins with analyzing what can be done given the current state of knowledge and what potential improvements in the decision can result from additional knowledge. In 2003, a risk assessor working at the Region 4 offices of EPA advised the project manager (PM) for the Barber's Orchard Superfund site in Waynesville, NC, to obtain site-specific bioavailability measurements for arsenic. The site was an old apple orchard at which arsenical pesticides were used. The land was subsequently sold to a developer who built luxury homes worth over $200K. A decision document for the site was written in 2004 with a cleanup cost over $30M. The decision was not approved by the Superfund Remedy Review Board.[23] The PM did eventually obtain site-specific information about arsenic bioavailability from an in vivo bioavailability study using monkeys conducted at the University of Florida.[24] The use of these bioavailability data reduced the cleanup costs from $32M to about $15M.

EXPERT ELICITATION

Formal EE is one of the means toward a structured and transparent way to address such uncertainties. When insufficient knowledge for decision making is available, this process provides a structured approach to seek the published and unpublished knowledge of experts for the purpose of developing quantitative estimates for use in risk assessment. Hence, EE is a process to synthesize the limited available information before conclusive scientific evidence is developed. A formal systematic method for elicitation improves the transparency and reproducibility of the information.

In the 1980s, EPA's Office of Air Quality Planning used EE to assess exposure response relationships for lead and ozone.[25] The advantage of EE is that the process combines knowledge from different disciplines that likely have differing views on the issue being considered. When data limitations or incomplete understanding of the problem at hand prevent conventional approaches to uncertainty analysis, EE may be helpful as a formal process to quantify expert judgments in terms of probability. Both the Red Book and Blue Book provide support for EE.[1,2] Circular A-4 from the Office of Management and Budget (OMB) concerning quantitative uncertainty analysis for regulatory decisions over $1B also supports EE.[26]

Methods for Expert Elicitation

One of the earliest structured methods for EE was the Delphi method developed in the 1950s at the RAND corporation and used by the US Air Force to address cold war issues. The Delphi technique seeks to obtain the most reliable consensus of opinion from a group of experts using both questionnaires and feedback.[27]

In many EEs, what is done is to suggest to the experts that they describe their estimates using gambling analogies as a means of encoding probabilities. Two kinds of experts are selected for participation in formal EE: substantive experts who possess knowledge of the subject matter being considered and normative experts with expertise in decision analysis, psychology, and group facilitation.

Using a consensus technique to develop information usually can be viewed as circumventing the scientific method; nonetheless, EE is generally accepted as a reasonable tool for developing increased certainty about the information needed for risk assessment. EE is not often used—it is usually not practical or necessary for most decisions based on environmental risk assessment because of resource limitations and concerns about public and stakeholder acceptance of the results.

COMPARISON BETWEEN SITE AND BACKGROUND

Before 2002, EPA used comparison to background as part of COPC selection. In fact, this was performed in the first example in the following text. Selection of the background dataset is critical to this. However, a guidance document on background comparison released in 2002 indicated in an appendix that background comparison should occur after risk characterization and that site-related risks should be compared to background risks.[28] Many risk assessors and PMs at EPA believed this policy would result in a confusing message—there would be at least two risk estimates and risk comparisons have been shown to be difficult for the lay public to understand fully.[19,29]

In a number of instances, many EPA PMs and risk assessors disagreed with the advice in the guidance and continued to use comparison to background as part of COPC selection.

EXAMPLES OF A FULLRISK CHARACTERIZATION

The remainder of this chapter will present two risk assessments including the risk characterizations. These examples were developed from real-world situations. The presentation has been tailored to illustrate both qualitative and quantitative uncertainty, and other aspects of the risk characterization.

In both examples, some narrative that might occur in a typical risk assessment is provided. Conclusions about the estimated level of risk are presented in a way to highlight the uncertainties. In addition, interactions between the risk assessor and the decision maker are presented to provide examples of how the risk assessment can aid decision making without compromising its scientific basis.

EXAMPLE #1: FORMER GOLD MINE SITE BEING CONSIDERED AS A HISTORIC PARK

Gold mining and milling at the site began after an 1899 discovery of gold and continued until approximately 1941, with the heaviest use in the 1930s.

The initial discovery of gold at the site in 1899 spurred a series of claims that were allowed to lapse after somewhat small yields of gold. In 1925, discovery of a rich vein of ore prompted the formation of a private mining corporation in 1929. Mining activities continued through the 1930s and the scale of mining operations was expanded to extract gold from 54 tons of ore a day. The minerals in the ore consisted mainly of pyrite, magnetite, and pyrrhotite, minerals containing arsenic and other heavy metals.

Mining was conducted by open underhand chiseling methods employed to break ore out of the ore shoots. The mined rock was carried by an aerial tram system to a nearby mill. Gold was extracted by mercury amalgamation and cyanide leaching. After gold and other valuable metals were extracted at the mill, mill tailings, which are the waste left over after extraction, were deposited over an area approximately 9 acres near the north fork of Trout Creek. The total depth of tailings generally varies from 0 to 2 ft.

The tailings were deposited in their current locations due to the failure of a tailings dam. The iron-rich surface of the tailings has oxidized and appears as a bright orange and red packed crust, with some localized gray areas of unoxidized pyrite in areas of greater disturbance. A layer of hardpan is present on the surface of the tailings and extends in depth to between 1 and 6 in. below the surface of the tailings. The hardpan is not easily mobilized due to its physical structure; the tailings do not support vegetation—except where they have been covered by soil or other organic matter. The tailings contain high levels of metals and are acidic in water. Sporadic mining and extraction attempts occurred throughout the 1980s; several partially empty, rusted drums of tailings are located on the site as a result of an attempt to pack and transport the tailings off-site to a separate mill.

Environmental sampling data for metals for both the site and background locations are provided in Tables 5.1 through 5.8.

The site is being developed as a historic park, and therefore, the following receptors are considered in this risk assessment:

- Adult park visitor
- Child park visitor
- Adult park worker
- Construction/excavation worker

TABLE 5.1

Surface Soil Sampling Results for Metals at the Former Gold Mine Site

Al	Sb	As	Ba	Be	Cd	Cr	Co	Cu	Fe	Pb	Mn	Hg	Ni	Se	Ag	Tl	Tin	Va	Zn
		389			1.6	14	16.8	601	84,500	2960	220	3.3	12.2	3.2	26.6			34	292
		61.5			0.72	10.9	6.7	385	63,500	627	171	0.39	7.7	1.1	7			23.6	81.6
		972			7.6	8.1	34.6	872	200,000	717	1,560	0.54	6.6	2.1	16.1			14.3	765
		3460			32.2	6	149	436	267,000	443	661	1.2	17.4	8.4	13.3			11.8	2450
		224			1.8	3.1	8.6	173	37,600	335	254	0.16	4.8	1.2	6.5			22.3	189
		6.1			0.46	7.7	185	55.1	22,300	22	16,200	0.079	42.7	5.3	0.66			35.8	44.6
		504			0.16	4.7	3.3	48.4	72,800	300	112	0.98	3.8	3.2	6.2			35.3	118
		20.6			0.16	4.7	5.3	40.4	12,900	31.5	256	0.018	6.7	5	0.5			32.8	40.1
1.9		510	300	<1	9	14	110	340	25	140	7300		12		2		<5	57	1000
3.7		430	200	<1	<2	50	8	540	9.3	1900	380		12		44		<5	70	140
8	NA	57	420	3	<2	33	16	83	5.5	720	920	0.026	20	0.45	0.4		<5	92	160
1.5	NA	2100	230	<1	<2	18	13	53	25	75	3,400	0.06	8	0.4	0.1		7	40	340
2,960	<4.0	49	37.6	0.07	0.71	1.6	6.41	12.9	18,500	6.81	1,350	<0.02	4.2	<0.2	0.38	<0.25		17.7	72.3
1	3	760	27	<1	5	6	44	2,500	14	820	2,000	0.11	14	3	23		<5	110	1100
1.1	3	670	23	<1	8	5	39	3,200	16	770	2,100	0.12	13	3	24		<5	120	1300
6.8	<1	71	500	<1	<2	22	7	72	5.3	22	690	<0.005	18	0.8	<2		24	140	65

0.91	3	510	12	<1	<2	5	48	1,400	13	700	3,000	0.25	14	3.2	15		<5	90	890
1460	<40	1780	6.82	<0.50	11.3	<5.0	19	267	157,000	1040	270	0.74	<10	7.49	24.8	0.94		16.5	1600
2,710	<40	1190	2.5	<0.50	8.2	<5.0	70.4	146	125,000	163	265	0.806	11	5.77	7.4	0.57		18.4	942
1610	<20	948	8.14	<0.25	7.22	<2.5	9.4	111	67,300	939	187	2.47	<5.0	9.34	19.8	2.83		17.3	661
	200	500	50	2	50	20	200	7,000	20	500	100		30		1,000		20	20	500
	200	1500	50	2	50	20	200	5,000	50	700	500		30		100		20	30	500
	200	500	70	2	50	30	20	15,000	7	100	500		15		700		20	100	500
	200	700	100	2	50	100	70	2,000	10	100	200		20		10		20	50	500
0.46	<15	960	13	<2	6	8	77	1,700	21	1300	400	0.7	10	3.5	45		<10	9	390
0.84	<15	1100	39	<2	11	16	33	1,700	13	910	1,100	0.27	7	1.3	26		<10	20	810
1.7	<15	960	110	<2	21	21	36	1,300	14	5000	240	1.5	6	3.4	110		<10	28	300
0.35	<15	435	18	<2	<4	5	12	795	9.8	6100	150	0.505	<4	1.95	48.5		<10	9	150
1	<15	870	15	<2	13	13	41	3,400	15	890	1,400	0.15	8	0.9	27		<10	18	720
1.4	17	1500	23	<2	9	9	20	300	13	880	400	1.1	<4	3.3	14		<10	26	980

Notes: A number preceded by "<" indicates the chemical was not detected and the number is the analytical reporting limit from the laboratory. Units are mg/kg soil. Chemical symbols were used to save space and represent the various metals as follows: Al = aluminum; Sb = antimony; As = arsenic; Ba = barium; Be = beryllium; Cd = cadmium; Cr = chromium; Co = cobalt; Cu = copper; Fe = iron; Pb = lead; Mn = manganese; Hg = mercury; Ni = nickel; Se = selenium; Ag = silver; Tl = thallium; Sn = tin; Va = vanadium; Zn = zinc.

TABLE 5.2

Background Concentrations of Surface Soil at the Former Gold Mine Site

Al	Sb	As	Ba	Be	Cd	Cr	Co	Cu	Fe	Pb	Mn	Hg	Ni	Se	Ag	Tl	Tin	Va	Zn
11,000	<4.0	9.32	190	0.32	0.21	23.5	11	37.8	23,500	7.38	443	0.0692	26.3	0.45	0.34		<0.25	54.1	57.2
0.09	<15	3,100	7	<2	4	9	20	560	45	12	230	<0.02	<4		<4	<10		21	25
0.03	<15	31	<2	<2	<4	<2	130	4,400	47	77	140	<0.02	<4	0.6	11	<10		<4	140
0.97	<15	500	9	<2	<4	87	<2	300	14	1,600	110	0.03	6	0.7	25	<10		72	130
4.5	<15	<20	97	<2	6	90	59	12,000	26	<8	6100	<0.02	12		7	37		95	900
1.6	<15	1,200	6	<2	95	12	33	4,100	29	2,900	2900	0.12	12	1.4	16	<10		52	7300
1,860	<40	21.1	24.5	<0.50	8.6	6.9	19	3,430	211,000	44	1070	0.14	<10	28.7	109	<10	<0.75	23.5	2200
		4			0.098	5.1	5.1	22	11,800	2.6	213	0.015	7.5	<5	1			37.7	28.4
		58.2			0.21	110	7.7	141	27,000	5	1190	0.046	50.1	<5.3	0.81			14.9	36.8
		24.7			1.7	67.5	11.5	122	27,300	2.3	1280	0.034	36	<5.3	0.32			28	363
		55.5			1.9	26.5	5.6	533	19,400	83.4	525	0.064	15.8	5.1	1.7			14.5	140
		1.3			<1	12.9	3.9	17.8	9,270	1.1	150	0.017	10	<5.1	<1			31.8	23.1
		130			1.3	33.8	10.4	527	30,400	68.2	762	0.085	19.3	<5.2	2.1			17.3	132

Notes: A number preceded by "<" indicates the chemical was not detected and the number is the analytical reporting limit from the laboratory. Units are mg/kg soil. Continuation of sampling results for the former gold mine site. Chemical symbols were used to save space and represent the various metals as follows: Al = aluminum; Sb = antimony; As = arsenic; Ba = barium; Be = beryllium; Cd = cadmium; Cr = chromium; Co = cobalt; Cu = copper; Fe = iron; Pb = lead; Mn = manganese; Hg = mercury; Ni = nickel; Se = selenium; Ag = silver; Tl = thallium; Sn = tin; Va = vanadium; Zn = zinc.

TABLE 5.3
Mixed Surface and Subsurface Soil Sampling Results for Metals at the Former Gold Mine Site, Part 1

Al	Sb	As	Ba	Be	Cd	Cr	Co	Cu	Fe	Pb	Mn	Hg	Ni	Se	Ag	Tl	Sn	Va	Zn
		389			1.6	14	16.8	601	84,500	2,960	220	3.3	12.2	3.2	26.6			34	292
		61.5			0.72	10.9	6.7	385	63,500	627	171	0.39	7.7	1.1	7			23.6	81.6
		972			7.6	8.1	34.6	872	200,000	717	1,560	0.54	6.6	2.1	16.1			14.3	765
		3460			32.2	6	149	436	267,000	443	661	1.2	17.4	8.4	13.3			11.8	2,450
		224			1.8	3.1	8.6	173	37,600	335	254	0.16	4.8	1.2	6.5			22.3	189
		6.1			0.46	7.7	185	55.1	22,300	22	16,200	0.079	42.7	5.3	0.66			35.8	44.6
		504			0.16	4.7	3.3	48.4	72,800	300	112	0.98	3.8	3.2	6.2			35.3	118
		20.6			0.16	4.7	5.3	40.4	12,900	31.5	256	0.018	6.7	5	0.5			32.8	40.1
1.9		510	300	<1	9	14	110	340	25	140	7,300		12		2		<5	57	1,000
3.7		430	200	<1	<2	50	8	540	9.3	1,900	380		12		44		<5	70	140
8		57	420	3	<2	33	16	83	5.5	720	920	0.026	20	0.45	0.4		<5	92	160
1.5		2100	230	<1	<2	18	13	53	25	75	3,400	0.06	8	0.4	0.1		7	40	340
2960	<4.0	49	37.6	0.068	0.71	1.6	6.41	12.9	18,500	6.81	1,350	<0.02	4.2	<0.2	0.38	<0.25		17.7	72.3
1	3	760	27	<1	5	6	44	2,500	14	820	2,000	0.11	14	3	23		<5	110	1,100
1.1	3	670	23	<1	8	5	39	3,200	16	770	2,100	0.12	13	3	24		<5	120	1,300
6.8	<1	71	500	<1	<2	22	7	72	5.3	22	690	<0.005	18	0.8	<2		24	140	65
0.91	3	510	12	<1	<2	5	48	1,400	13	700	3,000	0.25	14	3.2	15		<5	90	890

(continued)

TABLE 5.3 (continued)
Mixed Surface and Subsurface Soil Sampling Results for Metals at the Former Gold Mine Site, Part 1

Al	Sb	As	Ba	Be	Cd	Cr	Co	Cu	Fe	Pb	Mn	Hg	Ni	Se	Ag	Tl	Sn	Va	Zn
1460	<40	1780	6.82	<0.50	11.3	<5.0	19	267	157,000	1,040	270	0.74	<10	7.49	24.8	0.94		16.5	1,600
2710	<40	1190	2.5	<0.50	8.2	<5.0	70.4	146	125,000	163	265	0.806	11	5.77	7.4	0.57		18.4	942
1610	<20	948	8.14	<0.25	7.22	<2.5	9.4	111	67,300	939	187	2.47	<5.0	9.34	19.8	2.83		17.3	661
	200	500	50	2	50	20	200	7,000	20	500	100		30		1,000		20	20	500
	200	1500	50	2	50	20	200	5,000	50	700	500		30		100		20	30	500
	200	500	70	2	50	30	20	15,000	7	100	500		15		700		20	100	500
	200	700	100	2	50	100	70	2,000	10	100	200		20		10		20	50	500
0.46		960	13	<2	6	8	77	1,700	21	1,300	400	0.7	10	3.5	45		<10	9	390
0.84		1100	39	<2	11	16	33	1,700	13	910	1,100	0.27	7	1.3	26		<10	20	810
1.7		960	110	<2	<4	21	36	1,300	14	5,000	240	1.5	6	3.4	110		<10	28	300
0.35		435	18	<2	<4	5	12	795	9.8	6,100	150	0.505	<4	1.95	48.5		<10	9	150
1		870	15	<2	10	13	41	3,400	15	890	1,400	0.15	8	0.9	27		<10	18	720
1.4	17	1500	23	<2	9	9	20	300	13	880	400	1.1	<4	3.3	14		<10	26	980
0.74		1000	<2	<2	58	10	110	4,800	19	40	2,100	0.02	16	0.9	<4		<10	<4	2,900
0.38		2000	5	<2	50	6	69	2,800	23	570	1,100	0.38	9	1.9	15		<10	5	2,400
0.35		970	15	<2	4	<2	130	1,200	29	1,500	150	1.2	12	4.1	35		<10	<4	200
1.1		760	26	<2	7	6	6	180	5.7	710	190	0.56	<4	2	12		<10	22	420
2210	<20	739	20.2	<0.25	3.3	4	<5.0	722	94,200	825	98.5	0.241	<5.0	2.47	20.8	<0.25	<10	19.1	152

Note: A number preceded by "<" indicates the chemical was not detected and the number is the analytical reporting limit from the laboratory.

TABLE 5.4

Mixed Surface and Subsurface Soil Sampling Results for Metals at the Former Gold Mine Site, Part 2

Al	Sb	As	Ba	Be	Cd	Cr	Co	Cu	Fe	Pb	Mn	Hg	Ni	Se	Ag	Tl	Sn	Va	Zn
3780	<20	960	59.1	<0.25	3.4	7.6	7	228	106,000	57	71.8	0.105	<5.0	1.6	2.2	<0.25		39	107
2590	<40	1295	9.63	<0.50	11.65	70.05	20.5	308	142,500	807.5	216.5	0.9735	49.35	4.525	18.05	0.6		16.2	2,015
		622							78,900										
		845							88,900										
		702							83,000										
		210							48,000										
		389							52,400										
		396							51,600										
		208							66,500										
		21.7							24,300										
		3.91							10,000										
		1670							69,800										
		96.4							50,300										
1560	<20	1,170	3.64	<0.25	5.32	3.2	<5.0	1710	96,200	1330	147	0.442	<5.0	3.43	36.7	<0.25		14.8	225
411	<8.0	205	4.76	<0.10	1.5	3.6	4.6	1420	42,700	4190	65.4	3.87	10.8	10.6	73.8	4.78		5.89	139
1930	<20	6.92	19.5	<0.25	<1.0	7.4	<5.0	86.3	4,720	38.3	145	0.317	6.7	<0.4	<1.5	<0.25		5.59	35.5
1760	<40	449	32.1	<0.5	6.5	117	21	2000	329,000	2470	449	0.944	242	<0.05	35.5	<1.0		26	2,320
9660	<4.0	56.8	47.5	0.16	1.26	91.8	8.31	146	20,500	483	611	3	57.3	<0.4	10.1	<0.25		30.1	165

Note: A number preceded by "<" indicates the chemical was not detected and the number is the analytical reporting limit from the laboratory. Units are mg/kg soil.

TABLE 5.5

Background Mixed Surface and Subsurface Soil Sampling Results at the Former Gold Mine Site

Al	Sb	As	Ba	Be	Cd	Cr	Co	Cu	Fe	Pb	Mn	Hg	Ni	Se	Ag	Sn	Tl	Va	Zn
11,000	<4.0	9.32	190	0.32	0.21	23.5	11	37.8	23,500	7.38	443	0.0692	26.3	0.45	0.34		<0.25	54.1	57.2
0.09	<15	3100	7	<2	4	9	20	560	45	12	230	<0.02	<4		<4	<10		21	25
0.03	<15	31	<2	<2	<4	<2	130	4,400	47	77	140	<0.02	<4	0.6	11	<10		<4	140
0.97	<15	500	9	<2	<4	87	<2	300	14	1600	110	0.03	6	0.7	25	<10		72	130
4.5	<15	<20	97	<2	6	90	59	12,000	26	<8	6100	<0.02	12		7	37		95	900
1.6	<15	1200	6	<2	95	12	33	4,100	29	2900	2900	0.12	12	1.4	16	<10		52	7300
1,860	<40	21.1	24.5	<0.50	8.6	6.9	19	3,430	211,000	44	1070	0.14	<10	28.7	109	<10	<0.75	23.5	2200
		4			0.098	5.1	5.1	22	11,800	2.6	213	0.015	7.5	<5	1			37.7	28.4
		58.2			0.21	110	7.7	141	27,000	5	1190	0.046	50.1	<5.3	0.81			14.9	36.8
		24.7			1.7	67.5	11.5	122	27,300	2.3	1280	0.034	36	<5.3	0.32			28	363
		55.5			1.9	26.5	5.6	533	19,400	83.4	525	0.064	15.8	5.1	1.7			14.5	140
		1.3			<1	12.9	3.9	17.8	9,270	1.1	150	0.017	10	<5.1	<1			31.8	23.1
		130			1.3	33.8	10.4	527	30,400	68.2	762	0.085	19.3	<5.2	2.1			17.3	132
12,100	<4.0	6.64	238	0.35	0.21	23.4	12.4	43.1	25,400	5.36	509	0.153	27.4	0.44	0.34		<0.25	61.2	60.3
2.1	<15	50	5	<2	7	76	59	5,800	26	11	5200	<0.02	9	1.1	<4	49		43	820
0.14	<15	59	5	<2	5	5	31	160	49	<8	470	<0.02	4	2.4	<4	<10		7	20
0.07	<15	23	20	<2	<4	<2	2	97	1.1	<8	38	0.03	<4	0.1	<4	<10		5	53
0.02	<15	<20	4	<2	<4	<2	24	1,700	43	51	100	<0.02	<4	0.7	35	<10		<4	40
0.13	<15	98	5	<2	<4	<2	24	3,000	21	450	37	0.03	<4	1.8	41	<10		4	57

Note: A number preceded by "<" indicates the chemical was not detected and the number is the analytical reporting limit from the laboratory. Units are mg/kg soil.

TABLE 5.6
Surface Water Sampling Results at the Former Gold Mine Site

Al	Sb	As	Ba	Be	Cd	Cr	Co	Cu	Fe	Pb	Mn	Hg	Ni	Se	Ag	Sn	Tl	Va	Zn
0.12	2.4	<5	<10	23	<10	<10	<10	<10	<50	<50	<0.05	0.021	<10	<5	<0.05	<0.1	<1	<10	<10
0.06	1.8	6	<10	12	<10	<10	<10	<10	280	<50	<0.05	0.016	<10	<5	<0.05	<0.1	2	<10	<10
0.05	2	<5	<10	13	<10	<10	<10	<10	120	<50		0.021	<10	0.5	<0.01	<0.05	<0.05	<10	<10
<0.02	2	<5	<10	<10	<10	<10	<10	<10	310	<50		0.019	<10	<0.2	<0.01	<0.05	<0.05	<10	<10
0.1	5.6	19	<10	26	<10	<10	<10	28	2500	<50		0.024	<10	0.7	<0.01	<0.05	0.06	<10	150
	1.06	<10			<1	<1	<1	3.66	314	<1		0.032	2.23					17.2	<10
	3.6	4.3							81.1	<0.5		0.018						3.4	<4.0
	1.4	21.8							67	<0.5		0.016						<3.0	<4.0
		4.5			<0.33		<1	0.85	<50			0.016						4.1	4.7
		11.2			<0.33		0.25	1.4	120			0.018						<20	11.5
		6			<0.33		0.094	2.1	65.3			0.016						0.44	5.1
		4.3			<0.33		0.074	1.3	<50									3.1	2.9
		31.4			<0.33		<1	0.88	<50									1.3	2.6
		4.4			<0.33		<1	1.1	32.3									<20	4
		10.7			<0.33		0.28	1.2	297									0.53	3.5

Note: A number preceded by "<" indicates the chemical was not detected and the number is the analytical reporting limit from the laboratory. Units are µg/L.

TABLE 5.7
Background Surface Water Sampling Results at the Former Gold Mine Site

Al	Sb	As	Ba	Be	Cd	Cr	Co	Cu	Fe	Pb	Mn	Hg	Ni	Se	Ag	Sn	Tl	Va	Zn
<40	0.4	4.4	18	<1	190	<1	<2	<2	<4	30	<5	0.023	<4		<0.1	<0.4	<0.5	<2	<2
<40	<0.2	<0.8	3	<1	<50	<1	<2	<2	<4	<20	<5	0.021	<4		<0.1	<0.4	<0.5	3	<2
<40	<0.2	<0.8	12	<1	<50	<1	<2	<2	10	30	<5	0.013	<4		<0.1	<0.4	<0.5	<2	<2
45	0.35	3.3	4	<1	<50	<1	<2	3.5	19.5	1,050	<5	0.013	<4		<0.1	<0.4	<0.5	2	<2
<40	<0.2	1	12	<1	<50	<1	<2	<2	7	60	<5		<4		<0.1	<0.4	<0.5	<2	<2
<40	<0.2	<0.8	18	<1	<50	<1	<2	<2	5	60	<5		<4		<0.1	<0.4	<0.5	<2	<2
<40	0.56	<0.8	9	<1	<50	<1	<2	<2	5	30	<5		<4		<0.1	<0.4	<0.5	<2	<2
<40	<0.2	<0.8	5	<1	<50	<1	<2	<2	<4	200	<5		<4		<0.1	<0.4	<0.5	<2	<2
<40	<0.2	0.8	5	<1	<50	<1	<2	<2	<4	<20	<5		<4		<0.1	<0.4	<0.5	3	<2
120	0.07	<0.5	22	<10	67	<10	<10	<10	<10	170	<50		<10	<5	<0.05	<0.1	<1	<10	<10
56	<0.02	0.4	<5	<10	<10	<10	<10	<10	<10	100	<50		<10	0.3	<0.01	<0.05	<0.05	<10	<10
17	<0.02	1	7.7	<10	<10	<10	<10	<10	<10	420	<50		<10	0.4	<0.01	<0.05	<0.05	<10	<10
71	0.03	2	5.1	<10	11	<10	<10	<10	<10	60	<50		<10	0.8	<0.01	<0.05	<0.05	<10	<10
56.5	<0.02	0.75	<5	<10	<10	<10	<10	<10	<10	<50	<50		<10	0.4	<0.01	<0.05	<0.05	<10	<10
250	<0.02	0.7	24	<10	64	<10	<10	<10	<10	420	<50		<10	0.8	<0.01	<0.05	<0.05	<10	<10
32.6			1.1			<0.33		<1	1.2	<50								2.8	5.4
19.6			1.8			<0.33		<1	0.96	<50								2.3	4.7
<50			2.5			<0.33		<1	1.1	<50								2.4	3.1
229			8			0.24		0.36	8.8	390								1.7	33
<50			4.1			<0.33		<1	1	<50								2.8	6.3

Note: A number preceded by "<" indicates the chemical was not detected and the number is the analytical reporting limit from the laboratory. Units are μg/L.

TABLE 5.8
Groundwater Sampling Results at the Former Gold Mine Site

As	Ba	Cd	Cr	Cu	Fe	Pb	Hg	Ni	Se	Ag	Va	Zn
0.0015	0.014	<0.001	0.0037	0.0041	0.22	<0.001	<0.0002	0.0044	0.0025	0.0019	0.0013	0.007
0.0077	0.065	<0.001	0.0098	0.0088	10	0.0026	<0.0002	0.0069	<0.005	0.00084	0.0082	0.029
0.0046	0.033	<0.001	0.0029	0.0007	6.8	<0.001	<0.0002	0.0021	<0.005	0.00062	<0.02	0.0041
0.00059	0.016	<0.001	<0.002	0.004	0.38	0.00028	<0.0002	0.0021	0.0013	<0.001	0.0018	0.0036
0.00195	0.017	<0.001	<0.002	0.00305	1.045	0.00047	<0.0002	0.00101	<0.005	<0.001	0.001005	0.00345
0.0048	0.014	<0.001	0.0019	0.0078	1.5	0.0013	<0.0002	0.0023	<0.005	0.00038	0.0043	0.0099
0.0037	0.18	<0.001	0.0015	0.004	0.56	0.00023	<0.0002	0.0017	<0.005	0.00053	0.0055	0.0044
0.0025	0.028	0.0037	0.0074	0.0074	0.72	0.0003	<0.0002	0.0091	0.0083	0.0037	0.0024	0.038

Note: A number preceded by "<" indicates the chemical was not detected and the number is the analytical reporting limit from the laboratory. Units are μg/L.

Conceptual Site Model

The conceptual site model provides the framework of the risk assessment and is shown in Figure 5.1. It characterizes the primary and secondary potential sources and release mechanisms and identifies the primary exposure points, receptors, and exposure routes. Receptors include humans who contact environmental media at the site. This risk assessment focuses on potential human exposure to COPCs

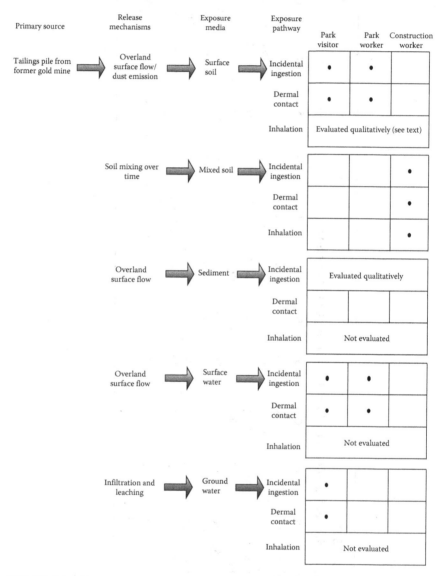

FIGURE 5.1 Conceptual site model for the former gold mine site showing sources of contamination, release mechanisms, receptors, and exposure routes. The bullets show complete exposure pathways considered in the quantitative risk assessment.

detected in soil, surface water, sediment, and groundwater at, and adjacent to, the site. Exposure points are places or "points" where exposure could potentially occur, and exposure routes include the basic pathways through which COPCs may potentially be taken up by the receptor.

Inhalation of fugitive dust was considered but not addressed quantitatively for the park visitors and workers. The reason is that all areas at the site were heavily vegetated except for the nine-acre tailings pile. However, as noted, the tailings pile was covered with a layer of hardpan that effectively prevents the generation of fugitive dust. Hence, inhalation of dust as an exposure pathway at the site was considered negligible for park visitors and park workers.

For the construction worker, vehicle traffic on contaminated unpaved roads typically accounts for the majority of emissions, with wind erosion, excavation soil dumping, dozing, grading, and filling operations contributing lesser emissions.[30] EPA's supplemental soil screening guidance provides a value of 4.4E + 08 m^3/kg was this was used as the RME value of construction-related particulate emission factor (PEF). A value representative of dust emissions in nonconstruction scenarios of 1.32E + 09 m^3/kg was used as the central tendency exposure (CTE) value. Dividing this value into a soil concentration expressed in mg/kg provides the air concentration of a given chemical.

Exposure to surface water was considered to occur by dermal contact only. Purposeful ingestion would likely not occur due to the potential for *Giardia* in water. Park visitors are made aware of this in an orientation session and cautioned not to drink from streams. Incidental ingestion would likely not occur because the water is too cold for swimming or even wading, remaining most of the year below 60°F.

In general, sediment exposure is thought to be minimal—most sediment on the skin washes off as a receptor moves through the water. For this reason, sediment was not evaluated quantitatively. The site is located in the northern United States, and even in the summer, the water temperature in Trout Creek, a small stream draining the site, remains below 60°F. Hence, it is unlikely that prolonged exposure to either surface water or sediment would occur. However, for surface water, park visitors were assumed to put their hands in the surface water during two daily events and thus some dermal contact might occur.

Groundwater was not included in the CSM but was evaluated based on a hypothetical domestic use. While this exposure scenario is not realistic for the site, it provides a protective risk evaluation.

Exposure Units

An EU is defined as the geographic area within which a receptor comes in contact with a contaminated medium during the ED. The EU is defined based on the receptor, exposure medium, and the nature of contact.[20] The current and future receptors at the site include park personnel, park visitors, and construction workers. These are atypical receptors and the entire area around the site was considered the EU. While concentrations of metals are somewhat higher in samples from the tailings pile, many of the metals appear to be elevated in both soil and background samples.

Data Analysis and Selection of Chemicals of Potential Concern (COPCs)

Screening levels were obtained from EPA's most recent RSLTs.[31] COPC screening was performed with a cancer risk of 1E-06 and an HQ value of 0.1. A value lower than 1 for the HQ is used to account for exposure to multiple chemicals.

Data were selected to represent both the site and exposure units appropriate to the medium considered. As the north fork of Trout Creek flows through the site, the creek goes underground and then reemerges as springs. The portion of the creek near the site is dry for part of the year. Hence, sediment in these locations is considered as surface soil and sampling results from the dry creek area were included as surface soil. The park worker and adult and child park visitors were assumed to be exposed to surface soil.

Surface water as potential dermal exposure media for workers and visitors were assumed to occur in the north fork of Trout Creek between its emergence as springs between the mine site and the tailings pile and the confluence with the south fork of Trout Creek.

A construction/excavation worker would be exposed to a mixture of surface and subsurface soil. Groundwater was evaluated as if it could be a source of water for domestic use. The site and background datasets are provided in Tables 5.1 though 5.8.

The selection of COPCs generally uses four screening steps:

1. Elimination of five inorganic constituents (calcium, chloride, magnesium, potassium, and sodium) that are considered essential human nutrients
2. Elimination of constituents for which the maximum detect did not exceed the screening benchmarks based on a cancer risk of 1E-06 and an HQ value of 0.1 (benchmarks were obtained from EPA's RSLT, November 2012 version)
3. Elimination of constituents that were detected in fewer than 5% of the relevant samples
4. Elimination of constituents that were shown to be higher than the background sample using the Wilcoxon–Mann–Whitney test[28]

EPA's ProUCL software v. 4.1 was used for the statistical comparison in step 4 and for calculation of the values for the 95% UCL of the mean that is recommended for use as the exposure point concentration.[32,33] ProUCL is a mature software and provides recommendations for which UCL value to use. Guidance documents for ProUCL provide details of the statistical methods and UCL calculation methods.[34–36]

The results of COPC screening for various media are shown in Tables 5.9 through 5.12. Table 5.9 shows the selection of COPCs in surface soil. Residential soil screening levels were used. The COPCs in surface soil are arsenic, cadmium, lead, mercury, silver, and zinc. Table 5.10 shows the selection of COPCs in mixed surface and subsurface soil. Commercial/industrial screening levels were used. The COPCs are arsenic, lead, manganese, and silver. Table 5.11 shows the selection of COPCs in surface water. Residential tap water screening levels were used. The COPCs in surface water are beryllium and iron. Table 5.12 shows the selection of

TABLE 5.9

Selection of Chemicals of Potential Concern for Surface Soil at the Former Gold Mine Site

Chemical	N	#D	Freq. of Detect (%)	Min. ND	Max. ND	Min. Detect	Max. Detect	Mean of Detects	SD Detects	RSLs Cancer	RSLs NC	Background Comparison	COPC (Y/N)	UCL95 EPC	Method
Surface soil (mg/kg) park visitors and workers															
Aluminum	23	23	100	—	—	0.35	9,660	1,047	2,114	—	7,700	<=BG	N	—	—
Antimony	23	8	35	1	40	3	200	103.3	103.5	—	3.1	<=BG	N	—	—
Arsenic	35	35	100	—	—	6.1	3,460	732.1	714.5	0.39	2.2	—	Y	1034	95% approx. gamma
Barium	27	27	100	—	—	2.5	500	87.5	130.7	—	1,500	—	N	—	—
Beryllium	27	8	30	0.1	2	0.07	3	1.65	1.01	1,400	16	—	N	—	—
Cadmium	35	28	80	1	4	0.16	50	12.38	15.82	1,800	7	—	Y	36.6	99% KM (Chebyshev)
Chromium	35	32	91	2.5	5	1.6	117	21.21	28.82	—	12,000	—	N	—	—
Cobalt	35	33	94	5	5	3.3	200	46.74	57.6	370	2.3	<=BG	N	—	—
Copper	35	35	100	—	—	12.9	15,000	1,568	2,798	—	310	<=BG	N	—	—
Iron	35	35	100	—	—	5.3	329,000	46,337	80,366	—	5,500	<=BG	N	—	—
Lead	35	35	100	—	—	6.81	6,100	1,078	1,435	—	400	—	Y	1,363	Mean for IEUBK or ALM[a]
Manganese	35	35	100	—	—	65.4	16,200	1,363	2,925	—	180	<=BG	N	—	—
Mercury	29	27	92	0.005	0.02	0.018	3.87	0.891	1.06	—	1	—	Y	1.678	95% KM (Chebyshev)
Nickel	35	30	85	4	10	3.8	242	22.36	43.12	1,300	150	<=BG	N	—	—
Selenium	29	24	86	0.2	0.6	0.4	10.6	3.653	2.803	—	39	—	N	—	—

(continued)

TABLE 5.9 (continued)

Selection of Chemicals of Potential Concern for Surface Soil at the Former Gold Mine Site

Chemical	N	#D	Freq. of Detect (%)	Min. ND	Max. ND	Min. Detect	Max. Detect	Mean of Detects	SD Detects	RSLs Cancer	RSLs NC	Background Comparison	COPC (Y/N)	UCL95 EPC	Method
Silver	35	33	94	1.5	2	0.1	1,000	74.98	205.1	—	39	—	Y	407	99% KM (Chebyshev)
Thallium	9	4	44	0.25	1	0.57	4.78	2.28	1.938	—	0.078	<=BG	N	—	—
Tin	18	6	33	5	10	7	24	18.5	5.857	—	4,700	—	N	—	—
Vanadium	35	35	100	—	—	5.59	140	39.75	35.36	—	39	<=BG	N	—	—
Zinc	35	35	100	—	—	35.5	2,450	585.3	603.2	—	2,300	—	Y	797.6	95% approx. gamma

[a] Integrated exposure/uptake/biokinetic model for lead and Adult Lead Model. See http://www.epa.gov/superfund/health/contaminants/lead/pbrisk.htm

TABLE 5.10

Selection of Chemicals of Potential Concern for Mixed Surface and Subsurface Soil at the Former Gold Mine Site

Chemical	N	#D	Freq. of Detect (%)	Min. ND	Max. ND	Min. Detect	Max. Detect	Mean of Detects	SD Detects	RSLs Cancer	RSLs NC	Background Comparison	COPC (Y/N)	UCL95 EPC	Method
Mixed surface and subsurface soil (mg/kg) construction/excavation worker															
Aluminum	30	30	100	—	—	0.35	9,660	1,089	1,981	—	99,000	—	N		
Antimony	30	8	27	1	40	3	200	103.3	103.5	—	41	<=BG	N		
Arsenic	53	53	100	—	—	3.91	3,460	726.7	659	1.6	26	—	Y	1,121	95% Chebyshev
Barium	34	33	97	2	2	2.5	500	75.68	120.8	—	19,000	—	N		
Beryllium	34	7	21	0.1	2	0.068	3	1.604	1.082	6,900	200	—	N		
Cadmium	42	34	81	1	4	0.16	58	14.18	18.41	9,300	80	—	N		
Chromium	42	38	91	2	5	1.6	117	20.59	28.08	—	150,000	—	N		
Cobalt	42	39	93	5	5	3.3	200	48.33	56.53	1,900	30	<=BG	N		
Copper	42	42	100	—	—	12.9	15,000	1,551	2,633	—	4,100	<=BG	N		
Iron	53	53	100	—	—	5.3	329,000	48,836	69,338	—	72,000	<=BG	N		
Lead	42	42	100	—	—	6.81	6,100	1,006	1,331	—	800	—	Y	1,006	Mean for ALM[a]
Manganese	42	42	100	—	—	65.4	16,200	1,229	2,697	—	2,300	—	Y	3,043	95% Chebyshev

(continued)

TABLE 5.10 (continued)

Selection of Chemicals of Potential Concern for Mixed Surface and Subsurface Soil at the Former Gold Mine Site

Chemical	N	#D	Freq. of Detect (%)	Min. ND	Max. ND	Min. Detect	Max. Detect	Mean of Detects	SD Detects	RSLs Cancer	RSLs NC	Background Comparison	COPC (Y/N)	UCL95 EPC	Method
Mercury	36	34	94	0.005	0.02	0.018	3.87	0.81	0.973	—	4.3	—	N		
Nickel	42	35	83	4	10	3.8	242	21.75	40.33	64,000	2,000	—	N		
Selenium	36	32	89	0.05	0.4	0.4	10.6	3.4	2.581	—	510	—	N		
Silver	42	39	93	1.5	4	0.1	1,000	66.09	189.5	—	510	—	Y	238.1	95% KM (Chebyshev)
Thallium	12	5	42	0.25	1	0.57	4.78	1.944	1.839	—	0.61	<=BG	N		
Tin	22	6	27	5	10	7	24	18.5	5.857	—	61,000	—	N		
Vanadium	42	40	95	4	4	5	140	37.31	33.88	—	520	—	N		
Zinc	42	42	100	—	—	35.5	2,900	682.8	753.2	—	31,000	—	N		

[a] Adult lead model. See http://www.epa.gov/superfund/health/contaminants/lead/pbrisk.htm

TABLE 5.11

Selection of Chemicals of Potential Concern for Surface Water at the Former Gold Mine Site

Chemical	N	#D	Freq. of Detect (%)	Min. ND	Max. ND	Min. Detect	Max. Detect	Mean of Detects	SD Detects	RSLs Cancer	RSLs NC	Background Comparison	COPC (Y/N)	UCL95 EPC	Method
Surface water (µg/L) park visitors and workers															
Aluminum	5	4	80	0.02	0.02	0.05	0.12	0.0825	0.033		1600		N		
Antimony	8	8	100	—	—	1.06	5.6	2.483	1.469		6.0[b]		N		
Arsenic	15	11	73	5	10	4.3	31.4	11.24	9.066	0.045	0.47	<=BG	N		
Barium	4	0	0	10	10	—	—	—	—		290		N		
Beryllium	5	4	80	10	10	12	26	18.5	7.047		1.6		Y	23.86	95% KM (t) UCL
Cadmium	13	0	0	0.33	10	—	—	—	—		5.0[b]		N		
Chromium	6	0	0	1	10	—	—	—	—		100[b]		N		
Cobalt	13	4	31	1	10	0.074	0.28	0.175	0.106		0.47		N		
Copper	13	9	69	10	10	0.85	28	4.499	8.856		62		N		
Iron	15	11	74	50	50	32.3	2500	380.6	711.5		1100		Y	1304	97.5% KM (Chebyshev)
Lead	8	0	0	0.5	50	—	—	—	—		15[b]		N		
Manganese	2	0	0	0.05	0.05	—	—	—	—		32		N		
Mercury	11	11	100	—	—	0.016	0.032	0.0197	0.0048		0.43[a]		N		
Nickel	6	1	17	10	10	2.23	2.23	2.23	—	—	30		N		
Selenium	5	2	40	0.2	5	0.5	0.7	0.6	0.141		7.8		N		
Silver	5	0	0	0.01	0.05	—	—	—	—		7.1		N		
Thallium	5	2	40	0.05	1	0.06	2	1.03	1.372		0.016	<=BG	N		
Tin	5	0	0	0.05	0.1	—	—	—	—		930		N		
Vanadium	15	7	47	3	20	0.44	17.2	4.296	5.871		7.8	<=BG	N		
Zinc	15	8	53	4	10	2.6	150	23.04	51.38		470		N		

[a] As mercuric chloride or other mercury salts.

[b] Denotes the federal MCL.

218 Environmental Risk Assessment

TABLE 5.12

Selection of Chemicals of Potential Concern for Groundwater at the Former Gold Mine Site

Chemical	N	#D	Freq. of Detect (%)	Min. ND	Max. ND	Min. Detect	Max. Detect	Mean of Detects	SD Detects	RSLs Cancer	RSLs NC	COPC (Y/N)	UCL95 EPC	Method
Groundwater (μg/L) potential residents														
Arsenic	8	8	100	—	—	0.59	7.7	3.418	2.28	0.045	0.47	Y	4.945	95% Student's-t
Barium	8	8	100	—	—	14	180	45.88	56.82		290	N		
Cadmium	8	1	12	1	1	3.7	3.7	3.7	—		5.0[b]	N		
Chromium	8	2	75	2	2	1.5	9.8	4.533	3.33		100[b]	N		
Copper	8	8	100	—	—	0.7	8.8	4.981	2.755		62	N		
Iron	8	8	100	—	—	220	10,000	2,653	3,670		1,100	Y	7,341	95% approx. gamma
Lead	8	6	75	1	1	0.23	2.6	0.863	0.94		15[b]	N		
Mercury	8	0	0	0.2	0.2	—	—	—	—		0.43[a]	N		
Nickel	8	8	100	—	—	1.01	9.1	3.701	2.884		30	N		
Selenium	8	3	38	5	5	1.3	8.3	4.033	3.743		7.8	Y	4.638	95% KM(t)
Silver	8	6	75	1	1	0.38	3.7	1.328	1.283		7.1	N		
Vanadium	8	7	88	20	20	1.005	8.2	3.501	2.642		7.8	Y	5.393	95% KM(t)
Zinc	8	8	100	—	—	3.45	38	12.43	13.4		470	N		

[a] As mercuric chloride or other mercury salts.
[b] Denotes the federal MCL used if lower than the concentration at HQ = 0.1.

COPCs in groundwater. No background data were available. Residential tap water screening levels were used. The COPCs are also beryllium and iron.

Exposure Assessment

An exposure assessment was conducted as part of the health risk assessment to evaluate the potential exposure pathways at the site. An exposure pathway is defined by the following four elements: (1) a source and mechanism of constituent release to the environment, (2) an environmental transport medium for the released constituent, (3) a point of potential contact with the contaminated medium (the exposure point), and (4) an exposure route at the exposure point. These pathways are shown in the conceptual site model (Figure 5.1). The purpose of the exposure assessment is to estimate the way a population may potentially be exposed to constituents at a site. Typically, exposure assessment involves projecting concentrations along potential pathways between sources and receptors. The projection usually is accomplished using site-specific data and, when necessary, mathematical modeling. Exposure can occur only when the potential exists for a receptor to experience direct contact with an environmental medium containing released constituents or if a mechanism exists for released constituents to be transported to a receptor. Without exposure, there is no risk; therefore, the exposure assessment is a critical component of the risk assessment.

Exposure Assumptions

To provide some understanding of the range of exposures and consequent risks, scenarios based on both RME and CTE were evaluated. Standard default values for assessing risk that generally lead to the RME risk estimates were used.[37–40]

The values for soil adherence factors were obtained from Exhibit 3-3 in EPA's *RAGS: Volume 1: Human Health Evaluation Manual (Part E, Supplemental Guidance for Dermal Risk Assessment) Final.*[4] The values for drinking water ingestion were from EPA's 2011 *Exposure Factors Handbook.*[40] The values for soil ingestion rate were from EPA's *Standard Default Exposure Assumptions.*[37]

For dermal contact with surface water, the values for skin surface area representing arms, forearms, and lower legs for both adults and children were from Exhibit CI in EPA's *RAGS, Part E, Supplemental Guidance for Dermal Risk Assessment.*[4] For park workers and construction workers, just the hands and forearms were assumed to contact soil and just the hands for surface water contact.

The concept of RME was envisioned to provide an estimate of the highest reasonable exposure possible to an individual. Such an individual is defined as the RME receptor and is considered to be at the 90th percentile of the exposure distribution or higher. The NCP indicates that site decisions should be based on the RME receptor and RME assumptions are shown for all four receptors in Table 5.13.[41] EPA has indicated that the RME approach is incomplete by presenting only a point estimate of risk with no indication of where it falls within the risk distribution and that CTE risk estimates should also be presented.[17,42]

TABLE 5.13
Exposure Assumptions Used in the Risk Assessment of the Former Gold Mine Site

Parameter	Abbr.	Units	Child Park Visitor		Adult Park Visitor		Park Worker		Construction Worker	
			RME Value	CTE Value	RME Value	CTE Value	RME Value	CTE Value	RME Value	CTE Value
Body weight	BW	kg	15	15	70	70	70	70	70	70
Averaging time—C	AT-C	Days	25,550	25,550	25,550	25,550	25,550	25,550	25,550	25,550
Averaging time—NC	AT-NC	Days	ED * 365	ED * 365	ED * 365	ED * 365	ED * 365	ED * 365	ED * 365	ED * 365
Exposure frequency	EF	Days/year	24	7	24	7	200	120	60	10
Exposure duration	ED	Years	6	2	30	7	25	15	1	1
Drinking water ingestion	IRW	L/day	1	0.5	2	1	2	1	1	1
Incidental water ingestion	IRWI	L/event	Not evaluated		Not evaluated quantitatively (see conceptual site model; Figure 5.1)				2	1
Inhalation rate	IN	m³/day							21	17
Soil ingestion rate	IR	mg/day	200	50	100	20	100	20	330	200
Skin surface area—hands, forearms, and lower legs for soil contact	SA	cm²	1,400	1,400	4,500	4,500	2,100	2,100	2,100	2,100
Skin surface area—hands for surface water contact	SAw	cm²	400	400	900	900	900	900	900	900
Soil adherence factor	SAF	mg/cm²	0.4	0.04	0.3	0.07	0.1	0.02	0.3	0.1
Relative bioavailability (As)	RBA	Percent	31%	4.1%	31%	4.1%	31%	4.1%	31%	4.1%
Particulate emission factor	PEF	m³/kg							1.32E09	1.32E09
Exposure time	ET	Hour							8	8

Exposure Pathways and Receptors

The most likely route of potential human exposure to constituents detected is through direct contact with soil. The potential for exposure to fugitive dust generated at the site is low due to the presence of hardpan covering the tailings pile, snow in the winter, and vegetative cover in the summer. Notwithstanding, this pathway was considered quantitatively for construction workers.

The exposure dose for oral and dermal exposure to soil or sediment was estimated for both carcinogens and noncarcinogens as follows:

$$
\begin{aligned}
&ADD\,(mg/kg\text{-}day) \\
&= \frac{C_{soil} \times IR_{soil} \times EF \times ED \times CF}{BW \times AT} + \frac{C_{soil} \times CF \times SAF \times ABS_{dermal} \times EF \times ED \times SSA}{BW \times AT}
\end{aligned}
\tag{5.3}
$$

where
    ADD is the average daily dose (mg/kg/day)
    C_{soil} is the concentration in soil (mg/kg)
    IR_{soil} is the soil ingestion rate (mg/day)
    EF is the exposure frequency (days/year)
    ED is the exposure duration (year)
    BW is the body weight (kg)
    AT is the averaging time (days)
    CF is the conversion factor (kg/mg) = 1E-06
    SAF is the soil adherence factor (mg/cm^2)
    ABS_{dermal} is the dermal absorption (chemical specific)
    SSA is the skin surface area (cm^2)
    PEF is the particulate emission factor (m^3/kg)

For the construction/excavation worker, the exposure concentration for inhalation exposure was estimated as follows:

$$
EC_{inhalation} = \frac{C_{soil} \times (1/PEF) \times ET \times EF \times ED}{(AT \times 24\,h/day)}
\tag{5.4}
$$

Equation 5.4 provides an exposure concentration in air that can be used with inhalation RfCs in units of mg/m^3 or IURs in units of per μg/m^3:

$$
Risk = EC_{inhalation} \times 1000 \times IUR
\tag{5.5}
$$

$$
Hazard\,quotient = \frac{EC_{inhalation}}{RfC}
\tag{5.6}
$$

Equations 5.5 and 5.6 are, respectively, used to estimate inhalation risk and inhalation hazard. Both IUR values and RfC values need to represent continuous exposure as a time-weighted average of exposure in the study used to derive the toxicity factor.[5] Because the values for soil concentration were in mg/kg,

$EC_{inhalation}$ was multiplied by 1000 to convert to $\mu g/m^3$ for assessing carcinogenic risk. The value of the PEF was the default value recommended by EPA.[43]

The intake or dose equation shown earlier is provided with two separate terms to show the separation of ingestion and dermal routes for soil contact. For both carcinogens and noncarcinogens, Equation 5.3 was applied to adults and children separately.

Regarding inhalation, the metals at the site are not volatile and exposure to fugitive dust generation will be very low. During the winter, the climate and snow cover will prevent any exposure to dust. During the summer, the extensive vegetation and the presence of the hardpan covering on the tailings pile will mitigate the generation of dust. Equations 5.4 through 5.6 were developed to be able to use a toxicity criterion expressed as an IUR value rather than an inhalation slope factor. The application of unit risks to different life stages (i.e., children and adults) requires understanding of whether the chemical acts by a potentially mutagenic MOA. This MOA is not likely for arsenic. A full discussion of this issue is presented in EPA's *RAGS, Part F, Supplemental Guidance for Inhalation Risk Assessment, Final*.[5,44]

Bioavailability of Arsenic

Recently, EPA recommended a default relative bioavailability (RBA) value of 60% for arsenic in soil.[45] What RBA measures is the difference in both bioaccessibility and GI absorption. Bioaccessibility is the proportion of the arsenic that dissolves in the gut lumen and GI absorption is the proportion of dissolved arsenic that moves from the gut lumen to the bloodstream. In bioavailability studies of arsenic in juvenile swine and monkeys, RBA values ranged from 4.1% to 78% with an arithmetic mean of 31 ± 16%.[46] In more recent studies on mice, soils containing arsenopyrite slag similar to the mine tailings at the site showed the lowest bioavailability—around 7%, similar to that observed in monkeys. In addition, in vivo bioavailability was found to correlate well with bioaccessibility measured in vitro.[47] As noted, bioaccessibility is essentially the solubility in the gastrointestinal tract. In vitro bioaccessibility measurements are considerably less costly than in vivo studies.

A site-specific value for bioavailability for the former gold mine site can be estimated using a comparison between results obtained from the soil leaching calculator at EPA's website at http://epa-prgs.ornl.gov/cgi-bin/chemicals/csl_search and actual soil and groundwater concentrations at the site. The calculator provides a default value of soil concentration protective of groundwater at the federal maximum contaminant level (MCL) of 10 µg/L. This value is 0.292 mg/kg. These values represent the potential ability of rainwater to dissolve arsenic in soil.

At the site, the arithmetic mean value for arsenic in mixed surface and subsurface soils at the site is 726.7 mg/kg and in groundwater is 3.4 µg/L (Tables 5.10 and 5.12). Comparing the soil/groundwater concentration ratio from the calculator and the site indicates there is a five order of magnitude difference in these estimates of arsenic solubility. Based on this comparison, the bioavailability of arsenic would be 0.001%. Hence, the RME value used in the risk calculation was 31%; the average from EPA's database and the CTE value was 4.1%, representing a low but still observable value.[46]

Toxicity Assessment

This section discusses the two general categories of toxic effects (noncarcinogenic and carcinogenic) evaluated in risk assessments and the toxicity values used to calculate potential risks. Toxicity values for potential noncarcinogenic and carcinogenic effects are determined from available databases. For this risk assessment, toxicity values were first obtained from either EPA's IRIS database or the Risk Assessment Information System (RAIS) at http://rais.ornl.gov.

Whenever possible, route-specific toxicity values have been used. However, toxicity values for dermal exposures have not yet been developed by USEPA; therefore, the oral toxicity values were used to derive adjusted toxicity values for use in assessing dermal exposure. The adjusted toxicity values represent the theoretical toxicity of the orally absorbed dose of the constituent based on the oral toxicity value and the assumed or measured gastrointestinal absorption (ABS_o) in the study underlying the NOAEL or LOAEL:

$$RfD_a = RfD_o \times ABS_o \qquad (5.7)$$

$$CSF_a = \frac{CSF_o}{ABS_o} \qquad (5.8)$$

Toxicity values were generally obtained from the RAIS at http://rais.ornl.gov/. This is a joint venture between USEPA and the Department of Energy. The hierarchy of sources of toxicity values recommended by EPA was used.[48] Table 5.14 shows the toxicity criteria and absorption factors used.

Risk Characterization

Potential risks to human health can be evaluated quantitatively by combining potential exposure and toxicity data. A distinction is made between noncarcinogenic and carcinogenic endpoints, and two general criteria are used to describe risk: the HQ for noncarcinogenic effects and ELCR for constituents thought to be potential human carcinogens.

To evaluate noncarcinogenic effects, exposure doses are averaged only over the expected exposure period, the exposure duration or ED in the calculation. The HQ is the ratio of the estimated exposure dose and the RfD for oral and dermal exposures and estimated inhalation concentration and the RfC for inhalation exposures. For the park visitor scenario, adults and children are considered separately for noncancer effects. Inhalation was assessed quantitatively for the construction/excavation worker only. Inhalation exposures leading to noncancer effects are assessed using RfCs as toxicity factors.[5,44,49]

TABLE 5.14

Toxicity Criteria and Absorption Factors

COPC	Noncancer Effects					Cancer Effects					Absorption Factors		
	Oral RfD (mg/kg-day)	Source	Dermal RfD^a (mg/kg-day)	Inhalation RfC (mg/m³)	Source	Oral CSF (mg/kg-day)$^{-1}$	Source	Dermal CSF^b (mg/kg-day)$^{-1}$	Inhalation IUR (mg/m³)$^{-1}$	Source	GI Abs. (Unitless)	Dermal Abs. (Unitless)	Skin Permeability (Kp cm/h)
Arsenic, inorganic	0.0003	IRIS	0.0003	1.5E-05	CALEPA	1.5	IRIS	1.5	0.0043	IRIS	1	0.03	0.001
Beryllium	0.002	IRIS	1.4E-05	2.0E-05	IRIS	—	—	—	0.0024	IRIS	0.007	—	0.001
Cadmium (diet)	0.001	IRIS	2.5E-05	2.0E-05	CALEPA	—	—	—	0.0018	IRIS	0.025	0.001	0.001
Cadmium (water)	0.0005	IRIS	2.5E-05	2.0E-05	CALEPA	—	—	—	0.0018	IRIS	0.05	0.001	0.001
Iron	0.7	PPRTV	0.7	—	—	—	—	—	—	—	1	—	0.001
Manganese (diet)	0.14	IRIS	0.14	5.0E-05	IRIS	—	—	—	—	—	1	—	0.001
Manganese (water)	0.024	IRIS	0.00096	5.0E-05	IRIS	—	—	—	—	—	0.04	—	0.001

Mercury (elemental)	—	—	—	0.0003	IRIS	—	—	—	—	—	1	—	0.001
Mercury, inorganic salts	0.0003	IRIS	2.1E-05	—	—	—	—	—	—	—	0.07	—	0.001
Selenium	0.005	IRIS	0.005	0.02	CALEPA	—	—	—	—	—	1	—	0.001
Silver	0.005	IRIS	0.0002	—	—	—	—	—	—	—	0.04	—	0.0006
Vanadium	0.005	IRIS	0.005	—	—	—	—	—	—	—	1	—	0.001
Zinc	0.3	IRIS	0.3	—	—	—	—	—	—	—	1	—	0.0006

Notes: RfD, reference dose; CSF, cancer slope factor; IRIS, Integrated Risk Information System (USEPA database of toxicity criteria); PPRTV, provisional peer-reviewed toxicity values (USEPA values not on IRIS); CALEPA, California Environmental Protection Agency; ATSDR, Agency for Toxic Substance and Disease Registry.

[a] Dermal RfD = oral RfD × GI absorption

[b] Dermal CSF = oral CSF/oral absorption

HQ values greater than 1 indicate that the estimated dose exceeds the RfD or RfC and could potentially lead to adverse effects. This ratio does not provide the probability of an adverse effect, but does reflect the concept of a highly protective threshold for the adverse effects. Although an HQ less than 1 indicates that health effects should not occur, an HQ that exceeds 1 does not mean that health effects will occur. RfDs should be considered protective because they are developed with UFs and thus have a margin of safety included. Hence, the RfD is a very good tool for CERCLA-type risk assessments that are ultimately used to develop a cleanup level with a high expectation of health protection. However, as noted earlier, the RfD is a poor tool for determining whether adverse human effects will occur—the use of the unadjusted NOAEL or LOAEL would be a better tool for assessing the actual potential for adverse effects. The sum of the HQs is the hazard index (HI) with the same acronym but different meaning than hazard identification.

Another limitation with the HI approach is that the assumption of dose additivity is applied to compounds that induce different effects by different mechanisms of action. Consequently, the summing of hazard indices for a number of compounds that are not expected to induce the same type of effects or that do not act by the same mechanism may overestimates the potential for effects.[3] Consistent with USEPA risk assessment guidelines for chemical mixtures, in the event that a total HI exceeds 1, HQs should be segregated HQs by target organ.[3]

The excess lifetime cancer risk (ELCR) is an estimate of the potential increased risk of cancer resulting from lifetime exposure to constituents detected in media at the facility. Estimated doses, or intakes, for each constituent are averaged over the hypothesized lifetime of 70 years. The multiplicative risk calculation assumes that a large dose received over a short period is equal to a smaller dose received over a longer period, as long as the total dose is the same. The ELCR for a particular chemical is equal to the product of the exposure dose and the CSF. For inhalation exposures, the ELCR is the product of the time-weighted average inhalation concentration and the unit risk value. Usually, these are expressed in units of $\mu g/m^3$ and $(\mu g/m^3)^{-1}$, respectively. The estimated risk values are a highly health-protective indication of the potential for increased cancer risk from contact with site media under the residential exposure scenario. Similar to RfDs, the CSF is an excellent and useful tool to develop protective cleanup levels but a poor predictor of the actual occurrence of cancer in humans. Because ELCRs are probabilities, they can be summed across routes of exposure and COPCs to derive a "total site risk."[3] Estimated ELCRs between 1/10,000 and 1/1,000,000 are theoretical as they cannot be observed in the US population with a background cancer rate of 30%–50%.[50]

Risk Assessment Results

The RME and CTE risk assessment calculations and results for park visitors, park workers, and excavation workers are shown in Tables 5.15 through 5.20, in the succeeding text.

TABLE 5.15

Quantitative Characterization of Risk and Hazard for Park Visitors (RME)

Chemical	Surface Soil EPC (mg/kg)	Oral Intake (mg/kg/day)	DA_{event} (mg/cm³)	Dermal Intake (mg/kg/day)	Dermal Absorption (ABS_d) (Unitless)	GI Absorption (ABS_{GI}) (Unitless)	Oral Bioavailability (Unitless)	Toxicity Values CSF_{oral}	CSF_{dermal}	Oral Risk	Dermal Risk	Total Risk
Adult cancer risks												
Arsenic	1034	8.33E-06	9.31E-06	3.37E-06	0.03	0.95	31%	1.5	1.6	3.87E-06	1.67E-06	6E-06
Child's cancer risk												
Arsenic	1034	3.89E-04	1.24E-05	3.26E-05	0.03	0.95	31%	1.5	1.6	1.81E-04	1.62E-05	2E-04

RME lifetime receptor

Summation of adult and child RME cancer risks = 2E-04

Chemical	Surface Soil EPC (mg/kg)	Oral Intake (mg/kg/day)	DA_{event} (mg/cm³)	Dermal Intake (mg/kg/day)	Dermal Absorption (ABS_d) (Unitless)	GI Absorption (ABS_{GI}) (Unitless)	Oral Bioavailability (Unitless)	Toxicity Values RfD_{oral}	RfD_{dermal}	Oral HQ	Dermal HQ	Total HI
Adult RME noncancer hazard from soil												
Arsenic	1034	9.71E-05	9.31E-06	3.93E-05	0.03	0.95	31%	3.00E-04	2.90E-04	1.00E-01	4.20E-02	0.14
Cadmium	36.6	3.44E-06	1.10E-08	4.64E-08	0.001	0.025	100%	0.001	2.50E-05	3.44E-03	1.86E-03	0.01
Lead	1363						Assessed with the adult lead model					
Mercury	1.68	1.58E-07	5.03E-09	2.13E-08	0.01	0.07	100%	0.0003	2.10E-05	5.25E-04	1.01E-03	0.002
Silver	407	3.82E-05	1.22E-06	5.16E-06	0.01	0.04	100%	0.005	0.0002	7.65E-03	2.58E-02	0.03
Zinc	798	7.49E-05	2.39E-06	1.01E-05	0.01	1	100%	0.3	0.3	2.50E-04	3.37E-05	0.0003

(continued)

TABLE 5.15 (continued)

Quantitative Characterization of Risk and Hazard for Park Visitors (RME)

Chemical	Surface Soil EPC (mg/kg)	Oral Intake (mg/kg/day)	Dermal Intake (mg/kg/day)	DA_{event} (mg/cm³)	Dermal Absorption (ABS_d) (Unitless)	GI Absorption (ABS_{GI}) (Unitless)	Oral Bioavailability (Unitless)	Toxicity Values RfD_{oral}	RfD_{dermal}	Oral HQ	Dermal HQ	Total HI
Adult RME noncancer hazard from surface water												
Beryllium	23.9	3.82E-06	7.18E-09			0.007	100%		1.4E-05		5.13E-04	0.0005
Iron	1304	2.09E-04	3.92E-07			1	100%		0.7		5.60E-07	0.000001
										Total RME hazard for adult park visitor = 0.2		
Child RME noncancer hazard from soil												
Arsenic	1034	9.07E-04	1.24E-05	7.61E-05	0.03	0.95	31%	3.00E-04	2.90E-04	9.37E-01	8.14E-02	1
Cadmium	36.6	3.21E-05	1.46E-08	8.98E-08	0.001	0.025	100%	0.001	2.50E-05	3.21E-02	3.59E-03	0.04
Lead	1363					Assessed with the IEUBK model						
Mercury	1.68	1.47E-06	6.72E-09	4.12E-08	0.01	0.07	100%	0.0003	2.10E-05	4.91E-03	1.96E-03	0.007
Silver	407	3.57E-04	1.63E-06	9.99E-06	0.01	0.04	100%	0.005	0.0002	7.14E-02	5.00E-02	0.1
Zinc	798	7.00E-04	3.19E-06	1.96E-05	0.01	1	100%	0.3	0.3	2.33E-03	6.53E-05	0.002
Child RME noncancer hazard from surface water												
Beryllium	23.9	3.82E-06	7.18E-09			0.007	100%		1.45E-05		5.13E-04	0.0005
Iron	1304	2.09E-04	3.92E-07			1	100%		0.7		5.60E-07	1E-06
										Total RME hazard for child park visitor = 1		

TABLE 5.16

Quantitative Characterization of Risk and Hazard for Park Visitors (CTE)

Chemical	Surface Soil EPC (mg/kg)	Oral Intake (mg/kg/day)	DA_{event} (mg/cm³)	Dermal Intake (mg/kg/day)	Dermal Absorption (ABS_d) (Unitless)	GI Absorption (ABS_{GI}) (Unitless)	Oral Bioavailability (Unitless)	Toxicity Values CSF_{oral}	CSF_{dermal}	Oral Risk/HI	Dermal Risk/HI	Total Risk/HI
Adult CTE cancer risks												
Arsenic	1034	1.6E-07	6.9E-05	2.4E-06	0.03	0.95	4.1%	1.5	1.6	1.0E-09	1.6E-07	1.7E-07
Child CTE cancer risk												
Arsenic	1034	1.89E-06	3.9E-05	2.0E-06	0.03	0.95	4.1%	1.5	1.6	1.2E-07	1.3E-07	2.5E-07

CTE lifetime receptor

Summation of adult and child CTE cancer risks = 4E-07

Chemical	Surface Soil EPC (mg/kg)	Oral Intake (mg/kg/day)	DA_{event} (mg/cm³)	Dermal Intake (mg/kg/day)	Dermal Absorption (ABS_d) (Unitless)	GI Absorption (ABS_{GI}) (Unitless)	Oral Bioavailability (Unitless)	Toxicity Values RfD_{oral}	RfD_{dermal}	Oral HQ	Dermal HQ	Total HI
Adult CTE noncancer hazard from surface soil												
Arsenic	1034	5.67E-06	2.17E-06	2.68E-06	1	0.95	4.1%	3E-04	2.9E-04	8E-04	4E-04	0.001
Cadmium	36.6	2.01E-07	2.56E-09	3.16E-09	0.03	0.025	100%	0.001	2.5E-05	2.0E-04	1.3E-04	0.0003
Lead	1363						Assessed with the adult lead model					
Mercury	1.68	9.21E-09	1.18E-09	1.45E-09	1	0.07	100%	0.0003	2.1E-05	3.1E-05	6.9E-06	4E-05
Silver	407	2.23E-06	2.85E-07	3.51E-07	1	0.04	100%	0.005	0.0002	4.5E-04	1.8E-03	0.0001
Zinc	798	4.37E-06	5.59E-07	6.89E-07	1	1	100%	0.3	0.3	1.5E-05	2.3E-06	0.00002

(continued)

TABLE 5.16 (continued)
Quantitative Characterization of Risk and Hazard for Park Visitors (CTE)

Chemical	Surface Soil EPC (mg/kg)	Oral Intake (mg/kg/day)	DA_{event} (mg/cm³)	Dermal Intake (mg/kg/day)	Dermal Absorption (ABS_d) (Unitless)	GI Absorption (ABS_{GI}) (Unitless)	Oral Bioavailability (Unitless)	Toxicity Values RfD_{oral}	Toxicity Values RfD_{dermal}	Oral HQ	Dermal HQ	Total HI
Adult CTE noncancer hazard from surface water												
Beryllium	23.9		3.82E-06	2.10E-09	1	0.007	100%		1.4E-05		0.003	0.003
Iron	1304		2.09E-04	1.14E-07	1	1	100%		0.7		1.6E-07	2E-07
									Total CTE hazard for adult park visitor = **0.004**			
Child CTE noncancer hazard from soil												
Arsenic	1034	6.6E-05	4.14E-05	7.40E-05	1	0.95	4.1%	3E-04	2.9E-04	0.2	0.007	0.02
Cadmium	36.6	2.3E-06	4.39E-08	7.86E-08	0.03	0.025	100%	0.001	2.5E-05	0.002	0.0004	0.005
Lead	1363					Assessed with the IEUBK model						
Mercury	1.68	1.1E-07	6.72E-08	1.20E-07	1	0.07	100%	0.0003	2.1E-05	0.0004	6E-06	0.006
Silver	407	2.6E-05	1.63E-05	2.91E-05	1	0.04	100%	0.005	0.0002	0.004	0.0001	0.2
Zinc	798	5.1E-05	3.19E-05	5.71E-05	1	1	100%	0.3	0.3	0.0002	2E-07	4E-04
Child CTE noncancer hazard from surface water												
Beryllium	23.9		1.91E-06	1.05E-09	1	0.007	100%		1.45E-05		7E-05	7E-05
Iron	1304		1.04E-04	5.72E-08	1	1	100%		0.7		8E-08	8E-08
									Total CTE hazard for child park visitor = **0.2**			

TABLE 5.17

Quantitative Characterization of Risk and Hazard for Park Workers (RME)

Park worker RME cancer risks

Chemical	Surface Soil EPC (mg/kg)	Oral Intake (mg/kg/day)	DA_{event} (mg/cm³)	Dermal Intake (mg/kg/day)	Dermal Absorption (ABS_d) (Unitless)	GI Absorption (ABS_{GI}) (Unitless)	Oral Bioavailability (Unitless)	Toxicity Values CSF_{oral}	Toxicity Values CSF_{dermal}	Oral Risk/HI	Dermal Risk/HI	Total Risk/HI
Arsenic	1034	2.89E-04	9.31E-06	5.46E-05	0.03	0.95	31%	1.5	1.6	1.3E-04	2.7E-05	2E-04

Park worker RME cancer risk = 2E-04

Park worker RME noncancer hazard from surface soil

Chemical	Surface Soil EPC (mg/kg)	Oral Intake (mg/kg/day)	DA_{event} (mg/cm³)	Dermal Intake (mg/kg/day)	Dermal Absorption (ABS_d) (Unitless)	GI Absorption (ABS_{GI}) (Unitless)	Oral Bioavailability (Unitless)	Toxicity Values RfD_{oral}	Toxicity Values RfD_{dermal}	Oral HQ	Dermal HQ	Total HI
Arsenic	1034	8.09E-04	3.10E-06	5.10E-05	1	0.95	31%	3E-04	2.9E-04	8.4E-01	5.5E-02	0.9
Cadmium	36.6	2.86E-05	3.66E-09	6.02E-08	0.03	0.025	100%	0.001	2.5E-05	2.9E-02	2.4E-03	0.03
Lead	1363					Assessed with the adult lead model						
Mercury	1.68	1.31E-06	1.68E-09	2.76E-08	1	0.07	100%	0.0003	2.1E-05	4.4E-03	1.3E-03	0.006
Silver	407	3.19E-04	4.07E-07	6.69E-06	1	0.04	100%	0.005	0.0002	6.4E-02	3.4E-02	0.1
Zinc	798	6.24E-04	7.98E-07	1.31E-05	1	1	100%	0.3	0.3	2.1E-03	4.4E-05	0.002

Park worker RME noncancer hazard from surface water

Chemical	Surface Soil EPC (mg/kg)	Oral Intake (mg/kg/day)	DA_{event} (mg/cm³)	Dermal Intake (mg/kg/day)	Dermal Absorption (ABS_d) (Unitless)	GI Absorption (ABS_{GI}) (Unitless)	Oral Bioavailability (Unitless)	Toxicity Values RfD_{oral}	Toxicity Values RfD_{dermal}	Oral HQ	Dermal HQ	Total HI
Beryllium	23.9	3.82E-06		5.99E-08	1	0.007	100%		1.4E-05		4.3E-03	0.004
Iron	1304	2.09E-04		3.27E-06	1	1	100%	0.7			4.67E-06	5E-06

Total RME hazard for park worker = 1

TABLE 5.18

Quantitative Characterization of Risk and Hazard for Park Workers (CTE)

Chemical	Surface Soil EPC (mg/kg)	Oral Intake (mg/kg/day)	DA_{event} (mg/cm³)	Dermal Intake (mg/kg/day)	Dermal Absorption (ABS_d) (Unitless)	GI Absorption (ABS_{GI}) (Unitless)	Oral Bioavailability (Unitless)	Toxicity Values CSF_{oral}	Toxicity Values CSF_{dermal}	Oral Risk/HI	Dermal Risk/HI	Total Risk/HI
Park worker RME cancer risks												
Arsenic	1034	2.08E-05	9.31E-06	1.97E-05	0.03	0.95	4.1%	1.5	1.6	1.3E-06	1.3E-06	2.6E-06
											Park worker CTE cancer risk = 3E-06	

Chemical	Surface Soil EPC (mg/kg)	Oral Intake (mg/kg/day)	DA_{event} (mg/cm³)	Dermal Intake (mg/kg/day)	Dermal Absorption (ABS_d) (Unitless)	GI Absorption (ABS_{GI}) (Unitless)	Oral Bioavailability (Unitless)	Toxicity Values RfD_{oral}	Toxicity Values RfD_{dermal}	Oral HQ	Dermal HQ	Total HI
Park worker RME noncancer hazard from surface soil												
Arsenic	1034	9.71E-05	3.10E-06	3.06E-05	1	0.95	4.1%	3E-04	2.9E-04	1.3E-02	4.3E-03	0.02
Cadmium	36.6	3.44E-06	3.66E-09	3.61E-08	0.03	0.025	100%	0.001	2.5E-05	3.4E-03	1.4E-03	0.005
Lead	1363						Assessed with the adult lead model					
Mercury	1.68	1.58E-07	1.68E-09	1.66E-08	1	0.07	100%	0.0003	2.1E-05	5.3E-04	7.9E-04	0.001
Silver	407	3.82E-05	4.07E-07	4.01E-06	1	0.04	100%	0.005	0.0002	7.7E-03	2.0E-02	0.03
Zinc	798	7.49E-05	7.98E-07	7.87E-06	1	1	100%	0.3	0.3	2.5E-04	2.6E-05	0.0003
Park worker RME noncancer hazard from surface water												
Beryllium	23.9		3.82E-06	3.59E-08	1	0.007	100%		1.4E-05		2.6E-03	0.003
Iron	1304		2.09E-04	1.96E-06	1	1	100%		0.7		2.8E-06	3E-06
										Total CTE hazard for park worker = 0.1		

TABLE 5.19
Quantitative Risk and Hazard Characterization for Construction/Excavation Workers (RME)

Construction Worker RME Cancer Risks (Oral, Dermal, and Inhalation)

Chemical	Surface Soil EPC (mg/kg)	Oral Intake (mg/kg/day)	DA_{event} (mg/cm³)	Dermal Intake (mg/kg/day)	Dermal Absorption (ABS_d) (Unitless)	GI Absorption (ABS_{GI}) (Unitless)	Oral Bioavailability (Unitless)	CSF_{oral}	CSF_{dermal}	Oral Risk/HI	Dermal Risk/HI	Total Risk/HI
Arsenic	1121	1.24E-05 / Inh. conc. 6.65E-10	1.01E-05	7.11E-07	0.03	0.95	31%	1.5	1.6	5.8E-06 / IUR 4.3E-03	3.5E-07	6.1E-06 / 2.9E-12

Construction worker RME cancer risk = 6E-06

Construction worker RME noncancer hazard from mixed soil (oral and dermal)

Chemical	Surface Soil EPC (mg/kg)	Oral Intake (mg/kg/day)	DA_{event} (mg/cm³)	Dermal Intake (mg/kg/day)	Dermal Absorption (ABS_d) (Unitless)	GI Absorption (ABS_{GI}) (Unitless)	Oral Bioavailability (Unitless)	RfD_{oral}	RfD_{dermal}	Oral HQ	Dermal HQ	Total HI
Arsenic	1121	8.69E-04	1.01E-05	4.9E-05	1	0.95	31%	3E-04	2.9E-04	8.98E-01	5.41E-02	1
Lead	1006					Assessed with the adult lead model							
Manganese	3043	2.36E-03	9.13E-06	4.5E-05	1	1	100%	0.14	0.14	1.68E-02	3.22E-04	0.02
Silver	238.1	1.85E-04	7.14E-07	3.52E-06	1	0.04	100%	0.005	0.0002	3.69E-02	1.76E-02	0.1

Construction worker RME noncancer hazard from mixed soil (inhalation)

Chemical	Surface Soil EPC (mg/kg)	Inh. conc.	RfC	Inh. HQ	Total HI
Arsenic	1121	4.65E-08	1.50E-05	3.10E-03	0.003
Lead	1006		Assessed with the adult lead model		
Manganese	3043	1.26E-07	5.00E-05	2.53E-03	0.003
Silver	238.1	9.88E-09	NA		

Total RME hazard for construction worker = 1

TABLE 5.20

Quantitative Risk and Hazard Characterization for Construction/Excavation Workers (CTE)

Construction worker CTE cancer risks (oral, dermal, and inhalation)

Chemical	Surface Soil EPC (mg/kg)	Oral Intake (mg/kg/day)	DA_{event} (mg/cm³)	Dermal Intake (mg/kg/day)	Dermal Absorption (ABS_d) (Unitless)	GI Absorption (ABS_{GI}) (Unitless)	Oral Bioavailability (Unitless)	Toxicity Values CSF_{oral}	CSF_{dermal}	Oral Risk/HI	Dermal Risk/HI	Total Risk/HI
Arsenic	1121	1.25E-06	3.36E-06	3.95E-08	0.03	0.95	4.1%	1.5	1.6	7.71E-08	2.59E-09	7.97E-08
		Inh. conc. 6.65E-10								IUR 4.30E-03		4.76E-13

Construction worker CTE cancer risk = 8E-08

Construction worker CTE noncancer hazard from mixed soil (oral and dermal)

Chemical	Surface Soil EPC (mg/kg)	Oral Intake (mg/kg/day)	DA_{event} (mg/cm³)	Dermal Intake (mg/kg/day)	Dermal Absorption (ABS_d) (Unitless)	GI Absorption (ABS_{GI}) (Unitless)	Oral Bioavailability (Unitless)	Toxicity Values RfD_{oral}	RfD_{dermal}	Oral HQ	Dermal HQ	Total HI
Arsenic	1121	8.77E-05	3.36E-06	2.76E-06	1	0.95	4.1%	3E-04	2.9E-04	1.20E-02	3.98E-04	0.01
Lead	1006	Assessed with the adult lead model										
Manganese	3043	2.38E-04	3.04E-06	2.50E-06	1	1	100%	0.14	0.14	1.70E-03	1.79E-05	0.002
Silver	238.1	1.86E-05	2.38E-07	1.96E-07	1	0.04	100%	0.005	0.0002	3.73E-03	9.78E-04	0.005

Construction worker CTE noncancer hazard from mixed soil (inhalation)

Chemical	Surface Soil EPC (mg/kg)	Inh. conc.	Toxicity Values RfC	Inh. HQ	Total HI
Arsenic	1121	7.76E-09	1.50E-05	5.17E-04	0.0005
Lead	1006	Assessed with the adult lead model			
Manganese	3043	2.11E-08	5.00E-05	4.21E-04	0.0004
Silver	238.1	1.65E-09	NA		

Total CTE hazard for construction worker = 0.02

The RME risk estimates for the various receptors were as follows:

Receptor	Cancer Risk	Hazard Index
Adult park visitor	RME = 6E-06; CTE = 2E-07	RME = 0.2; CTE = 0.004
Child park visitor	RME = 2E-04; CTE = 2E-07	RME = 1; CTE = 0.2
Park service worker	RME = 2E-04; CTE = 3E-06	RME = 1; CTE = 0.1
Construction worker	RME = 6E-06; CTE = 8E-08	RME = 1; CTE = 0.02

The major contributors to hazard indices were arsenic and silver. The critical toxic effects of arsenic are hyperpigmentation, keratosis, and possible vascular complication; hence, the vascular system is the target of the noncancer effects of arsenic. The critical toxic effect of silver is argyria, a medically benign but permanent bluish-gray discoloration of the skin. Although the deposition of silver is permanent, it is not associated with any adverse health effects. Hence, the HI could be segregated, and when arsenic only was included, all HIs were less than unity.

Risk from Lead Exposure to Children

Because exposure to lead occurs from multiple media, the IEUBK model is used to assess the risks of lead to children.[51] The endpoint in the IEUBK model is the proportion of a hypothetical population of children 6 years old and under with blood lead concentrations greater than 10 μg/dL. The regulatory target is to have 95% or more of the hypothetical population with blood lead concentrations less than 10 μg/dL.

Lead exposure to children 6 years or less from the site was assessed as a time-weighted average between the site and their home.[52] The IEUBK model was executed using default values for all inputs save two. Children were assumed to visit the site for 24/365 days out of the year (Table 5.13). The outdoor soil lead concentration was 277 mg/kg based on a weighted average between 200 mg/kg, the default value used in the model, and 1363 mg/kg, the average in surface soil at the site. A second source of lead in the "multiple source analysis" option used was "second home dust." The default value of the conversion factor for soil lead to indoor dust lead is 0.7 and the second home dust concentration used was 0.7 × 1363 mg/kg or 954 mg/kg. Exposure to this second source would occur 24/365 days or 6.6% of the time. The average is used because all sources of uncertainty and variability—including that in the exposure point concentration—are included in the GSD, which is included in the current implementation of the IEUBK model.[52]

The percentage of the hypothetical population of children modeled this way that had blood lead concentrations less than 10 μg/dL is 98.3%. Hence, lead concentrations in soil at the site are below levels of regulatory concern for children.

Risk from Lead Exposure to Adults

USEPA's Adult Lead Model was used along with a lead soil concentration of 1363 mg/kg. This value was the higher of the two average concentrations from either surface soil or mixed soil and, thus, would be protective of any receptors evaluated.

The model spreadsheet and guidance can be obtained at http://www.epa.gov/ superfund/health/contaminants/lead/products.htm#alm. Updated values of the GM blood lead concentration in women of child-bearing age and the GSD were used in the model.[53,54] There was a 98.8% probability that fetal blood lead concentration would be less than the target of 10 μg/dL. Hence, lead concentrations in soil at the site are not of concern for adults.

As with children, all sources of uncertainty and variability are considered by EPA to be addressed in the GSD.[52,54]

Characterization of Uncertainty

The risk estimates presented here are conservative estimates of potential risks associated with potential exposure to constituents detected in media at the former gold mine site. Uncertainty is inherent in the risk assessment process, and a brief discussion of these uncertainties is presented in this section. Each of the three basic building blocks for risk assessment (monitoring data, exposure scenarios, and toxicity values) and for the exposure assessment (parameters, models, and scenarios) contribute to the overall uncertainty.

Samples collected during site investigations were intended to characterize the nature and extent of potential contamination at the site. Subsequently, most of the samples were collected from locations selected in a directed manner to accomplish this goal. Sampling locations selected in this way provide considerable information about the site but tend to be concentrated in areas of higher levels of contamination. Therefore, data from sampling locations selected in this manner tend to overestimate constituent concentrations representative of the potential exposure area. The samples were obtained to support contaminant characterization and ultimately, remediation, and not to estimate human exposure. Hence, this risk assessment is based on the assumption that the available monitoring data adequately describe the occurrence of constituents in media at the site.

Environmental sampling itself introduces uncertainty. This source of uncertainty can be reduced through a well-designed sampling plan, use of appropriate sampling techniques, and implementation of laboratory data validation and quality assurance/ quality control (QA/QC). The most likely source of uncertainty regarding concentration is the possible mismatch between the assumptions about receptor behavior.

The toxicity values and other toxicological information used in this report are likewise associated with significant uncertainty. In addition, humans are different than laboratory animals. In addition, the effects shown by the animals in the high-dose studies are often very different than the effects reported by humans in parallel epidemiology studies. As indicated, arsenic is the risk driver for both the child visitor and the park service worker, both exposure scenarios with hazard indices above one. The RfD for arsenic was estimated from human data on circulatory effects.

Range of Uncertainty Based on Bioavailability

As an exercise, risk estimates were calculated using arsenic bioavailability estimates of 1% and 100% for those receptors showing risk or hazard above regulatory thresholds. For the park service worker, the range of cumulative risk estimates based on this bioavailability range is 4E-06 to 3E-04 and the range of cumulative hazard indices is 1–5. For the child visitor, the range of cumulative hazard indices is 0.9–4. Because the arsenic at the site is very likely in the form of arsenopyrite, 1% is more likely as an accurate value of the bioavailability of arsenic at the site. Hence, if this value of 1% were used rather than 31%, then all receptors in all use scenarios would have risk or hazard estimates below regulatory criteria.

Comparison to Background Risks

As an additional exercise, exposure point concentrations were calculated for arsenic and cobalt from background surface soil samples. Arsenic and cobalt show the highest HQs in surface soil. The background samples used are shown in Table 5.2. It should also be noted that the background locations are described as "iron-rich" or as "pyrite." Hence, it is likely that the arsenic in the general area of the former gold mine site exists as arsenopyrite and has very low bioaccessibility.

Exposure point concentrations were calculated for COPCs in surface soil. The 95% UCL concentrations for arsenic, cadmium, and zinc were higher than site concentrations. It is left as an exercise at the end of the chapter for the reader to calculate risks from background soil for all receptors.

Conclusions

The child park visitor and the park service worker were the only receptors showing unacceptable risks. The medium of concern is surface soil. Target organ separation according to USEPA (1989) indicated that only the park service worker would experience noncancer risks from arsenic above levels of concern.

Receptor	HI (Total)	HI (By Target Organ, as Only)
Child park visitor	1.7	0.8
Park service worker	2.3	1.3

Because the arsenic at the site appears to be predominantly in the form of arsenopyrite, which has a very low bioaccessibility, it is likely that these risks are overestimated and risks at the site are very likely below regulatory criteria.

Risk Characterization and Risk Management

When the results of the risk assessment were presented to the PM, she immediately noted that for the child visitor and the park worker, there was a huge difference in RME and CTE cancer risk estimates from arsenic. The explanation was given that the largest contributors to this difference were the RME and CTE bioavailability estimates, which were approximately an order of magnitude apart. The risk assessor had come to this meeting prepared and showed his spreadsheet calculations with the overhead projector. The PM asked what the RME cancer risk would be for 4.1% bioavailability and the risk assessor performed this calculation by changing a single value in the spreadsheet.* The risk dropped to 4E-05, now well within the target risk range of 10^{-6} to 10^{-4}.[41]

The PM asked about the reason for the low bioavailability value. The risk assessor was able to show her the high values of arsenic in soil and the low values in surface water and groundwater. He noted that the same low bioavailability values had been measured for other gold mine soils and was likely due to poor solubility of the arsenic. The risk assessor explained that the knowledge base about arsenic bioavailability included historical knowledge of the bioavailability of metals at mining sites, guidance from USEPA, and two laboratory animal studies—all indicated that arsenic in mining soils had very low bioavailability. The risk assessor also noted that a relatively inexpensive validated in vitro assay for bioavailability was available.[55]

"Makes sense to me," was the PM's comment. She then asked if 4.1% was appropriate to use for the RME calculation. The risk assessor said that such use likely was appropriate. Her last question was about the solubility and the risk assessor was able to explain his estimation of the five order of magnitude difference in solubility that was shown earlier in this chapter.

"Based on that," said the PM, "4.1% is probably an overestimate of the bioavailability." The risk assessor agreed.

"I can sell this," she said. "The park workers were complaining because of their perception of the arsenic risk. They wouldn't get out of their trucks anywhere near the tailings pile. Bunch o' whiners!"

"The animals don't seem to mind it," said the risk assessor. "There's a photo in the ecological risk assessment of a ground squirrel on the tailings pile. Some of them have built burrows in the tailings. The arsenic levels in those tailings is over 7500 parts per million and it sure doesn't seem to have hurt the animals."

"Can I get the photo?" asked the PM. "I can show it at the next meeting. Maybe it'll stop those slackers from whining."

"In my day," said the risk assessor, "we would have called them goldbricks, which seems kind of fitting for a gold mine."

RISK ASSESSMENT FOR CONSUMPTION OF CONTAMINATED FISH IN A COASTAL COMMUNITY

In this example, marine fish contain bioaccumulative chemicals produced by legacy industrial processes. These fish are consumed by coastal anglers and others. The conceptual model for the site is shown in Figure 5.2. Because only a single exposure pathway is considered, this conceptual model is much simpler than that for the former gold mine example.

Chlorine has been a commodity chemical for over 100 years. Most often, chlorine is produced by the chlor-alkali process by electrolytic decomposition of brine

---

* These spreadsheets are provided for you and are available at the publisher's website (http://www. crcpress.com/product/isbn/9781466598294). For an exercise at the end of this chapter, you will repeat this calculation.

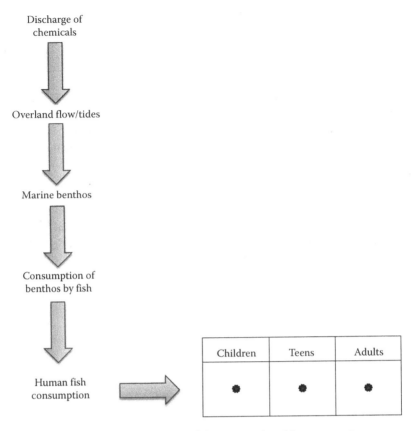

	Children	Teens	Adults
	●	●	●

FIGURE 5.2 Conceptual site model for fish consumption risk assessment.

or seawater. One of the past uses of chlorine was the production of pesticides, specifically the pesticide toxaphene, first introduced in 1945. Toxaphene was widely used as an insecticide on cotton, soybeans, and corn.[56]

Toxaphene was produced by passing chlorine gas through camphene derived from pine stumps in the presence of UV radiation.[57] This process resulted in the nonselective addition of chlorine to mainly bornane molecules, leading to a highly complex technical mixture of chlorinated camphenes and bornanes.[58] The total worldwide use of toxaphene is estimated to be 1.3M tons of which about 40% is used in the United States, primarily in the southeast.[58] Toxaphene was deregistered in the United States in 1982 and banned in 1990. The most persistent toxaphene congeners in humans are known as p-26, p-50, and p-62. The sum of the concentrations of these three congeners is referred to as Σ3PC and provided the basis for a toxicity criterion developed in 2006.[59]

The chlorine used to produce toxaphene was produced at a nearby chlor-alkali plant. Both PCBs and mercury are used in the electrolytic cells needed for the chlor-alkali process. Toxaphene, mercury, and Aroclor 1268, a high molecular weight PCB mixture, were released during chlorine production and toxaphene manufacture.

All three chemicals bioaccumulate in fish and this example demonstrates the risk assessment of consumption of fish contaminated with these three chemicals. The last number in the Aroclor designation represents the weight percentage of chlorine in the PCB mixture. The most highly chlorinated mixture is Aroclor 1268 with 68% chlorine by weight.

The high molecular weight congeners tend to be less toxic than other PCB congeners. EPA does not have toxicity criteria for Aroclor 1268; because the toxicity of this mixture is lower than other PCB mixtures; both a CSF and RfD specific to Aroclor 1268 have been developed and are published in the scientific literature.[59,60]

Selection of COPCs

The site where toxaphene was produced is located on the Atlantic coast in the southeastern United States. Manufacture of the pesticide occurred from 1948 until 1980. In 1997, 2001, and 2005, fish were sampled and toxaphene tissue residues analyzed. Full datasets for toxaphene congeners in fish, including p-26, p-50, and p-62, were available from samples obtained in 1997 and 2005.

Fish samples are analyzed about every 2 years for mercury and PCBs as part of routine monitoring for fish advisories; samples are available from 2002 forward. Datasets for Aroclor 1268 and mercury were available from samples obtained from 2002 through 2006 (Tables 5.21 through 5.23).

Because the site is in a relatively rural location with no other industry around, toxaphene, Aroclor 1268, and mercury were the only chemicals of potential concern. Hence, in this example, the exposure and toxicity assessments will be emphasized rather than COPC selection.

Reduction in Total Toxaphene Concentrations in All Fish

From data obtained in 1997, prior to any remediation, the concentration for total toxaphene residues from fish was 5250.8 ± 6531.3 µg/kg (mean ± standard deviation).[62] In fish collected in 2001, the concentration of total toxaphene residues was 1400 ± 3500 µg/kg.[63] During the time between these two sampling events, dredging had been performed in a canal to remove sediment with the highest toxaphene concentrations—the likely reason for these reductions.

In addition to the reduction in total toxaphene (TTX) residues, Σ3PC concentrations in all fish decreased from a mean of 222.0 µg/kg in 1997 to a mean of 21.0 µg/kg in 2005, about a 10-fold reduction. The average percentage of Σ3PC in TTX in the 1997 fish samples was 4.14%. The average percentage of Σ3PC in TTX in fish in the 2005 samples was 1.14%, an approximate 75% reduction. Hence, both the concentration of TTX and the percentage of persistent congeners decreased. Had single congener analysis been performed on the 2001 samples, a trend analysis might have revealed a decreasing concentration term; however, such analysis was not performed.[63]

Because p-26, p-50, and p-62 are the congeners most resistant to metabolism, biotransformation, or abiotic weathering, the reduction in both the concentration of Σ3PC in fish and the percentage of Σ3PC in total toxaphene residues from 1997 to 2005 is somewhat surprising.[59] The expectation is that the percentage of Σ3PC

TABLE 5.21
Concentrations of Aroclor 1268, Methylmercury, and Toxaphene as Σ3PC in Fish

	Atlantic Croaker							Red Drum					
Aroclor 1268 (mg/kg)		Methylmercury (mg/kg)		Toxaphene (µg/kg)			Aroclor 1268 (mg/kg)		Methylmercury (mg/kg)		Toxaphene (µg/kg)		
					TTX	Σ3PC						TTX	Σ3PC
2002 0.17		2002 0.067		1997	5,300	163.16	2002 0.16		2002 0.226		1997	490	25.93
2002 0.20		2002 0.462		1997	11,000	334.19	2002 0.33		2002 0.550		1997	1400	57.68
2002 0.96		2002 0.156		1997	11,000	296.69	2002 0.32		2002 0.230		1997	79	2.25
2002 0.60		2002 0.208		1997	6,300	207.69	2002 <0.05		2002 0.111		2005	596.0	8.81
2002 1.56		2002 0.168		2005	350	1.0	2002 0.19		2002 0.440		2005	955.5	25.09
2002 1.02		2002 0.112		2005	419	1.0	2002 <0.05		2002 0.312		2005	1774.5	38.32
2002 0.64		2002 0.195		2005	1,085	3.0	2002 <0.05		2002 0.156		2005	361.0	3.71
2002 0.49		2002 0.351					2002 <0.05		2002 0.242		2005	221.0	0.45
2002 0.58		2002 0.139					2002 0.17		2002 0.195		2005	937.0	15.61
2002 0.71		2002 0.174					2002 <0.05		2002 0.276		2005	159.0	4.93
2002 0.90		2002 0.149					2002 <0.05		2002 0.165		2005	159.5	8.12
2002 0.81		2002 0.170					2002 <0.05		2002 0.136		2005	209.0	4.22
2002 2.24		2002 0.228					2002 <0.05		2002 0.062		2005	85.0	5.88
2002 0.87		2002 0.233					2002 0.46		2002 0.240		2005	75.5	3.28
2002 0.81		2002 0.155					2002 <0.05		2002 0.253		2005	59.5	6.17

(continued)

TABLE 5.21 (continued)
Concentrations of Aroclor 1268, Methylmercury, and Toxaphene as Σ3PC in Fish

Atlantic Croaker

Aroclor 1268 (mg/kg)		Methylmercury (mg/kg)		Toxaphene (µg/kg) TTX	Σ3PC
2002	0.39	2002	0.018		
2002	<0.05	2002	0.210		
2002	1.04	2002	0.096		
2002	1.06	2002	0.147		
2005	2.10	2005	0.522		
2006	0.58	2006	0.267		
2006	1.40	2006	0.163		
2006	0.36	2006	0.172		

Red Drum

Aroclor 1268 (mg/kg)		Methylmercury (mg/kg)		Toxaphene (µg/kg) TTX	Σ3PC
2005	0.10	2005	0.377		
2005	0.03	2005	0.173		
2005	0.11	2005	0.266		
2005	0.05	2005	0.173		
2005	0.24	2005	0.398		
2005	0.05	2005	0.146		
2005	0.07	2005	0.255		

Note: Please note the difference in units between the measurements.

TABLE 5.22

Concentrations of Aroclor 1268, Methylmercury, and Toxaphene as Σ3PC in Fish

Southern Kingfish						Spot					
Aroclor 1268	Mercury	Toxaphene (μg/kg)				Aroclor 1268	Mercury	Toxaphene (μg/kg)			
			TTX	Σ3PC					TTX	Σ3PC	
2002 0.50	2002 0.235	1997	2100	94.90		2002 0.56	2002 0.073	1997	23,000	1210.20	
2002 0.31	2002 0.097	1997	230	10.09		2002 0.12	2002 0.115	1997	13,000	749.09	
2002 0.68	2002 0.230	1997	370	16.19		2002 1.09	2002 0.050	1997	7,600	396.79	
2002 0.51	2002 0.368	1997	840	38.08		2002 0.69	2002 0.120	1997	5,700	252.43	
2002 0.68	2002 0.146	2005	546	5.18		2002 3.07	2002 0.166	1997	12,000	555.81	
2002 1.19	2002 0.297	2005	3038	45.40		2002 1.33	2002 0.136	1997	16,000	520.86	
2002 0.48	2002 0.430	2005	2417	84.47		2002 1.31	2002 0.186	1997	2,400	82.99	
2002 0.48	2002 0.225	2005	1032	0.53		2002 <0.05	2002 0.047	1997	760	29.68	
2002 0.12	2002 0.288	2005	370	3.67		2002 <0.05	2002 0.086	1997	1,300	32.09	
2002 0.18	2002 0.483	2005	402	2.13		2002 0.80	2002 0.077	2005	9,160	117.10	
2002 1.25	2002 0.260	2005	446	1.89		2002 0.05	2002 0.083	2005	9,396	127.63	
2002 0.59	2002 0.240	2005	196	3.73		2002 1.30	2002 0.084	2005	808	7.61	
2002 0.35	2002 0.350	2005	746	6.00		2002 0.96	2002 0.093	2005	910	11.41	
2002 0.53	2002 0.275	2005	85	0.55		2002 <0.05	2002 0.061	2005	191	1.25	
2002 0.39	2002 0.299					2002 0.93	2002 0.059	2005	260	0.50	

(continued)

TABLE 5.22 (continued)
Concentrations of Aroclor 1268, Methylmercury, and Toxaphene as Σ3PC in Fish

Southern Kingfish

Aroclor 1268		Mercury		Toxaphene (µg/kg)	
				TTX	Σ3PC
2002	0.70	2002	0.456		
2002	0.28	2002	0.625		
2002	<0.05	2002	0.275		
2002	1.34	2002	0.448		
2002	<0.05	2002	0.315		
2002	<0.05	2002	0.352		
2002	0.31	2002	0.308		
2002	0.30	2002	0.325		
2002	0.90	2002	0.506		
2002	0.80	2002	0.450		
2005	0.89	2005	0.975		
2005	0.15	2005	0.382		
2005	0.20	2005	0.189		
2005	0.18	2005	0.743		
2005	0.34	2005	0.252		

Spot

Aroclor 1268		Mercury		Toxaphene (µg/kg)		
					TTX	Σ3PC
2002	0.68	2002	0.056			
2002	<0.05	2002	0.090			
2002	0.32	2002	0.067			
2002	<0.05	2002	0.069			
2002	<0.05	2002	0.066			
2002	<0.05	2002	0.095			
2002	1.49	2002	0.125			
2002	0.21	2002	0.095			
2002	1.19	2002	0.042			
2002	<0.05	2002	0.059	2005	3,272	106.91
2002	<0.05	2002	0.059			
2002	<0.05	2002	0.128			
2005	0.63	2005	0.250			
2005	0.06	2005	0.300			

Note: Aroclor 1268 and methylmercury are in mg/kg and toxaphene Σ3PC is in µg/kg.

TABLE 5.23

Concentrations of Aroclor 1268, Methylmercury, and Toxaphene as Σ3PC in Fish

Spotted Sea Trout

Aroclor 1268		Mercury		Toxaphene (µg/kg)		
					TTX	Σ3PC
2002	0.32	2002	0.460	1997	3300	193.29
2002	0.98	2002	0.476	1997	4400	117.39
2002	<0.05	2002	0.220	1997	3600	166.76
2002	0.82	2002	0.288	1997	940	30.8
2002	0.50	2002	0.360	1997	1300	37.84
2002	0.36	2002	0.494	1997	1300	55.99
2002	0.28	2002	0.350	1997	110	5.4575
2002	0.50	2002	0.384	2005	1077	37.62
2002	0.15	2002	0.408	2005	1478	32.96
2002	0.17	2002	0.400	2005	686	31.50
2002	0.83	2002	0.312	2005	1249	22.66
2002	0.11	2002	0.418	2005	1134	9.92
2002	0.35	2002	0.350	2005	566	8.61
2002	0.55	2002	0.406	2005	179	7.77
2002	0.78	2002	0.416	2005	258	7.24
2002	<0.05	2002	0.252	2005	234	5.77
2002	<0.05	2002	0.207	2005	116	3.79
2002	<0.05	2002	0.252	2005	128	2.78
2002	<0.05	2002	0.288	2005	311	0.94

Striped Mullet

Aroclor 1268		Mercury		Toxaphene (µg/kg)		
					TTX	Σ3PC
2002	1.30	2002	0.032	1997	2,000	125.72
2002	1.18	2002	0.019	1997	1,300	59.65
2002	1.32	2002	0.011	1997	2,100	116.57
2002	2.70	2002	0.022	1997	8,300	299.52
2002	1.57	2002	0.021	1997	26,000	1514
2002	0.10	2002	<0.01	1997	17,000	375.105
2002	0.33	2002	0.019	2005	970	50.87
2002	1.05	2002	0.025	2005	1,800	95.695
2002	2.28	2002	0.018	2005	4,100	240.54
2002	1.98	2002	0.026	2005	4,511	40.56
2002	4.42	2002	0.019	2005	3,285	27.99
2002	2.02	2002	0.029	2005	2,227	37.56
2002	10.50	2002	0.030	2005	524	1.37
2002	0.24	2002	0.030	2005	231	0.50
2002	0.26	2002	0.030	2005	242	0.50
2002	<0.05	2002	0.014	2005	604	4.07
2002	<0.05	2002	0.022	2005	342	1.40
2002	<0.05	2002	0.014	2005	126	1.25
2002	<0.05	2002	0.012			

(continued)

TABLE 5.23 (continued)

Concentrations of Aroclor 1268, Methylmercury, and Toxaphene as Σ3PC in Fish

	Spotted Sea Trout						Striped Mullet				
Aroclor 1268		Mercury		Toxaphene (µg/kg)		Aroclor 1268		Mercury		Toxaphene (µg/kg)	
				TTX	Σ3PC					TTX	Σ3PC
2002	<0.05	2002	0.288			2002	<0.05	2002	0.012		
2002	<0.05	2002	0.235			2002	0.50	2002	<0.01		
2002	<0.05	2002	0.288			2002	0.29	2002	<001		
2002	0.19	2002	0.336			2002	<0.05	2002	0.017		
2002	<0.05	2002	0.220			2002	1.29	2002	0.006		
2002	<0.05	2002	0.280			2002	1.00	2002	0.014		
2002	<0.05	2002	0.484			2002	3.78	2002	0.023		
2002	<0.05	2002	0.624			2002	0.19	2002	0.022		
2002	<0.05	2002	0.460			2002	0.27	2002	0.020		
2005	0.18	2005	0.428			2005	3.40	2005	0.041		
2005	0.19	2005	0.362			2005	2.20	2005	0.049		
2005	0.16	2005	0.641			2005	2.00	2005	0.051		
2005	0.43	2005	0.759			2005	1.70	2005	0.026		
2005	0.61	2005	0.941			2005	0.71	2005	0.026		
2005	0.20	2005	0.564			2005	0.03	2005	0.013		
2005	0.11	2005	0.300			2005	0.43	2005	0.014		

Note: Units as specified earlier.

in total toxaphene residues would increase because other congeners would be metabolized or degraded.[64–66]

One possible interpretation of this result is that the dredging perturbed the ecosystem in such a way that the fish sought prey items from other locations that were less contaminated. Supporting the hypothesis that the reduction in Σ3PC is due to changes in feeding strategies or locations of some of the fish species are concentrations in red drum changed very little. Red drum move in and out with the tides and feed over a wide area. This type of feeding strategy will tend to dilute the amount of Σ3PC in the prey items with high concentrations of toxaphene. Large differences in individual concentrations were observed in fish species will smaller home ranges, such as striped mullet and spot (Tables 5.21 through 5.23). Had congener analysis been available for more than two instances, the reduction in Σ3PC might have been taken into account in the risk assessment in a quantitative way; this was not possible with only two time points.

Exposure Assessment: Concentrations in Fish

Tables 5.21 through 5.23 show the data for the three COPCs in the various species of fish.

Table 5.24 shows COPC selection with the statistics and exposure point concentrations for the various species.

Exposure Assessment: Human Factors

This portion of the exposure assessment deals with the choice to eat fish, which species are consumed, the portion size of fish, and how often people consume self-caught fish.

Proportion of Species Consumed

The Marine Recreational Fisheries Statistics Program of the Office of Science and Technology within NOAA conducts the Marine Recreational Fisheries Statistics Survey (MRFSS) to produce catch, effort, and participation estimates and to provide biological, social, and economic data. USEPA made use of these data obtained from 1986 to 1993 to determine estimates of consumption of marine fish presented in the *Exposure Factors Handbook*.[38,40]

The data (Tables 5.21 through 5.23) consist of analytical results from fish species likely to be consumed by humans (e.g., red drum, spotted sea trout) as well as those less likely to be consumed (e.g., spot croaker, striped mullet). The likelihood of consumption of a given species is based on a relative species harvest analysis of the MRFSS data from 2001 through 2005 (Table 5.25).

The MRFSS consists of a telephone survey and an intercept or creel survey conducted on 2-month intervals. These 2-month intervals are called waves. The period of two months was chosen because it was the maximum time for easy recall of past fishing trips. The intercept data from 2001 through 2005 was used here.

A recent study by the NAS revealed that the MRFSS was considerably flawed in its execution and the data generated are inaccurate and biased.[67] The criticisms by the NAS were several: (1) sampling and statistical issues, such as failure to

TABLE 5.24
COPC Selection for Contaminants in Fish Tissue

Constituent/ Species	Frequency Det. / Total	Percent Detects (%)	Range of SQLs Min – Max	Range of Detects Min – Max	Average Detect	Screening Level*	Alternate Screening Level	COPC (Y/N)?	UCL	Method (PROUCL)
Atlantic Croaker										
Aroclor 1268	22 / 23	96	0.05–0.05	0.166–2.244	0.885	0.0016	0.056	Y	1.044	95% KM (BCA) UCL
Total toxaphene	7 / 7	100		0.35–11	5.065	0.0029	NA	Y	8.506	95% Student's t UCL
Toxaphene as Σ3PC	7 / 7	100		0.001–0.334	0.144	NA	0.027	Y/Y	0.25	95% Student's t UCL
Mercury	23 / 23	100		0.018–0.522	0.198	0.014	NA	Y	0.248	95% approx. gamma UCL
Red Drum										
Aroclor 1268	13 / 22	59	0.05–0.05	0.03–0.456	0.175	0.0016	0.056	Y	0.164	95% KM (t) UCL
Total toxaphene	15 / 15	100		0.0595–1.775	0.504	0.0029	NA	Y	0.851	95% approx. gamma UCL
Toxaphene as Σ3PC	15 / 15	100		0.00045–0.058	0.014	NA	0.027	Y/N	0.0244	95% approx. gamma UCL
Mercury	22 / 22	100		0.062–0.55	0.245	0.014	NA	Y	0.287	95% Student's t UCL
Southern Kingfish										
Aroclor 1268	30 / 33	91	0.05–0.05	0.12–1.4	0.579	0.0016	0.056	Y	0.649	95% KM (BCA) UCL
Total toxaphene	14 / 14	100		0.0845–3.038	0.915	0.0029	NA	Y	1.504	95% approx. gamma UCL
Toxaphene as Σ3PC	14 / 14	100		0.00053–0.0949	0.0223	NA	0.027	Y/Y	0.0483	95% approx. gamma UCL

Analyte									Method
Mercury	100		0.097–1.13	0.386	0.014	NA	Y	0.452	95% approx. gamma UCL
Spot									
Aroclor 1268	62	0.05–0.05	0.062–3.07	0.93	0.0016	0.056	Y	0.838	95% KM (% bootstrap) UCL
Total toxaphene	100		0.191–23	6.61	0.0029	NA	Y	11.88	95% approx. gamma UCL
Toxaphene as Σ3PC	100		0.0005–1.21	0.263	0.0029	0.027	Y/Y	0.623	95% Adjusted gamma UCL
Mercury	100		0.042–0.3	0.101	0.014	NA	Y	0.12	95% Student's t UCL
Spotted Sea Trout									
Aroclor 1268	62	0.05–0.05	0.089–0.98	0.383	0.0016	0.056	Y	0.342	95% KM (% bootstrap) UCL
Total toxaphene	100		0.11–4.4	1.177	0.0029	NA	Y	1.855	95% approx. Gamma UCL
Toxaphene as Σ3PC	100		0.00094–0.193	0.041	0.0029	0.027	Y/Y	0.0713	95% approx. gamma UCL
Mercury	100		0.207–0.941	0.39	0.014	NA	Y	0.434	95% approx. gamma UCL
Striped Mullet									
Aroclor 1268	84	0.05–0.05	0.027–10.5	1.615	0.0016	0.056	Y	2.737	95% KM (Chebyshev) UCL
Total toxaphene	100		0.126–26	4.203	0.0029	NA	Y	7.732	95% approx. gamma UCL
Toxaphene as Σ3PC	100		0.0005–1.514	0.166	0.0029	0.027	Y/Y	0.415	95% Adjusted gamma UCL
Mercury	92	0.01–0.01	0.0063–0.0514	0.0231	0.014	NA	Y	0.0251	95% KM (BCA) UCL

Note: All units are mg/kg.

TABLE 5.25

Proportion of Species Caught as Percentage of Total Recreational Catch

Year	Wave	Atlantic Croaker (%)	Red Drum (%)	Southern Kingfish (%)	Spot Croaker (%)	Spotted Sea Trout (%)	Striped Mullet (%)
2001	1	0.00	0.00	0.00	0.00	0.00	0.00
2001	2	1.84	19.14	14.14	0.17	35.06	0.00
2001	3	0.00	0.00	42.36	0.00	20.08	0.00
2001	4	0.34	6.38	13.18	0.12	39.93	0.00
2001	5	0.05	37.40	19.05	0.04	30.22	0.00
2001	6	0.00	26.15	5.60	0.00	45.40	0.00
2002	1	0.00	0.00	0.00	0.00	0.00	0.00
2002	2	0.43	15.66	15.80	0.00	33.57	0.00
2002	3	0.00	2.13	13.13	0.00	31.02	0.00
2002	4	0.07	19.93	26.61	0.00	34.27	0.51
2002	5	0.63	31.13	12.23	0.06	43.52	0.00
2002	6	0.00	25.03	0.86	0.00	59.72	0.00
2003	1	0.00	25.64	9.27	0.00	52.03	0.00
2003	2	0.40	30.53	44.89	0.00	17.44	0.00
2003	3	5.84	7.70	10.26	0.12	21.77	22.10
2003	4	11.35	8.06	10.72	0.00	19.18	0.54
2003	5	1.58	37.60	8.05	0.00	39.01	0.00
2003	6	0.31	12.35	3.51	0.01	81.13	0.05
2004	1	0.00	0.00	0.00	0.00	0.00	0.00
2004	2	0.00	12.20	44.42	0.00	22.59	0.00
2004	3	2.44	11.23	24.19	0.00	20.43	0.00
2004	4	0.61	2.22	36.92	0.00	43.15	0.00
2004	5	0.00	33.55	20.59	0.15	21.61	0.00
2004	6	0.00	25.64	9.27	0.00	52.03	0.00
2005	1	0.00	0.00	0.00	0.00	0.00	0.00
2005	2	0.00	1.03	81.62	0.00	3.75	0.00
2005	3	0.00	3.05	29.72	0.00	30.45	0.00
2005	4	1.84	19.14	14.14	0.17	35.06	0.00
2005	5	0.00	43.54	8.10	0.07	38.90	0.00
2005	6	0.00	40.51	3.04	0.20	44.62	0.00
Mean		0.96	17.14	17.99	0.04	31.59	0.80
Median		0.00	14.01	12.68	0.00	32.29	0.00

include anglers with access to private property and the use of different survey methods in different states; (2) lack of reliable human dimensions data, such as social, behavioral, attitudinal, and economic data; (3) lack of coordination between federal and state personnel and "balkanization" of the survey methods and designs; and (4) lack of communication and outreach with anglers.

Even if the MRFSS data were reliable, its use would entail an estimation of consumption from the harvest—with considerable uncertainty in the results.[67,68] If MRFSS data from a sufficiently large area are included, it is appropriate to use MRFSS data to obtain the relative abundance of species in the overall catch. The proportion of various species in the MRFSS data would reflect both the relative abundance of various species and angler success. Table 5.25 shows the average percentage of the various species of fish caught by coastal Georgia anglers between 2001 and 2005 developed from the MRFSS data. The MRFSS data are available from the NOAA Fisheries website (http://www.st.nmfs.noaa.gov/st1/recreational/downloads.html) as SAS export files. However, data from 2004 and 2005 are no longer available, possibly because of the critique from the NAS.

Because the concentrations of COPCs are different in different species of fish, likely due to their feeding strategies, weighting the species-specific exposure point concentrations according to angler success and preferences is necessary for a more accurate exposure estimate. Inclusion of this information in the exposure calculation is made quite simple by the use of a fraction ingested (FI) term applied to individual fish species.[3]

Fish Consumption Rates

Site-specific information from the Brunswick area obtained by ATSDR and the Glynn County Health Department in 1997 was used to develop exposure assumptions for subsistence fish consumers.[69] The monthly frequency of self-caught fish meals was estimated using three categories: <1/week, about 1/week, and >1/week. These categories were considered to represent 3, 5, and 7 meals per month, respectively. Fish meal sizes were obtained from Table 16-111 in EPA's 2011 *Exposure Factors Handbook*.[40] The value of seven meals per month was multiplied by the 75th percentile meal size to represent the RME fish consumption rate; these values are 18, 30, and 31 g/day for children, adolescents, and adults, respectively. The value of three meals per month was combined with the 50th percentile meal size for children, adolescents, and adults to represent the central tendency fish consumption rate; these values are 5.8, 7.2, and 9.3 g/day for children, adolescents, and adults, respectively (Table 5.26).

TABLE 5.26
Exposure Assumptions

Receptor	Meal Size (Table 10-123, EFH 2011)		Meals per Month		Fish Consumption Rate (g/day)	
	RME	CTE	RME	CTE	RME	CTE
Child (2–5)	77	58	7	3	18	5.8
Adolescent (6–19)	127	72	7	3	30	7.2
Adult (20–60+)	134	93	7	3	31	9.3

Toxicity Assessment

Toxaphene

Once in the environment, toxaphene undergoes weathering by both biotic and abiotic means, and a reduction in the number of congeners occurs. The three most persistent congeners observed in fish, marine mammals, and humans are p26, p-50, and p-62 and their structures are shown in Figure 5.3.[59] In the late 1990, due to concern about human exposure to weathered toxaphene via fish consumption, the EU commissioned the MATT study; MATT stands for "Monitoring, Analysis, and Toxicity of Toxaphene."[70] EPA had previously developed an oral CSF for technical toxaphene of 1.1 per mg/kg/day based on liver tumor occurrence in mice. The toxicity assessment on IRIS was conducted prior to any focus on MOA. In addition, weathering produces changes in composition of toxaphene and the toxicity of weathered toxaphene was unknown.

Both technical toxaphene and weathered toxaphene produce rodent liver tumors via the same MOA as phenobarbital—activation of the CAR and resulting

p-26	2-endo, 3-exo, 5-endo, 6-exo, 8,8,10,10 octachlorobornane or 2-exo, 3-endo, 5-exo, 6-endo, 8,8,10,10 octachlorobornane	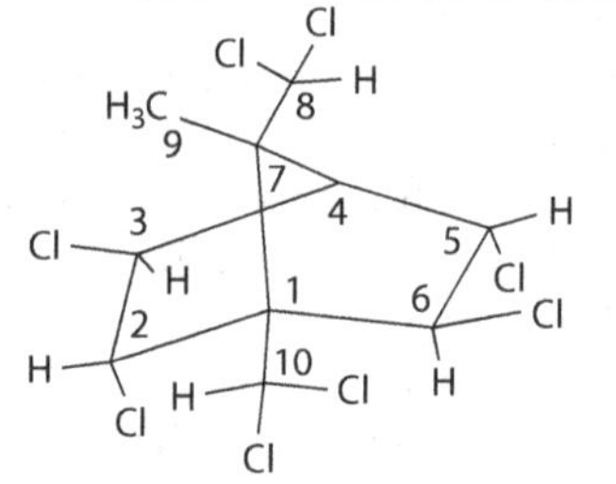
p-50	2-endo, 3-exo, 5-endo, 6-exo, 8,8,9,10,10 nonachlorobornane or 2-exo, 3-endo, 5-exo, 6-endo, 8,8,9,10,10 nonachlorobornane	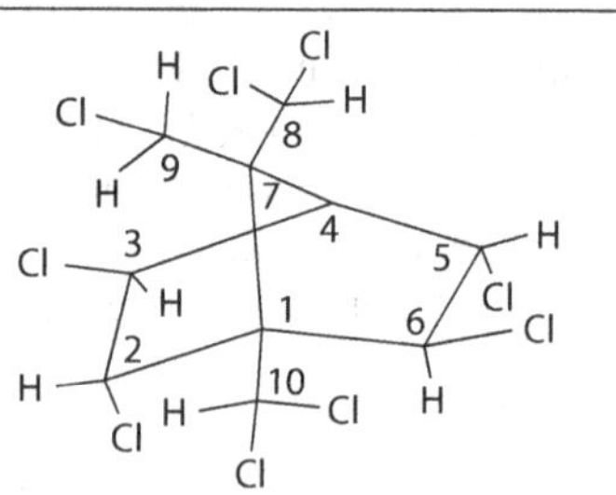
p-62	2,2,5,5,8,9,9,10,10 nonachlorobornane	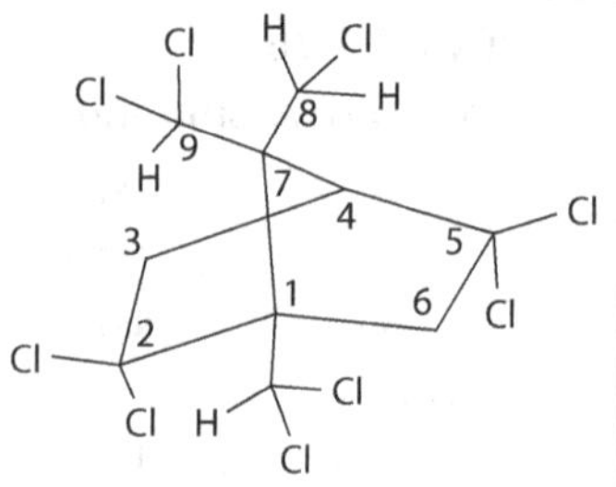

FIGURE 5.3 Structures of the three persistent toxaphene congeners.

induction of CYP enzymes leading to cytotoxicity with regenerative proliferation, a MOA not relevant to humans.[71,72]

The MATT report misinterpreted this CSF as an "RfD for carcinogenicity" and indicated that the US TDI value for a 60 kg individual would be 66 mg. The MATT derived a TDI of 0.41 mg/day for a 60 kg individual, equivalent to 0.007 mg/kg/day.[73,74]

In response to a request from a group of environmental activists living near the facility that produced toxaphene, EPA's Office of the Inspector General issued a memo indicating that the EPA regional office needed a way to conduct a risk assessment of weathered toxaphene—all that was needed was a toxicity criterion.[75] At the urging of the PM, one of the regional risk assessors teamed with Dr. Randy Manning, the state toxicologist for Georgia to whom this book is dedicated, to develop a toxicity criterion from the experiments used in the MATT report. An RfD with a value of 2E-05 mg/kg-day for Σ3PC was published soon thereafter.[59]

Aroclor 1268

PCBs were produced for use as dielectric fluids during the twentieth century. There are 209 different PCB congeners depending on the position and level of chlorination. A general PCB structure and a specific PCB congener are shown in Figure 5.4. The various PCB congeners produce various effects at the cellular and biochemical level; whether PCBs produce health effects in humans remain both controversial and unconfirmed.[76,77]

The congeners that produce the greatest toxicity are those in the middle range of chlorination, with four and five chlorines. Aroclor 1248 and Aroclor 1254 with 48% and 54% chlorine by weight are considered the most toxic of the PCB mixtures. Aroclor 1268 is 68% chlorine by weight and contains only a small percent-

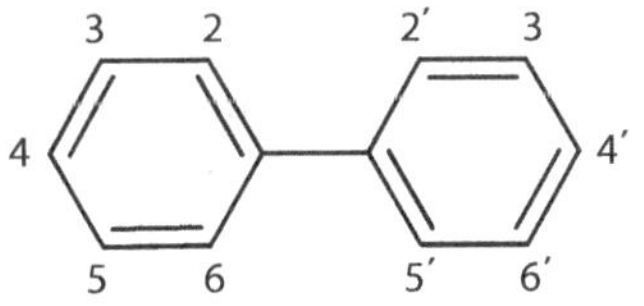

Chlorination positions on biphenyl for nomenclature of PCBs
(a)

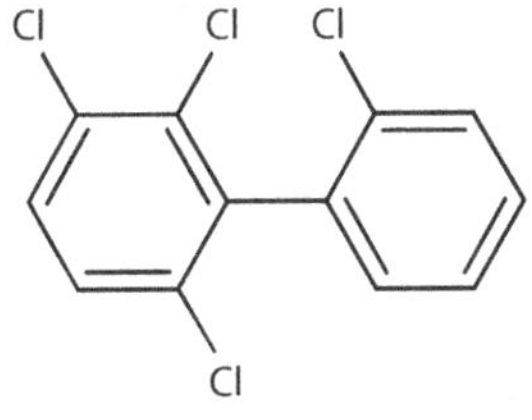

2,2′,3,6-tetraCB or PCB-45
(b)

FIGURE 5.4 (a) PCB structure and (b) nomenclature.

age of congeners with less than six chlorines—hence, Aroclor 1268 has lower toxicity than the other Aroclor mixtures.[60,78]

The CSF for PCBs is based on liver tumors occurring in rodents dosed with PCBs.[79] The MOA is similar to that of dioxin-like chemicals and is likely not applicable to humans.[80] For mixtures of high-risk, persistent congeners such as would occur in fish, the CSF is 2.0 per mg/kg/day. A CSF specific for Aroclor 1268 has been developed with a value of 0.27 per mg/kg/day.[60]

PCB mixtures have noncancer effects and RfDs exist in EPA's IRIS database for Aroclor 1016 and Aroclor 1254. The RfD value for Aroclor 1016 is 7E-05 mg/kg/day based on neurodevelopmental effects in monkeys. The RfD value for Aroclor 1254 is 2E-05 mg/kg/day based on dermal effects that appear similar to chloracne in humans. An RfD value for Aroclor 1268 has been developed based on comparison of congener composition between Aroclor 1016 and Aroclor 1254 with values of 1E-03 and 4E-04.[78]

Mercury

In fish, mercury exists as methylmercury (MeHg or CH_3Hg^+). Inorganic mercury is deposited from the air by rainfall in significant amounts and is transported in watersheds to water bodies. Methylmercury is formed in freshwater and estuarine ecosystems, primarily by sulfate-reducing bacteria in sediments. MeHg may become demethylated and inorganic mercury may return to the atmosphere by volatilization.[81,82] A reservoir of mercury exists in the atmosphere and contributes to MeHg in fish and biota. Figure 5.5 provides a schematic diagram of mercury cycling.

Methylmercury produces neurodevelopmental effects and yet unborn children represent a sensitive subpopulation. The RfD for methylmercury on EPA's IRIS database has a value of 1E-04 mg/kg/day and is based on longitudinal epidemiologic studies conducted in the Faroe Islands, the Seychelles, and New Zealand.[83–85] The dose response analysis was conducted on the Faroese data using a one-compartment pharmacokinetic model.[86] Other values for this RfD are available in the scientific literature with a recommended value of 4E-04 mg/kg/day.[87,88]

Risk Assessment Results

The RME cancer risk to the lifetime receptor was 7E-04. Of that 2E-04 was attributable to Aroclor 1268 and 5E-04 attributable to toxaphene. The hazard indices for adults, adolescents, and children were 3, 6, and 10, respectively. These values can be calculated from Tables 5.27 through 5.29. Table 5.30 shows the RME risk results using alternate toxicity criteria.

Characterization of Uncertainty

One of the major uncertainties of the exposure assessment was the use of the MRFSS data, even just to assign percentages of the catch that was consumed. Other fish species were reported in the MRFSS—for example, southern flounder—but these species were not obtained during sampling and thus could not be analyzed

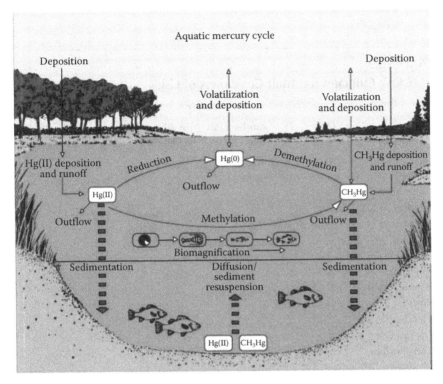

FIGURE 5.5 Diagram of aquatic mercury cycling. Changing concentrations of mercury in the marsh due to methylation/deposition/volatilization enabled the use of the entire mercury and methylmercury datasets for calculation of the exposure point concentrations.

for toxaphene. In addition, all the COPCs are lipophilic and bioaccumulative. Different cooking methods may remove some of these contaminants from fish. Last, the use of the site-specific fish consumption survey data along with the MRFSS data on species proportion of the catch raises the question of resource utilization.[40] The implicit assumption is that the proportion of fish caught represents the proportion of fish consumed. For example, striped mullet represented 0.8% of the catch and spot croaker represented 0.04%. Did fish consumers actually eat these small bony fish?

The question of resource utilization is likely the largest uncertainty in the exposure assessment. The site-specific fish consumption study did not ask what species of fish were consumed. The implicit assumption of using the MRFSS data is that the entire catch is consumed.

Trophic level weighting has been used to develop fish consumption guidelines for MeHg.[89] Fish at trophic level 2 (TL2) consist of herbivores, planktivores, and detritivores; fish at trophic level 3 (TL3) consist of secondary piscine omnivores with diets that include other fish and invertebrates, and fish at trophic level 4 (TL4) are high level carnivores, generally top predators that are exclusively or almost exclusively

TABLE 5.27

RME Risk Estimates for Adult Consumers of Fish

Fish Species	EPC (mg/kg)	FI (%)	Cancer — Aroclor 1268 (mg/kg-day)	Cancer — Toxaphene (mg/kg-day)	Non-cancer — Aroclor 1268 (mg/kg-day)	Non-cancer — Mercury (mg/kg-day)
Atlantic Croaker						
Aroclor 1268	1.044	0.96	1.9E-06		4.4E-06	
Total Toxaphene	8.506			1.5E-05		
Toxaphene as Σ3PC	0.25					
Mercury	0.248					1.1E-06
Red Drum						
Aroclor 1268	0.164	17.14	5.3E-06		1.2E-05	
Total Toxaphene	0.851			2.8E-05		
Toxaphene as Σ3PC	0.0244					
Mercury	0.287					2.2E-05
Southern Kingfish						
Aroclor 1268	0.649	17.99	2.2E-05		5.2E-05	
Total Toxaphene	1.504			5.1E-05		
Toxaphene as Σ3PC	0.0483					
Mercury	0.452					3.6E-05
Spot						
Aroclor 1268	0.838	0.04	6.4E-08		1.5E-07	
Total Toxaphene	11.88			9.0E-07		
Toxaphene as Σ3PC	0.623					
Mercury	0.12					2.1E-08
Spotted Seatrout						
Aroclor 1268	0.342	31.59	2.1E-05		4.8E-05	
Total Toxaphene	1.855			1.1E-04		
Toxaphene as Σ3PC	0.0713					
Mercury	0.434					6.1E-05
Striped Mullet						
Aroclor 1268	2.737	0.81	4.2E-06		9.8E-06	
Total Toxaphene	7.732			1.2E-05		
Toxaphene as Σ3PC	0.415					
Mercury	0.0251					8.9E-08
Total Intake from Fish			5.4E-05	2.2E-04	1.3E-04	1.2E-04
CSF/RfD			2	1.1	7.E-05	1E-04
Risk/Hazard			1.1E-04	2.4E-04	1.8	1.2

TABLE 5.28
RME Risk Estimates for Adolescent Consumers of Fish

Fish Species	EPC (mg/kg)	FI (%)	Cancer		Non-cancer	
			Aroclor 1268 (mg/kg-day)	Toxaphene (mg/kg-day)	Aroclor 1268 (mg/kg-day)	Mercury (mg/kg-day)
Atlantic Croaker						
Aroclor 1268	1.044	0.96	8.6E-07		6.7E-06	
Total Toxaphene	8.506			7.0E-06		
Toxaphene as Σ3PC	0.25					
Mercury	0.248					1.6E-06
Red Drum						
Aroclor 1268	0.164	17.14	2.4E-06		1.9E-05	
Total Toxaphene	0.851			1.3E-05		
Toxaphene as Σ3PC	0.0244					
Mercury	0.287					3.3E-05
Southern Kingfish						
Aroclor 1268	0.649	17.99	1.0E-05		7.8E-05	
Total Toxaphene	1.504			2.3E-05		
Toxaphene as Σ3PC	0.0483					
Mercury	0.452					5.4E-05
Spot						
Aroclor 1268	0.838	0.04	2.9E-08		2.2E-07	
Total Toxaphene	11.88			4.1E-07		
Toxaphene as Σ3PC	0.623					
Mercury	0.12					3.2E-08
Spotted Seatrout						
Aroclor 1268	0.342	31.59	9.3E-06		7.2E-05	
Total Toxaphene	1.855			5.0E-05		
Toxaphene as Σ3PC	0.0713					
Mercury	0.434					9.1E-05
Striped Mullet						
Aroclor 1268	2.737	0.81	1.9E-06		1.5E-05	
Total Toxaphene	7.732			5.4E-06		
Toxaphene as Σ3PC	0.415					
Mercury	0.0251					1.4E-07
Total Intake from Fish			2.4E-05	9.9E-05	1.9E-04	1.8E-04
CSF/RfD			2	1.1	7.E-05	7.E-05
Risk/Hazard			4.9E-05	1.1E-05	2.7	1.8

TABLE 5.29
RME Risk Estimates for Child Consumers of Fish

Fish Species	EPC (mg/kg)	FI (%)	Cancer Aroclor 1268 (mg/kg-day)	Cancer Toxaphene (mg/kg-day)	Non-cancer Aroclor 1268 (mg/kg-day)	Non-cancer Mercury (mg/kg-day)
Atlantic Croaker						
Aroclor 1268	1.044	0.96	8.6E-07		1.2E-05	
Total Toxaphene	8.506			7.0E-06		
Toxaphene as Σ3PC	0.25					
Mercury	0.248					2.9E-06
Red Drum						
Aroclor 1268	0.164	17.14	2.4E-06		3.4E-05	
Total Toxaphene	0.851			1.3E-06		
Toxaphene as Σ3PC	0.0244					
Mercury	0.287					5.9E-05
Southern Kingfish						
Aroclor 1268	0.649	17.99	1.0E-05		1.4E-04	
Total Toxaphene	1.504			2.3E-05		
Toxaphene as Σ3PC	0.0483					
Mercury	0.452					9.8E-05
Spot						
Aroclor 1268	0.838	0.04	2.9E-08		4.0E-07	
Total Toxaphene	11.88			4.1E-07		
Toxaphene as Σ3PC	0.623					
Mercury	0.12					5.8E-08
Spotted Seatrout						
Aroclor 1268	0.342	31.59	9.3E-06		1.3E-04	
Total Toxaphene	1.855			5.0E-05		
Toxaphene as Σ3PC	0.0713					
Mercury	0.434					1.6E-04
Striped Mullet						
Aroclor 1268	2.737	0.81	1.9E-06		2.7E-05	
Total Toxaphene	7.732			5.4E-06		
Toxaphene as Σ3PC	0.415					
Mercury	0.0251					2.4E-07
Total Intake from Fish			2.4E-05	9.9E-05	3.4E-04	3.2E-04
CSF/RfD			2	1.1	7.E-05	1E-04
Risk/Hazard			4.9E-05	1.1E-04	4.9	3.2

TABLE 5.30
RME Intakes, Showing Alternative Toxicity Criteria, Risks and Hazards

	Cancer		Non-cancer	
	Aroclor 1268 (mg/kg-day)	Toxaphene as Σ3PC (mg/kg-day)	Aroclor 1268 (mg/kg-day)	Mercury (mg/kg-day)
Adult				
Total Intake from Fish	5.4E-05	1.8E-05	1.3E-04	1.2E-05
CSF/RfD	0.27	2.E-05	1E-03	4.E-04
Risk/Hazard	1.5E-05	0.9	0.1	0.3
Adolescent				
Total Intake from Fish	2.4E-05	2.8E-05	1.9E-04	1.8E-04
CSF/RfD	0.27	2.E-05	1.E-03	4.E-04
Risk/Hazard	6.6E-06	1.4	0.2	0.5
Child				
Total Intake from Fish	2.4E-05	5.0E-05	3.4E-04	3.2E-04
CSF/RfD	0.27	2.E-05	1.E-03	4.E-04
Risk/Hazard	6.6E-06	2.5	0.3	0.8

piscivorous. Trophic level weighting factors are used for the MeHg levels observed at the various trophic levels for the purpose of fish consumption advisories. Application of EPA's trophic level weighting scheme for methylmercury yielded an overall fish concentration of 0.168 mg/kg for methylmercury. This is lower than the criterion value of 0.3 mg/kg at which fish consumption advisories would be put into place.[90] Note the EPCs vary widely.

The toxicity of Aroclor 1268 and that of toxaphene also contribute to the uncertainty. The alternative toxicity values suggest that these substances are much less toxic than indicated by the toxicity criteria on IRIS—by several orders of magnitude.

Risk Assessment and Risk Management

In preparation for a public meeting with the active community environmental group near the site, the risk manager sought information from the risk assessor. The issues about which she sought clarification on were the differences between the toxicity criteria in the IRIS database and the alternate criteria in the scientific literature. She was most concerned about the hazard from methylmercury. The risk assessor suggested that she present the differences in the risk assessment and the trophic level weighted intakes and a very brief discussion of the reasons for the differences in the toxicity evaluations. She asked if the removal of the sediments containing the largest amounts of toxaphene had had any effect on the fish concentrations. The RA indicated that there had been a notable reduction.[63]

"I'll attend the meeting as backup," he offered. She took him up on the offer.

At the meeting, the principal of the local elementary school offered to serve as a facilitator. They accepted his offer. At the meeting, there was no discussion of either toxaphene or Aroclor 1268—the interest of those attending the meeting was mercury. The risk assessor fielded a question from the head of the local environmental group.

"Why are the fish advisory levels that are considered safe higher than the levels that show a risk from mercury?" asked this man. "How does that possibly make sense?"

"Sir," responded the risk assessor, "that's a really good question. I wondered that myself when I noticed it. Here's why—when fish tissue concentrations are used for fish advisories, the outcome is a change in peoples' fish consumption behavior. However, when fish tissue concentrations are used in a risk assessment, the outcome is a cleanup level in sediment and high confidence is needed that this sediment cleanup level will result in fish concentrations that are not a health concern. So the risk assessment takes into account the uncertainty of relating environmental levels of mercury in sediment and other environmental media to mercury levels in fish tissue levels."

"Wow," said the man, "I never thought of that. So we're actually getting more bang for the buck in terms of health protection from the risk assessment."

"That's exactly right," chimed in the PM. The risk assessor looked around the room and saw many heads nodding. Before the PM could continue, a number of the attendees gathered their belongings and got up to leave.

EXERCISES FOR THOUGHT AND DISCUSSION

Simple Risk Assessment

Conduct a risk assessment for groundwater using the COPCs shown in Table 5.12.

Working with ProUCL

Go to EPA's website at http://www.epa.gov and enter the term "ProUCL" in the search box. Download the software and documentation and check that the statistics and UCL values in the tables are correct. The tables in this chapter provide all the data. In addition, spreadsheets are available on the publisher's website at http://www.crcpress.com/product/isbn/9781466598294 for you to work through as well.

Central Tendency Exposure Estimate for Fish Consumption

Recreate Table 5.30 using CTE fish consumption rates from Table 5.26. See how much lower the ingestion estimates and corresponding risks/hazards can be.

Risk Calculation Using Alternate Toxicity Criteria

Conduct a risk evaluation using the alternate toxicity criteria discussed in the narrative and found in Table 5.30. Remember you will have to treat toxaphene as Σ3PC and use the alternate toxicity criterion as an RfD. Suppose you used these alternate toxicity criteria and CTE exposure assumptions. How low would the risk/hazard estimates be?

USE OF ALTERNATE TOXICITY CRITERIA

Each of the examples in this chapter uses alternative criteria for bioavailability and toxicity, respectively. The use of these criteria is in the spirit of the Habicht memo discussed in Chapter 1. Develop convincing arguments both for and against the use of such nonstandard approaches. Set up a role-playing scenario in which one group, representing the regulated entity, attempts to convince others, representing the regulatory agency, that the use of new science is appropriate.

CALCULATION OF FISH ADVISORIES

Take a look at both of the EPA documents in the references (numbers 97 and 98) that relate to fish advisory levels for methylmercury. Using the data in Tables 5.21 through 5.23, reproduce the risk assessor's calculation using trophic level weighting to determine whether fish advisory levels are needed.

REFERENCES

1. National Research Council (NRC). *Risk Assessment in the Federal Government: Managing the Process.* Washington, DC: The National Academies Press, 1983. http://www.nap.edu/catalog.php?record_id=366
2. National Research Council (NRC). *Science and Judgement in Risk Assessment.* Washington, DC: The National Academies Press, 1994. http://www.nap.edu/catalog.php?record_id=2125
3. United States Environmental Protection Agency (USEPA). *Risk Assessment Guidance for Superfund: Volume 1: Human Health Evaluation Manual (Part A) (Interim Final) (RAGS).* EPA/540/1-89/002. Washington, DC, December, 1989. http://www.epa.gov/oswer/riskassessment/ragsa/
4. United States Environmental Protection Agency (USEPA). *Risk Assessment Guidance for Superfund: Volume 1: Human Health Evaluation Manual (Part E, Supplemental Guidance for Dermal Risk Assessment) Final.* EPA/540/R/99/005. OSWER 9285.7-02EP. PB99-963312. Washington, DC, July, 2004. http://www.epa.gov/oswer/riskassessment/ragse/ (accessed August 19, 2012).
5. United States Environmental Protection Agency (USEPA). *Risk Assessment Guidance for Superfund: Volume 1: Human Health Evaluation Manual (Part F, Supplemental Guidance for Inhalation Risk Assessment) Final.* EPA-540-R-070-002. OSWER 9285.7-82. Office of Superfund Remediation and Technical Innovation. Washington, DC, January, 2009. http://www.epa.gov/oswer/riskassessment/ragsf/index.htm (accessed September 8, 2012).
6. Morgan MG, Henrion M, and Small M. *Uncertainty: A Guide to Dealing with Uncertainty in Quantitative Risk and Policy Analysis.* Cambridge, U.K.: Cambridge University Press, 1990.
7. Cullen AC and Frey HC. *Probabilistic Techniques in Exposure Assessment: A Handbook for Dealing with Variability and Uncertainty in Models and Inputs.* New York: Plenum Press, 1999.
8. Bernstein PL. *Against the Gods: The Remarkable Story of Risk.* New York: John Wiley & Sons, Inc., 1998.
9. Ferson SS and Ginzburg LR. Different methods are needed to propagate ignorance and variability. *Reliab Eng Syst Safe.* 1996;54:133–144.

10. Cooper AR, Page AS, Wheeler BW, Hillsdon M, Griew P, and Jago R. Patterns of GPS measured time outdoors after school and objective physical activity in English children: The PEACH project. *Int J Behav Nutr Phys Act.* 2010;7:31.

11. Wheeler BW, Cooper AR, Page AS, and Jago R. Greenspace and children's physical activity: A GPS/GIS analysis of the PEACH project. *Prev Med.* 2010, August;51(2):148–152.

12. Ko S, Schaefer PD, Vicario CM, Binns HJ, and Safer Yards Project. Relationships of video assessments of touching and mouthing behaviors during outdoor play in urban residential yards to parental perceptions of child behaviors and blood lead levels. *J Expo Sci Environ Epidemiol.* 2007, January;17(1):47–57.

13. Dunton GF, Liao Y, Intille SS, Spruijt-Metz D, and Pentz M. Investigating children's physical activity and sedentary behavior using ecological momentary assessment with mobile phones. *Obesity (Silver Spring).* 2011, June;19(6):1205–1212.

14. Dunton GF, Liao Y, Intille S, Wolch J, and Pentz MA. Physical and social contextual influences on children's leisure-time physical activity: An ecological momentary assessment study. *J Phys Act Health.* 2011, January;8(Suppl 1):S103–S108.

15. Dunton GF, Kawabata K, Intille S, Wolch J, and Pentz MA. Assessing the social and physical contexts of children's leisure-time physical activity: An ecological momentary assessment study. *Am J Health Promot.* 2012;26(3):135–142.

16. Institute of Medicine IOM. *Environmental Decisions in the Face of Uncertainty.* Washington, DC: The National Academies Press, 2013.

17. United States Environmental Protection Agency (USEPA). Memorandum from F. Henry Habicht. *Guidance on Risk Characterization for Risk Managers and Risk Assessors.* Washington, DC, February 26, 1992. http://www.epa.gov/oswer/riskassessment/pdf/habicht.pdf

18. National Center for Educational Statistics (NCES). *Adult Literacy in America: A First Look at the Findings of the National Adult Literacy Survey,* April, 2002. http://nces.ed.gov/pubs93/93275.pdf (accessed February 14, 2012).

19. Peters E, Hibbard J, Slovic P, and Dieckmann N. Numeracy skill and the communication, comprehension, and use of risk-benefit information. *Health Aff (Millwood).* 2007;26(3):741–748.

20. United States Environmental Protection Agency (USEPA). *Risk Assessment Guidance for Superfund (RAGS) Volume III—Part A: Process for Conducting Probabilistic Risk Assessment.* EPA 540-R-02-002. OSWER 9285.7-45. PB2002 963302. Washington, DC, December, 2001. http://www.epa.gov/oswer/riskassessment/rags3adt/

21. National Research Council (NRC). *Science and Decisions: Advancing Risk Assessment.* Washington, DC: The National Academies Press, 2009. http://www.nap.edu/catalog.php?record_id=12209

22. Michaels D. *Doubt Is Their Product: How Industry's Assault on Science Threatens Your Health.* Oxford, U.K.: Oxford University Press, 2008.

23. United States Environmental Protection Agency (USEPA). Memorandum from Elliot Laws. *Formation of the Superfund Remedy Review Board.* Washington, DC, November 28, 1995. http://www.epa.gov/superfund/programs/nrrb/pdfs/11-28-95.pdf

24. Roberts SM, Munson JW, Lowney YW, and Ruby MV. Relative oral bioavailability of arsenic from contaminated soils measured in the cynomolgus monkey. *Toxicol Sci.* 2007, January;95(1):281–288.

25. United States Environmental Protection Agency (USEPA). *Expert Elicitation Task Force White Paper.* External Review Draft. Science Policy Council. Washington, DC, January 6, 2009. http://www.epa.gov/osa/pdfs/elicitation/Expert_Elicitation_White_Paper-January_06_2009.pdf

26. United States Office of Management and Budget (OMB). *Guidelines for Regulatory Analysis, Circular A-4.* Washington, DC, September 17, 2003. http://www.whitehouse.gov/sites/default/files/omb/assets/regulatory_matters_pdf/a-4.pdf

27. Rowe G and Wright G. The Delphi technique as a forecasting tool: Issues and analysis. *Int J Forecast.* 1999;15(4):353–375.

28. United States Environmental Protection Agency (USEPA). *Guidance for Comparing Background and Chemical Concentrations in Soil for CERCLA Sites.* EPA 540-R-01-003. OSWER 9285.7-41. Washington, DC, September, 2002. http://www.epa.gov/oswer/riskassessment/pdf/background.pdf

29. United States Environmental Protection Agency (USEPA). *Seven Cardinal Rules of Risk Communication.* OPA-87-020. Washington, DC, April, 1988. http://www.epa.gov/publicinvolvement/pdf/risk.pdf

30. United States Environmental Protection Agency (USEPA). *Supplemental Guidance for Developing Soil Screening Levels for Superfund Sites.* OSWER 9355.4-24. Washington, DC, December, 2002. http://www.epa.gov/superfund/health/conmedia/soil/ (accessed July 4, 2012).

31. United States Environmental Protection Agency USEPA. *Regional Screening Table—User's Guide.* Mid-Atlantic Region 3. Philadelphia, PA, November, 2012. http://www.epa.gov/reg3hwmd/risk/human/rb-concentration_table/

32. United States Environmental Protection Agency (USEPA). *Supplemental Guidance to RAGS: Calculating the Concentration Term.* OSWER 9285.7-08i. PB92-963373. Washington, DC, May, 1992. http://www.epa.gov/oswer/riskassessment/pdf/1992_0622_concentrationterm.pdf

33. United States Environmental Protection Agency (USEPA). *Calculating Upper Confidence Limits for Exposure Point Concentrations at Hazardous Waste Sites.* OSWER 9285.6-10. Washington, DC, December, 2002. http://www.epa.gov/oswer/riskassessment/pdf/ucl.pdf

34. United States Environmental Protection Agency (USEPA). *ProUCL Software.* Site Characterization and Monitoring Technical Support Center. Las Vegas, NV, n.d. http://www.epa.gov/osp/hstl/tsc/software.htm#about

35. United States Environmental Protection Agency (USEPA). *Data Quality Assessment: Statistical Methods for Practitioners,* EPA QA/G-9S. EPA/240/B-06/003. Washington, DC, February, 2006. http://www.epa.gov/quality/qs-docs/g9s-final.pdf

36. United States Environmental Protection Agency (USEPA). *On the Computation of a 95% Upper Confidence Limit of the Unknown Population Mean Based Upon Data Sets With Below Detection Limit Observations.* EPA/600/R-06/022. Las Vegas, NV, March, 2006. http://www.epa.gov/osp/hstl/tsc/Singh2006.pdf

37. United States Environmental Protection Agency (USEPA). *Risk Assessment Guidance for Superfund Volume I: Human Health Evaluation Manual. Supplemental Guidance. "Standard Default Exposure Factors."* PB91-921314. OSWER Directive 9285.6-03 Washington, DC, March 25, 1991. http://www.epa.gov/oswer/riskassessment/pdf/oswer_directive_9285_6-03.pdf

38. United States Environmental Protection Agency (USEPA). *Exposure Factors Handbook (1997 Final Report).* EPA/600/P-95/002F a-c, Washington, DC, 1997. http://cfpub.epa.gov/ncea/risk/recordisplay.cfm?deid=12464

39. United States Environmental Protection Agency (USEPA). *Child-Specific Exposure Factors Handbook (Final Report) 2–9.* EPA/600/R-06/096F, Washington, DC, 2008. http://cfpub.epa.gov/ncea/risk/recordisplay.cfm?deid=199243

40. United States Environmental Protection Agency (USEPA). *Exposure Factors Handbook 2011 Edition (Final).* EPA/600/R-09/052F. Washington, DC, September, 2011. http://cfpub.epa.gov/ncea/risk/recordisplay.cfm?deid=236252

41. United States Environmental Protection Agency (USEPA). *National Oil and Hazardous Substances Pollution Contingency Plan (NCP) Proposed Rule.* 53 Federal Register 51394. December 12, 1988. http://www.epa.gov/superfund/policy/remedy/sfremedy/pdfs/ncppropream.pdf (accessed September 30, 2012).
42. United States Environmental Protection Agency (USEPA). *Guidance for Risk Characterization.* Science Policy Council. Washington, DC, February, 1995. http://www.epa.gov/spc/pdfs/rcguide.pdf
43. United States Environmental Protection Agency (USEPA). *Soil Screening Guidance: Users' Guide.* EPA540/R-96/018. Washington, DC, July, 1996. http://www.epa.gov/superfund/health/conmedia/soil/ (accessed July 6, 2012).
44. United States Environmental Protection Agency (USEPA). *A Review of the Reference Dose and Reference Concentration Processes, Final Report.* EPA/630/P-02/002F. Risk Assessment Forum. Washington, DC, December, 2002. http://www.epa.gov/raf/publications/pdfs/rfd-final.pdf
45. United States Environmental Protection Agency (USEPA). *Recommendations for Default Value for Relative Bioavailability of Arsenic in Soil.* OSWER 9200.1-113. Washington, DC, December, 2012. http://www.epa.gov/superfund/bioavailability/pdfs/Arsenic%20Bioavailability%20POLICY%20Memorandum%2012-20-12.pdf
46. United States Environmental Protection Agency (USEPA). *Compilation and Review of Data on Relative Bioavailability of Arsenic in Soil.* OSWER 9200.1-113. Washington, DC, December, 2012. http://www.epa.gov/superfund/bioavailability/pdfs/Arsenic%20Bioavailability%20SCIENCE%20Report_SRC%2009-20-12.pdf
47. Bradham KD, Scheckel KG, Nelson CM, Seales PE, Lee GE, Hughes MF et al. Relative bioavailability and bioaccessibility and speciation of arsenic in contaminated soils. *Environ Health Perspect.* 2011, November;119(11):1629–1634.
48. United States Environmental Protection Agency (USEPA). Memorandum from Michael Cook. *Human Health Toxicity Values in Superfund Risk Assessments.* OSWER Directive 9285.7-53. Washington, DC, December 5, 2003. http://www.epa.gov/oswer/riskassessment/pdf/hhmemo.pdf
49. United States Environmental Protection Agency (USEPA). *Methods for Derivation of Inhalation Reference Concentrations and Application of Inhalation Dosimetry.* EPA/600/8-90/066F. Research Triangle Park, NC, October, 1994. http://cfpub.epa.gov/ncea/cfm/recordisplay.cfm?deid=71993 (accessed September 25, 2012).
50. American Cancer Society (ACS). *Cancer Facts and Statistics/Cancer Facts and Figures.* http://www.cancer.org/research/cancerfactsstatistics/ (accessed October 1, 2013).
51. United States Environmental Protection Agency (USEPA). *Reference Manual for the Integrated Exposure Uptake Biokinetic Model for Lead in Children (IEUBK) Windows® 32-bit Version.* OSWER #9285.7-44. Washington, DC, May, 2002. http://www.epa.gov/superfund/health/contaminants/lead/products/rmieubk32.pdf
52. United States Environmental Protection Agency (USEPA). *Technical Support Document: Parameters and Equations Used in the Integrated Exposure Uptake Biokinetic Model for Lead in Children (v0.99d).* EPA 540/R-94/040. PB94-963505. OSWER #9285.7-22. Washington, DC, December, 1994. http://www.epa.gov/superfund/health/contaminants/lead/products/tsd.pdf
53. United States Environmental Protection Agency (USEPA). *Recommendations of the Technical Review Workgroup for Lead for an Approach to Assessing Risks Associated With Adult Exposures to Lead in Soil, Final.* EPA-540-R-03-001. Washington, DC, January, 2003. http://www.epa.gov/superfund/health/contaminants/lead/products/adultpb.pdf

54. United States Environmental Protection Agency (USEPA). *Update of the Adult Lead Methodology's Default Baseline Blood Lead Concentration and Geometric Standard Deviation Parameter.* OSWER 9200.2-82. Washington, DC, June, 2009. http://www.epa.gov/superfund/health/contaminants/lead/products/almupdate.pdf

55. United States Environmental Protection Agency (USEPA). *Compilation and Review of Data on Relative Bioavailability of Arsenic in Soil.* OSWER 9200.1-113. Washington, DC, December, 2012. http://www.epa.gov/superfund/bioavailability/pdfs/Arsenic%20Bioavailability%20SCIENCE%20Report_SRC%2009-20-12.pdf

56. Kucklick JR and Helm PA. Advances in the environmental analysis of polychlorinated naphthalenes and toxaphene. *Anal Bioanal Chem.* 2006, October;386(4): 819–836.

57. de Geus HJ, Besselink H, Brouwer A, Klungsøyr J, McHugh B, Nixon E et al. Environmental occurrence, analysis, and toxicology of toxaphene compounds. *Environ Health Perspect.* 1999, February;107(Suppl 1):115–144.

58. de Geus HJ, Wester PG, Schelvis A, de Boer J, and Brinkman UA. Toxaphene: A challenging analytical problem. *J Environ Monit.* 2000, October;2(5):503–511.

59. Simon T and Manning R. Development of a reference dose for the persistent congeners of weathered toxaphene based on in vivo and in vitro effects related to tumor promotion. *Regul Toxicol Pharmacol.* 2006, April;44(3):268–281.

60. Warren DA, Kerger BD, Britt JK, and James RC. Development of an oral cancer slope factor for Aroclor 1268. *Regul Toxicol Pharmacol.* 2004, August;40(1):42–53.

61. Simon DL, Maynard EJ, and Thomas KD. Living in a sea of lead—Changes in blood—And hand-lead of infants living near a smelter. *J Expo Sci Environ Epidemiol.* 2007, May;17(3):248–259.

62. Maruya KA and Lee RF. Aroclor 1268 and toxaphene in fish from a southeastern U.S. estuary. *Environ Sci Technol.* 1998;32;1069–1075.

63. Maruya KA, Francendese L, and Manning RO. Residues of toxaphene decrease in estuarine fish after removal of contaminated sediments. *Estuaries.* 2005, October;28(5):786–793.

64. Angerhöfer D, Kimmel L, Koske G, Fingerling G, Burhenne J, and Parlar H. The role of biotic and abiotic degradation processes during the formation of typical toxaphene peak patterns in aquatic biota. *Chemosphere.* 1999, August;39(4):563–568.

65. Vetter W and Luckas B. Enantioselective determination of persistent and partly degradable toxaphene congeners in high trophic level biota. *Chemosphere.* 2000, August;41(4):499–506.

66. Vetter W, Klobes U, and Luckas B. Distribution and levels of eight toxaphene congeners in different tissues of marine mammals, birds and cod livers. *Chemosphere.* 2001;43(4–7):611–621.

67. National Academy of Sciences (NAS). *Review of Recreational Fisheries Survey Methods.* Washington, DC: The National Academies Press, 2006. http://www.nap.edu/catalog.php?record_id=11616

68. Rupp EM, Miller FI, and Baes CF. Some results of recent surveys of fish and shellfish consumption by age and region of U.S. residents. *Health Phys.* 1980, August;39(2):165–175.

69. Department of Health and Human Services (DHHS). Agency for Toxic Substances and Disease Registry. *Final report: Consumption of Seafood and Wildgame Contaminated with Mercury, Brunswick, Glynn County, GA.* Atlanta, GA, July, 1999. http://www.atsdr.cdc.gov/hac/PHA/ArcoQuarry/consumption_seafood_final_report.pdf

70. European Union (EU). Investigation into the monitoring, analysis and toxicity of toxaphene in marine foodstuffs, final report. FAIR CT PL.96.3131. Brussels, BE. https://www.marine.ie/NR/rdonlyres/2EFEB26A-DD5B-49BC-8106-9CC4657F942F/0/Toxaphene.pdf

71. Ross J, Plummer SM, Rode A, Scheer N, Bower CC, Vogel O et al. Human constitutive androstane receptor (CAR) and pregnane X receptor (PXR) support the hypertrophic but not the hyperplastic response to the murine nongenotoxic hepatocarcinogens phenobarbital and chlordane in vivo. *Toxicol Sci.* 2010, August;116(2):452–466.

72. Goodman JI, Brusick DJ, Busey WM, Cohen SM, Lamb JC, and Starr TB. Reevaluation of the cancer potency factor of toxaphene: Recommendations from a peer review panel. *Toxicol Sci.* 2000, May;55(1):3–16.

73. Besselink H, Nixon E, McHugh B, Klungsoyr J, and Brouwer A. In vitro and in vivo tumor promoting potency of technical toxaphene, uv-irradiated toxaphene, and biotransformed toxaphene. *Organohalogen Compd.* 2000;47:113–116.

74. Besselink H, Nixon E, McHugh B, Rimkus G, Klungsøyr J, Leonards P et al. Evaluation of tumour promoting potency of fish borne toxaphene residues, as compared to technical toxaphene and UV-irradiated toxaphene. *Food Chem Toxicol.* 2008, August;46(8):2629–238.

75. United States Environmental Protection Agency (USEPA). *Ombudsman Report: Appropriate Testing and Timely Reporting are Needed At the Hercules 009 Landfill Superfund Site.* 2005, September 9.

76. Golden R and Kimbrough R. Weight of evidence evaluation of potential human cancer risks from exposure to polychlorinated biphenyls: An update based on studies published since 2003. *Crit Rev Toxicol.* 2009;39(4):299–331.

77. Kimbrough RD and Krouskas CA. Human exposure to polychlorinated biphenyls and health effects: A critical synopsis. *Toxicol Rev.* 2003;22(4):217–233.

78. Simon T, Britt JK, and James RC. Development of a neurotoxic equivalence scheme of relative potency for assessing the risk of PCB mixtures. *Regul Toxicol Pharmacol.* 2007, July;48(2):148–170.

79. Mayes BA, McConnell EE, Neal BH, Brunner MJ, Hamilton SB, Sullivan TM et al. Comparative carcinogenicity in Sprague-Dawley rats of the polychlorinated biphenyl mixtures Aroclors 1016, 1242, 1254, and 1260. *Toxicol Sci.* 1998, January;41(1):62–76.

80. Guzelian P, Quattrochi L, Karch N, Aylward L, and Kaley R. Does dioxin exert toxic effects in humans at or near current background body levels? An evidence-based conclusion. *Hum Exp Toxicol.* 2006, February;25(2):99–105.

81. United States Environmental Protection Agency (USEPA). *South Florida Ecosystem Assessment: Phase I/II (Summary)—Everglades Stressor Interactions: Hydropatterns, Eutrophication, Habitat Alteration, and Mercury Contamination. Monitoring for Adaptive Management: Implications for Ecosystem Restoration.* EPA 904-R-01-002. Region 4 Science and Ecosystem Support Division. September, 2001. http://www.epa.gov/region4/sesd/reports/epa904r01002/epa904r01002.pdf

82. United States Environmental Protection Agency (USEPA). *Mercury Study Report to Congress. Volume V: Health Effects of Mercury and Mercury Compounds.* EPA-452/R-97-007. Washington, DC, December, 1997. http://www.epa.gov/mercury/report.htm

83. Crump KS, Kjellström T, Shipp AM, Silvers A, and Stewart A. Influence of prenatal mercury exposure upon scholastic and psychological test performance: Benchmark analysis of a New Zealand cohort. *Risk Anal.* 1998, December;18(6):701–713.

84. Myers GJ, Davidson PW, Shamlaye CF, Axtell CD, Cernichiari E, Choisy O et al. Effects of prenatal methylmercury exposure from a high fish diet on developmental milestones in the Seychelles Child Development Study. *Neurotoxicology.* 1997;18(3):819–829.

85. Grandjean P and White RF. Effects of methylmercury exposure on neurodevelopment. *JAMA.* 1999, March 3;281(10):896; author reply 897.

86. Swartout J and Rice G. Uncertainty analysis of the estimated ingestion rates used to derive the methylmercury reference dose. *Drug Chem Toxicol.* 2000, February;23(1):293–306.

87. Shipp AM, Gentry PR, Lawrence G, Van Landingham C, Covington T, Clewell HJ et al. Determination of a site-specific reference dose for methylmercury for fish-eating populations. *Toxicol Ind Health.* 2000, November;16(9–10):335–438.

88. Allen BC, Hack CE, and Clewell HJ. Use of Markov Chain Monte Carlo analysis with a physiologically-based pharmacokinetic model of methylmercury to estimate exposures in US women of childbearing age. *Risk Anal.* 2007, August;27(4):947–959.

89. United States Environmental Protection Agency (USEPA). *Guidance for Implementing the January 2001 Methylmercury Water Quality Criterion.* EPA 823-R-10-001. Washington, DC, April, 2010. http://water.epa.gov/scitech/swguidance/standards/criteria/aqlife/methylmercury/upload/mercury2010.pdf

90. United States Environmental Protection Agency (USEPA). *Water Quality Criterion for the Protection of Human Health: Methylmercury Final.* EPA-823-R-01-001. Washington, DC, January, 2001. http://water.epa.gov/scitech/swguidance/standards/criteria/aqlife/methylmercury/upload/2009_01_15_criteria_methylmercury_mercury-criterion.pdf

6 Ecological Risk Assessment

> A man is ethical only when life, as such, is sacred to him, that of plants and animals as that of his fellow men, and when he devotes himself helpfully to all life that is in need of help.
>
> **Albert Schweitzer**
> *Out of My Life and Thought, New American Library, 1964*

Ecological risk assessment addresses the parts of NEPA that require protection of the environment. Thus far, in this textbook, human health has been the sole focus. This chapter is devoted to ecological risk assessment—protection of biota in the wild and the ecosystems they inhabit.

Ecological risk assessment led the way in problem formulation, as discussed in Chapter 2, but is often given only cursory attention by decision makers. Concern for protection of natural areas, wild populations, and similar resources was envisioned explicitly in NEPA discussed in Chapter 1[1] (Figure 1.1).

Adverse ecological effects may result from exposure to one or more stressors. The lion's share of ecological risk assessment activity is predictive: often, data are combined with uncertain assumptions to predict the effect of stressors on ecological receptors both at the individual and population levels. In those cases, where data are collected to answer particular questions or test specific hypotheses as part of the ecological risk assessment, careful planning may reduce this uncertainty.

Ecological risk assessment has been developed with a focus on problem formulation and thus seems to follow the scientific method to a greater extent than does human health risk assessment. The reason for this is that the endpoints are less clear. For example, what does "a significant effect on a population" mean? Can the population withstand a reduction of 10% or 20%?

One can think of ecological risk assessment in two parts. The first part is theoretical and predictive. The second part consists of data gathering that seeks to provide evidence regarding any hypotheses developed in the first part; hence, this data collection is like an experiment.

Ecological risk assessments are anthropocentric in that they are designed to address and inform specific risk management decisions from a human perspective—the ecosystems and biota to be protected are those given value by societal consent, as pointed out in Chapter 1. Humans may not always know what assumptions or actions serve to improve the health of an ecosystem, community, or population; instead, the value is determined by a general societal consensus. The uncertainty of knowing the determinants of ecological health creates an ongoing tension in ecological risk assessment.

In the 1991 report of a colloquium to develop a set of inference guidelines for ecological risk assessment following the suggestion of the Red Book, EPA admitted that the development of standard methods for ecological risk assessment had lagged behind those for human health. The reason given was that ecological risk assessments address a variety of endpoints at different levels of biological organization—from individuals to communities to ecosystems—and choosing these endpoints presented a challenge.

This single chapter can do no more than provide an introduction to the field of ecological risk assessment. There are as many nuances and complexities in ecological risk assessment as in human health risk assessment. For those writing ecological risk assessments, familiarity with EPA guidance (discussed in the following text) is a necessary starting point.

Dr. Glen Suter is one of the foremost scientists in the field of ecological risk assessment (ERA). He worked as a staff scientist at the Oak Ridge National Laboratory (ORNL) until 1998 and then joined EPA. Dr. Suter has written a recent textbook on ecological risk assessment, also published by Taylor & Francis. As noted, this single chapter is not intended to be comprehensive, and those serious about wanting to learn more are encouraged to consult Dr. Suter's book.[2]

EPA GUIDANCE FOR ECOLOGICAL RISK ASSESSMENT

As part of the response to the recommendation by the NAS Red Book to develop inference guidelines for risk assessment, EPA's RAF held discussions in 1990 for the purpose of developing such guidelines specifically for ecological risk assessment.[3] These discussions resulted in *Framework for Ecological Risk Assessment* that underwent peer review and was published by the agency in 1992.[4]

The framework document was fairly general and started with the existing paradigm for human health risk assessment. The proposed framework consisted of three major phases: (1) problem formulation, (2) analysis, and (3) risk characterization.

The framework also recognized the need for ecological risk assessment to consider effects at the population, community, or ecosystem levels. The framework introduced flexibility in the choice of endpoints noting that "no single set of ecological values to be protected can be generally applied" and recommended that these values be selected based on both science and policy considerations.

In 1989, EPA's Superfund program released *RAGS, Volume II: Environmental Evaluation Manual, Interim Final* as a companion to RAGS, Volume I, for human health.[5] Although this document provided a thoughtful discussion of ecological risk issues, what was lacking was a prescriptive explanation of how to conduct and present an ecological risk assessment. The risk assessment community would have to wait until 1997 for such a document.

The first guidance document issued by EPA's RAF was the *Framework for Ecological Risk* Assessment.[4] In 1995, the RAF published draft *Proposed Guidelines for Ecological Risk Assessment.*[6] These were peer reviewed and finalized in 1998.

Concurrently, within the Superfund program, a much-needed revision of the 1989 guidance was being developed. This revision was published in 1997 as *Ecological Risk Assessment Guidance for Superfund: Process for Designing and Conducting Ecological Risk Assessments—Interim Final.*[7] This guidance is commonly referred to as the "Process Document."

The Process Document issued by the Superfund program differs in the number of steps and prescriptiveness of the process from the RAF's 1992 framework document and the 1998 *Guidelines for Ecological Risk Assessment*, also issued by the RAF.[8] Regardless of these differences, the Process Document is considered generally consistent with the framework and the guidelines.

The process described consisted of eight steps. A diagram is shown in Figure 6.1. In practice, the majority of ecological risk assessments conclude following step 3. The flexibility of this process is not shown in the diagram. In practice, step 3 has evolved to include a consideration of all available information, and most often, comparison to background concentrations or background risks, food chain modeling, bioavailability considerations, or other factors provide sufficient information to stop the process. Step 3 has been separated into parts A and B. A screening-level ecological risk assessment (SLERA) includes steps 1, 2, and 3a; a baseline ecological risk assessment (BERA) includes steps 1 through 7 and usually involves site-specific data gathering.

To fill in gaps in the guidance, EPA's Superfund program has occasionally issued ECO Update Bulletins. These can be found at http://www.epa.gov/oswer/riskassessment/ecoup/index.htm and provide a range of guidance on specific issues. Essentially what has happened is that rather than rewriting guidance documents, ecological risk assessors within the Superfund program have chosen to issue these periodic updates.

EPA's Office of Pesticide Programs has much less guidance and relies more heavily than on the framework and guidelines from the RAF. The Office of Pesticide Programs has issued specific guidance on endangered and threatened species.[9] In addition, there are a number of concerns about the use and permitting of rodenticides in the Office of Pesticide Programs.[10]

There are a number of new terms specific to ecological risk assessment. A selection of these terms is provided in Box 6.1.[7]

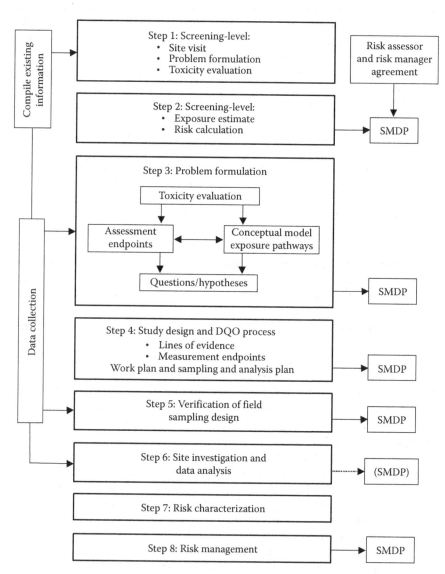

FIGURE 6.1 Eight-step process used in ecological risk assessment. The acronym "SMDP" means scientific/management decision point. (From United States Environmental Protection Agency (USEPA), *Ecological Risk Assessment Guidance for Superfund: Process for Designing and Conducting Ecological Risk Assessments (Interim Final)*, EPA 540-R-97-006, OSWER 9285.7-25, PB97-963211, Washington, DC, June, 1997.)

BOX 6.1   DEFINITION OF SELECTED TERMS
FOR ECOLOGICAL RISK ASSESSMENT[7]

Abiotic: Characterized by the absence of life. Abiotic materials or conditions include nonliving environmental media (e.g., water, sediment) or light, temperature, humidity, or other factors.

Area use factor: The ratio of an organism's foraging range to the area of contamination at the site. A value of 100% is the most protective.

Assessment endpoint: An explicit expression of the environmental or ecological value being investigated or sought to be protected by the risk assessment.

Baseline ecological risk assessment (BERA): The portion of the risk assessment that includes site-specific data collection.

Benthic community: The community of sediment-dwelling organisms at the bottom of a water body.

Bioaccumulation: A process by which chemicals are taken up by an organism due to exposure. Generally, bioaccumulative chemicals tend to increase in concentration moving up the food chain.

Biomagnification: The result of bioaccumulation by which tissue concentrations of chemicals at a higher trophic level exceed those in organisms at a lower trophic level.

Community: A group of populations of different species living at a specified location and time.

Biotic: Characterized as living. Biotic refers to the protists, plants, animals, and communities that comprise ecosystems.

Conceptual model: A series of working hypotheses of how contaminants or other stressors might affect ecosystems or communities. These hypotheses describe the relationships between exposure scenarios and assessment endpoints and between assessment endpoints and measurement endpoints.

Chemical of potential ecological concern (COPEC): A substance with the potential to affect ecological receptors adversely due to its concentration, distribution, and mode of toxicity.

Ecosystem: The biotic community and abiotic environment at a specific time and place, including the chemical, physical, and biological interactions.

Food chain transfer: The process by which higher-trophic-level organism substances are exposed to substances occurring in the tissues of lower-trophic-level prey organisms.

(continued)

**BOX 6.1 (continued)   DEFINITION OF SELECTED
TERMS FOR ECOLOGICAL RISK ASSESSMENT[7]**

Hazard index (HI): Sum of the HQs for multiple substances and/or multiple exposure pathways.

Hazard quotient (HQ): Ratio of an exposure level to a substance to a TRV selected for the risk assessment for that substance (e.g., LOAEL or NOAEL).

Home range: Area in which an organism lives.

Lowest observable adverse effect level (LOAEL): Lowest level of a stressor evaluated in a toxicity test or biological field survey that has a statistically significant adverse effect on the exposed organisms compared with unexposed organisms.

Measurement endpoint: Measurable characteristic reflecting the assessment endpoint.

No observed adverse effect level (NOAEL): Highest level of a stressor evaluated in a toxicity test or biological field survey that causes no statistically significant difference in effect compared with controls.

Scientific/Management decision point (SMDP): Point during the risk assessment process when stakeholder discussions occur to decide whether available information is sufficient to support risk management or whether additional information is needed.

Screening-level ecological risk assessment (SLERA): The initial portion of an ecological risk assessment that includes screening for COPECs, initial problem formulation, and conceptual site model.

Toxicity reference value (TRV): Numerical value expression the exposure–response relationship in an ERA. Can be a NOAEL, LOAEL, or BMD.

Trophic level: A classification of species within a community based on feeding or predator/prey relationships.

EIGHT STEPS OUTLINED IN THE PROCESS DOCUMENT

Here, each of the eight steps in the ecological risk assessment processes detailed in EPA's Process Document will be considered. This process is used by a large number of regulatory agencies and is the general method for proceeding. Most state agencies require a SLERA, which consists of the first three steps of the process. The BERA includes those steps that occur after the SLERA, should they be necessary. Following that, the last part of this chapter will be an example of ecological risk assessment similar to the case studies in Chapter 5.

STEP 1: SCREENING-LEVEL PROBLEM FORMULATION AND ECOLOGICAL EFFECTS EVALUATION

This first step is supposed to address five distinct issues:

1. Environmental setting and known contaminants
2. Fate-and-transport mechanisms
3. Mechanisms of ecotoxicity and categories of receptors
4. Exposure pathways
5. Screening endpoints for ecological risk

In many SLERAs, the first four issues are discussed in narrative, and the SLERA focuses on screening chemical concentrations measured in various environmental media. These considerations are used to develop a preliminary site conceptual model of how contaminants occur at the site and how they might affect the ecosystem on both individual and population levels.

SOURCES OF ECOLOGICAL SCREENING VALUES

Ecological screening benchmarks exist for biota, soil, and sediment. Criteria exist for surface water. Criteria are acceptable regulatory values, whereas benchmarks are intended for use as screening values. Air and groundwater are excluded because these are not usually considered in ecological risk assessment. A comprehensive source for these values is the RAIS at http://rais.ornl.gov maintained by the Oak Ridge National Laboratory. While at Oak Ridge, Dr. Glen Suter, mentioned previously, authored many of the documents that provide commonly used screening benchmarks.[11-16]

Ecological Screening Benchmarks for Surface Water

The list of National Ambient Water Quality Criteria (NAWQC) provides values for surface water. These criteria are generally protective but may not be applicable at every site. These criteria are based on three specific endpoints—human health, aquatic life, and organoleptic effects in humans such as smell and taste. These criteria are updated periodically and are currently available at http://water.epa.gov/scitech/swguidance/standards/criteria/current/index.cfm#altable. It should go without saying that the values based on aquatic life should be used in ecological risk assessment.

Ecological Screening Benchmarks for Soil

The first attempt at developing soil screening criteria for ecological risk assessment was begun in the 1980s by the Dutch government.[17] Ecological screening values for soil may be based on protection of endpoints in plants, soil invertebrates and the soil ecosystem, or vertebrates. The next attempt at a systematic collection of chemical-specific screening values was conducted at ORNL. For plants, soil benchmarks were determined based on a literature review and were based on laboratory experiments.[13] For invertebrates and soil bacteria, three specific endpoints were examined—toxicity to earthworms, toxicity to soil microbes and heterotrophic processes in the soil, and toxicity to invertebrates other than earthworms.[12] Earthworm toxicity was assessed from laboratory experiments reported in the scientific literature; toxicity to heterotrophic microbes was assessed based on the changes in the activities of various microbial enzymes, respiration, or the ability to fix nitrogen. Toxicity to invertebrates other than earthworms was assessed from laboratory experiments on mollusks, arthropods, or nematodes. Soil values were also developed for birds and mammals based on food chain modeling, for example, soil to earthworms to birds.[14]

In 1998, Gary Friday of the Westinghouse Savannah River Company compiled ecologically based soil screening values from a variety of sources including ORNL documents, US Fish and Wildlife Service benchmarks, the Dutch values, and values from the Canadian Council of Ministers of the Environment.[18] Over time, various sets of soil screening values have been assembled by different regulatory agencies in state governments with considerable overlap.

In 2003, EPA developed ecological soil screening levels (Eco-SSLs) for 24 chemicals. Eco-SSLs are concentrations of contaminants in soil protective of ecological receptors that contact or live in soil or ingest biota living in soil. Four groups of receptors are considered: plants, soil invertebrates, birds, and mammals. The 24 chemicals include metals and bioaccumulative organic chemicals.[19]

Ecological Screening Benchmarks for Sediment

The development and use of sediment screening benchmarks are controversial. Generally, sediment benchmarks are based on three methods: laboratory bioassays, field studies, and models of equilibrium partitioning.

The first method is laboratory bioassays conducted on contaminated sediments collected in the field or background sediments spiked in the laboratory with single chemicals or mixtures. The endpoint is generally mortality of a sediment-dwelling organism. Most often, the amphipod *Hyalella azteca* is used as the test organism. Other organisms used are the midge *Chironomus tentans* or the mayfly *Hexagenia* spp. Questions have been raised about whether these laboratory tests are applicable to field populations.[20]

The second method, field surveys, attempts to estimate the highest concentration of a particular contaminant that can be tolerated by 95% of the benthos. This screening-level concentration (SLC) approach uses field data from sites with different concentrations of contaminants in sediments and on the co-occurrence of benthic infaunal species in these sediments. At least 10 species and 10 different locations are required for each chemical. The frequency distribution of the concentrations of a contaminant at all sites where a given species is present is calculated, and the 90th percentile of this distribution is used as the SLC for that species. When these species-specific SLC values are developed for at least 10 species, the 5th percentile of the resulting distribution is thought to represent the concentration that 95% of the species can tolerate. This 5th percentile value becomes the SLC. This method assumes that the chemical concentration data used cover the full tolerance range of each species. Hence, a range of at least two orders of magnitude is needed for some validity. The concentration range is not always reported and the full tolerance range of most species remains unknown.[21]

The third method, equilibrium partitioning, is based on the idea that sediment-dwelling benthic organisms contact the interstitial pore water in sediment, and concentrations in pore water would thus reflect the most relevant exposure metric. This equilibrium-partitioning approach estimates the concentration of a chemical in pore water based on the concentration in bulk sediment. For nonionic organic chemicals, the partition coefficient is roughly equal to the organic carbon partition coefficient (K_{oc}) multiplied by the fraction of organic carbon in sediment.[22] For metals, the equilibrium-partitioning approach is more difficult and may not be possible because of a variety of factors; hence, for metals, measurement of the bioavailable pool of metal in sediment requires simultaneous measurement of acid-volatile sulfides.[23–25]

Clearly, there are areas of great uncertainty associated with all three approaches. The bulk of the work to integrate these three approaches has been conducted by Dr. Don MacDonald, a private consultant in Nanaimo on the island of Vancouver in British Columbia. The National Oceanic and Atmospheric Administration (NOAA) has adopted many of Dr. MacDonald's recommendations.[26–28] The state of Florida also depends on an evaluation by Dr. MacDonald for marine and estuarine sediment quality benchmarks.[29]

EPA's regional office in Chicago (Region 5) provides a compilation of sediment screening benchmarks on a single web page.[30]

Natural Resource Trustees

Communication with natural resource trustees and the decision to conduct a natural resource damage assessment (NRDA) may commence during step 1 of the ecological risk assessment process.

The National Oil and Hazardous Substances Pollution Contingency Plan (40 CFR 300) (NCP) and the CERCLA require that an NRDA be conducted as part of most hazardous waste site evaluations. Natural resource trustees include other federal agencies such as the NOAA; the US Fish and Wildlife Service; state officials, usually designated by the governor; and representatives from Native American tribes.

The requirements for completing an NRDA for the various trustees are slightly different. Risk assessors working on these will need to consult with both PMs and attorneys to make sure these requirements are met.

Step 2: Screening-Level Exposure Estimate and Risk Calculation

This step is analogous to screening for chemicals of potential concern (COPCs) in human health risk assessment. The acronym COPECs meaning "chemicals of potential ecological concern" has come into common usage. In this step, COPECs are determined by comparing the maximum detected level of a chemical in an environmental medium with the screening benchmark. If the maximum detect is greater than the benchmark, the chemical is considered a COPEC. The Process Document indicates that highly conservative exposure factors should be used. The other part of the step is the calculation of a screening-level HI as the ratio of the maximum detected concentration and the screening benchmark.

This calculation will produce a purposive overestimate of risk. In addition, because of uncertainty in the screening benchmarks, a number of these screening benchmarks are below naturally occurring background concentrations in soil. EPA's 2005 revision to the *Guidance for Developing Ecological Soil Screening Levels* indicates the following:

> It is EPA's policy to not screen against background levels. Background concentrations, the speciation of metals, and the effects of conservative modeling assumptions are generally taken into account in the initial steps of the baseline risk assessment.[19]

The upshot of this policy is that preliminary estimates of ecological risk are often reversed once background comparison is performed. However, the ecological risk assessment process does not include this comparison until step 3. These reversals are often difficult to communicate to stakeholders who may not be familiar with the process.

Regarding step 2, the Process Document indicates that ecological risk is estimated by comparing maximum concentrations detected with the ecotoxicity screening benchmarks collected in step 1. This results in a set of highly conservative (HQs) for the chemicals detected at the site. The Process Document also prescribes a scientific/management decision point (SMDP) at the conclusion of step 2. At this SMDP, the risk manager and risk assessment team will decide that either the SLERA is adequate to determine that ecological threats are negligible or that the process should continue to a more detailed ecological risk assessment.

This prescription for an SMDP following step 2 proved problematic. If an SMDP consisted of an informal conversation between a risk assessor and a PM, there would of course be no delay in the process. However, if the SMDP required the presence of stakeholders to attend a meeting in person, a situation involving airplane travel and other expenses, the difficulty that arose was that the information developed in steps 1 and 2 was almost always insufficient to decide whether the process should go forward.

Step 3: Baseline Risk Assessment Problem Formulation

Although a preliminary problem formulation had been developed in step 1, the baseline problem formulation used all other available information to develop more refined estimates of risk. The problem with putting this activity off until after the first SMDP was that the information needed to decide whether to proceed had not been developed. Step 3 included the following activities:

- Refinement of COPEC selection
- Review of information on contaminant fate and transport, exposure pathways, and ecosystems potentially at risk

- Comparison of site and background concentrations or risks
- Selection of assessment endpoints and refined estimation of exposure, likely using food chain modeling
- Development of a refined conceptual model and problem formulation with working hypotheses or questions that would be addressed by further site investigation

In 2000, the Department of Defense could not obtain funding for activities related to ecological risk assessment at military bases further into the future than the next SMDP. In response to this, regional toxicologists split step 3 into two parts—3a and 3b. In step 3a, information from the first four items in the previous list would be addressed. The SLERA would consist of a report or technical memorandum detailing the results of steps 1, 2, and 3a. An SMDP would then be conducted and would have sufficient information to make the decision whether to proceed. Step 3b would then include only the development of a refined site conceptual site model and problem formulation. Step 3b would occur only if the decision to move forward to a full BERA had been made in the SMDP.[31]

Step 3a

Once step 3 had been split into two parts, the development of the SLERA consisting of steps 1, 2, and 3a allowed ecological risk activities at many sites to be concluded much more quickly. Any available information that had bearing on the ecological risk at the site and did not require additional data gathering could be included. As noted, such information often consisted of a background comparison, assessment of bioavailability, and food chain modeling. For ecological risk assessment where significant work and resources might be involved if the assessment progressed, background comparison was almost always conducted—even though doing so was inconsistent with EPA's guidance on the use of background.[32]

Step 3b

Based on the information developed in steps 1, 2, and 3a and presented in the SLERA, a refined problem formulation was developed that included a detailed conceptual model and hypotheses regarding how the contaminants at the site are negatively impacting the ecosystem. These hypotheses should provide testable predictions that were subject to confirmation based on data that could be gathered at the site. Problem formulation would include consideration and refinement of the site conceptual model, exposure pathways, and selection of assessment endpoints and measurement endpoints. The exposure pathways must be linked to the assessment endpoints.

A particular strong line of evidence that ecological effects observed at the site are indeed site-related can be obtained by sampling along a biological gradient. For example, at a site where the occurrence of PAHs in sediment has been hypothesized to be toxic to benthic invertebrates, colocated sample points at which both sediment samples for laboratory PAH analysis and measures of number and diversity of benthic invertebrates (as a measure of the health of this ecological community) are obtained along a PAH concentration gradient.

The most useful time for the first comprehensive SMDP is after the SLERA has been completed. At this time, risk assessors, decision makers, and other stakeholders can discuss possible paths forward. The measurement endpoints need to reflect testable hypotheses; the choice of these hypotheses is based on the assessment endpoints and this choice needs agreement between all involved on these endpoints.

An excellent example of refined problem formulation is available in the literature. EPA worked with Dr. Don MacDonald (mentioned previously) on the ecological risk assessment for the Calcasieu Estuary near Lake Charles, Louisiana. In 1937, the Calcasieu ship channel allowed Lake Charles to become a deepwater port and enabled rapid industrial development. Because of this development, a portion of the estuary sediment became contaminated from industrial wastewater discharges, municipal wastewater discharge, spills associated with shipping activities, and likely other disposal processes. The SLERA was completed in 1999 and identified metals, PAHs, PCBs, pesticides, polychlorinated dibenzo-p-dioxins/polychlorinated dibenzofurans (PCDDs/PCDFs), chlorophenols, chlorinated benzenes, chlorinated ethanes, phthalates, cyanide, and acetone as COPECs. The refined problem formulation identified sediments, surface water, and surface microlayer at the air–water interface as relevant exposure media. A number of COPECs were bioaccumulative and others were directly toxic. The contaminants and relevant ecological receptors are shown in Table 6.1.

STEP 4: STUDY DESIGN AND DATA QUALITY OBJECTIVE PROCESS

The products of step 4 are the Work Plan (WP) and Sampling and Analysis Plan (SAP) for data collection activities related to the measurement endpoints. The WP should describe the following:

- Site conceptual model
- Exposure pathways
- Assessment endpoints
- Testable hypotheses
- Measurement endpoints
- Uncertainties and assumptions

The SAP provides details of the actual data collection and analysis procedures, including sampling techniques, data reduction, statistical analyses, QA, and QC. EPA has developed guidance on data quality objectives to ensure data collection results in interpretable results.[33–38]

Once the WP and SAP are developed, an SMDP should occur to ensure that all involved parties agree on the measurement endpoints and methods of data collection and interpretation. Rewriting the WP and SAP is often much more efficient than mobilizing a field sampling team on two or more separate occasions.

TABLE 6.1
Receptors and Contaminants Considered in the Calcasieu Estuary

Type of Toxicity	Receptors	Aquatic		Terrestrial	
		Contact	Ingestion	Contact	Ingestion
Bioaccumulative	Mercury	●	●		●
	PAHs				
	PCBs				
	Dioxins/furans				
	Hexachlorobenzene				
	Hexachlorobutadiene				
	Aldrin				
	Dieldrin				
Directly toxic in sediment	Copper	●	●		●
	Chromium				
	Lead				
	Mercury				
	Nickel				
	Zinc				
	PAHs				
	PCBs				
	HCB				
	HCBD				
	Carbon disulfide				
	Acetone				
	Ammonia				
	Hydrogen sulfide				
Directly toxic in surface water	Copper	●			●
	Mercury				
	1,2 Dichloroethane				
	Trichloroethane				

Step 5: Field Verification of Sampling Design

Before mobilizing a field sampling team, verification of the SAP should be performed to determine whether the proposed samples can actually be collected. Is the proposed sampling appropriate for the site and can the sampling approach be implemented?

Often, in step 4, as part of developing the WP and SAP, the ability and efficiency of the proposed data collection should be field-tested. The data quality objective process provides a means of specifying the number of samples needed to obtain sufficient statistical power to support a chosen level of confidence.[37,38] Can a sufficient number of samples be obtained? For example, if the SAP indicates collection of soil invertebrates, usually earthworms, it is necessary to ensure that earthworms are indeed present at the site and can be obtained in sufficient quantity for the proposed laboratory testing. Another example might be sediment sampling in a lake or river—is the water shallow enough that a team member in waders can collect the samples; or is a boat needed along with a dredge or grab sampler? Is the water deep enough to require a winch to retrieve the sampler?[39]

Reference areas are used to determine possible ecological impacts from site-related contaminants. Reference areas represent "background" conditions and should be selected to be as similar to the site as possible in all aspects except contamination. For example, should an ecological risk assessment be conducted at the former goldmine site discussed in Chapter 5, a reference creek might be found within the same drainage but topographically separated and thus likely unaffected by former mining activities at the site. The reference areas for soil and sediment comparisons should be evaluated for similarity in terms of slope, habitat, species potentially present, and other soil and sediment characteristics. The reference areas for surface water should be evaluated for similarity in terms of flow rates, substrate type, water depth, temperature, turbidity, oxygen levels, water hardness, pH, and possibly other water quality parameters.[7]

If fulfillment of the WP and SAP are not feasible, these will need to be rewritten. Hence, step 5 may be an iterative process and will require an SMDP to review any changes to the WP and SAP. As noted in the Process Document,

> In the worst cases, changes in the measurement endpoints could be necessary, with corresponding changes to the risk hypotheses and sampling design. Any new measurement endpoints must be evaluated according to their utility for inferring changes in the assessment endpoints and their compatibility with the site conceptual model (from Steps 3 and 4). Loss of the relationship between measurement endpoints and the assessment endpoints, the risk questions or testable hypothesis, and the site conceptual model will result in a failure to meet study objectives.[7]

STEP 6: SITE INVESTIGATION AND ANALYSIS PHASE

This step should be straightforward and follows the WP and SAP developed in steps 4 and 5. Despite careful planning, unexpected conditions may arise in the field. For example, a spring flood may change the course of a river once the water subsides. How will the new river course affect the chosen sampling locations?

Sampling along a range of contaminant concentrations will likely provide much useful information. Changing site conditions may require decisions in the field to ensure the measurement endpoints reflect the assessment endpoints and the data obtained are sufficient to address the hypotheses put forward in problem formulation.

In the field, initial sampling may reveal unexpected aspects of the nature and extent of contamination or about biological effects along contamination gradients. At times, the WP and SAP may need field modification, but it is important to ensure that the study objectives, both in terms of data quality and measurement endpoints, be met.

Exposure of organisms, communities, or populations can be defined as the co-occurrence in time and place of these ecological components and one or more stressors.[4] Spatial patterns or distributions of the stressor(s) and ecological components may appear to predict effects on ecological components. Sampling for biological effects along a gradient of contamination is especially important for demonstrating relationships between the spatial patterns of exposure and effects.

The biological gradient of effects can often be used to develop a site-specific TRV. This site-specific value will incorporate not only the intrinsic toxicity of the chemical or stressor but also site-specific aspects such as bioavailability. The site-specific TRV may be predictive of the spatial pattern of effects and this would constitute a strong piece of evidence that the stressor was indeed producing effects at the site. Likely other evidence will also emerge from the site investigation. This evidence should be evaluated for causal associations, potential confounding, and overall probative value.[4,40]

Site PMs or others in a risk management role may find such an exposure–response analysis especially useful for the evaluation of various cleanup strategies. Because the plan for site investigation has already been determined in steps 4 and 5, the hope is that the information developed in step 6 can be integrated into a credible risk characterization in step 7.

STEP 7: RISK CHARACTERIZATION

In step 7, information developed in step 6 on exposure and effects are integrated into a statement of risk regarding the assessment endpoints determined during problem formulation. Different studies or datasets can provide multiple lines of evidence to support the conclusions in step 6. All these lines of evidence should be factored into the risk characterization.

The risk characterization ideally provides the risk manager with contaminant concentration levels in various abiotic media that provide upper and lower bounds on the estimated threshold for adverse ecological effects. The upper bound would be developed using central tendency measures of exposure along with a TRV that represents a LOAEL value. The lower bound would be developed using more protective/conservative measures of exposure and a TRV that represents a NOAEL value. For higher-trophic-level organisms considered in assessment endpoints, actual sampling of tissue concentrations may not be possible for practical or ethical reasons; hence, the study design will necessarily have specified how trophic transfer/food chain modeling could be used to back-calculate the abiotic concentration representing the bounds on the effect threshold.

Risk characterization should also consider the uncertainties attendant in the risk characterization, both fate-and-transport and food chain models.

STEP 8: RISK MANAGEMENT

Although the Red Book and the Blue Book are explicit about the need for separation of risk assessment and risk management, both the risk manager and risk assessor must take into account the potential for ecological damage that might result from the implementation of risk management options. For example, how foolish it would be to bulldoze and clear a bottomland hardwood swamp to remove soil/sediment contamination because of predicted effects in avian populations if the adult females are feeding young birds in the nests and the resident population appears strong and healthy!

Obviously, the risk manager must understand the risk assessment, including the various assumptions and associated uncertainties. This final decision of selecting a risk management option will be greatly helped by a robust problem formulation in the earlier steps of the process.

SCREENING-LEVEL ECOLOGICAL RISK ASSESSMENT FOR A FORMER MANUFACTURED GAS PLANT IN A RAILROAD YARD

Manufactured gas was used throughout most of the nineteenth century and the first half of the twentieth century until the advent of natural gas production. The manufacturing process typically consisted of the gasification of combustible materials, almost always coal, but also wood and oil.

In the coal gasification process, coal gas was produced through the distillation of bituminous coal under anaerobic conditions. The gases produced were drawn off and some of the vapors were converted to liquids. The remaining vapor was coal gas.

In the carburetted water gas process, coal gas was produced and passed into a carburettor where oil was introduced into the vapor. This oil–gas mixture was superheated to thermally "crack" the oil. Carburetted water gas was, thus, a mixture of the gaseous products of coal and petroleum. Impurities were removed and the gas was passed through a scrubber that brought the vapor into direct contact with water. This process increased the thermal content of the fuel and the form of coal gas produces was known as "water gas." One of the scrubbing methods consisted of application of direct electrical current to precipitate the particles of coal tar. This electrical process required transformers and other components containing PCBs used as heat exchange and dielectric fluids. Often, railroads served manufactured gas plants to enable shipping of the gas to consumers.

In this example, outflow from a detention pond built to collect surface runoff from a manufactured gas plant enters a small unnamed stream that supports limited aquatic life (Figure 6.2). The stream flows about 1.5 river miles before its confluence with a river known internationally for its high-quality trout fishing. The contaminants at the former manufactured gas plant include metals from the spent coal, PAHs from the manufactured gas process, pesticides, PCBs from transformers, and polychlorinated dioxins and furans, occurring both as by-products of gas production and as contaminants of PCB mixtures.

The detention pond is large enough to support a fish population, but conversations with the plant manager indicated the pond had never been stocked with fish and that he had never seen fish in it. The chain-link fence surrounding the pond was old, and around the dam, there were gaps in the fence that could allow access by small terrestrial animals. Birds could, of course, access the pond from the air. The plant manager indicated the pond was dredged once a year to ensure it would have sufficient capacity to catch and hold runoff from the plant site. The dredge spoils were taken to a hazardous waste landfill about 400 miles away. Site sampling data from surface water and sediment in the detention pond for metals, PAHs, chlorinated pesticides, PCBs as Aroclors, and polychlorinated dibenzodioxins and furans (PCDD/Fs) are shown in Tables 6.2 through 6.10 in the following text. Sampling data for river sediment near the confluence with the tributary are shown in Tables 6.11 through 6.14.

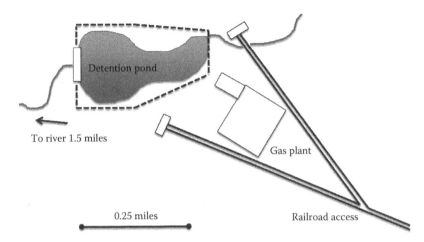

FIGURE 6.2 Simplified map of the manufactured gas plant. The dashed line around the pond shows the approximate location of a chain-link fence.

ECOLOGICAL RISK ASSESSMENT OF POLYCHLORINATED DIBENZODIOXINS AND FURANS

Most dioxins, furans, and dioxin-like compounds lack individual screening benchmarks. However, the congener-specific dioxin and furan data can be consolidated into a single measure, called the TEQ of the sample. The TEQ is calculated by multiplying the concentrations of each congener or congener containing chlorine at the 2,3,7, and 8 positions in a sample by a toxicity equivalence factor (TEF) and summing those products. The TEF normalizes the toxicity of those congeners to the toxicity of the 2,3,7,8-TCDD congener, generally considered to be the most toxic of the dioxin, furan, and dioxin-like compounds. A great deal of effort by internationally known scientists has gone into developing the TEF scheme for dioxin-like chemicals.[41–43] In effect, the TEQ indicates the concentration of TCDD that would have the same toxicity as the mixture of dioxins and furans being evaluated. The TEFs used here reference the WHO values for mammals, birds, and fish.[42] Figure 6.3 shows the structure of several PCDD/Fs.

ECOLOGICAL RISK ASSESSMENT OF POLYCYCLIC AROMATIC HYDROCARBONS AND POLYCHLORINATED BIPHENYLS

PAHs consist of many different chemicals, likely at least 100. PAHs occur from incomplete combustion, such as vehicle exhaust. PAHs are the carcinogenic molecules in cigarette smoke. Asphalt contains PAHs and they occur in sediment in most water bodies due to runoff. PAHs are a by-product of the manufactured gas process.

The bioavailability and toxicity of PAHs vary considerably. Higher molecular weight PAHs are less water soluble and have consequently low bioavailability. Although lower molecular weight PAHs have greater water solubility and are more bioavailable, they may be more subject to loss from evaporation and other

TABLE 6.2
Surface Water Sampling Results for Metals from the Detention Pond at the Manufactured Gas Plant

	Al	Sb	As	Ba	Be	Cd	Cr	Co	Cu	Fe	Pb	Mn	Hg	Ni	Se	Ag	Tl	Va	Zn
SW-1	22	<5	<5	34	<10	<0.05	0.27	<0.01	<0.5	248	<0.05	42	<0.01	<1	<0.2	<0.05	<1	<10	<10
SW-2	<0.02	<5	<5	36	<10	0.14	0.28	<0.01	1.6	580	<0.05	74	<0.01	5.3	<0.2	0.40	<1	<10	<10
SW-3	30	<5	<5	36	<10	0.22	<0.05	<0.01	2.2	220	<0.05	69	<0.01	16.6	<0.2	0.43	<1	<10	<10
SW-4	42	<5	<5	37	<10	<0.05	2.90	<0.01	2.1	526	0.95	104	<0.01	33.8	<0.2	0.57	<1	<10	<10
SW-5	87	<5	<5	39	<10	0.29	0.49	<0.01	1.6	273	<0.05	83	<0.01	1.5	<0.2	<0.05	<1	<10	<10
SW-6	70	<5	<5	40	<10	0.36	0.61	<0.01	3.6	488	<0.05	83	<0.01	33.8	<0.2	<0.05	<1	<10	<10
SW-7	<0.02	<5	<5	41	<10	0.44	<0.05	<0.01	3.9	294	<0.05	42	<0.01	200	<0.2	<0.05	<1	<10	<10
SW-8	107	<5	<5	43	<10	0.53	0.74	<0.01	3.0	458	<0.05	42	<0.01	1.5	1.1	<0.05	<1	<10	<10
SW-9	55	<5	<5	44	<10	0.63	1.09	<0.01	2.5	313	27	42	<0.01	2.7	<0.2	0.43	<1	<10	4.5
SW-10	133	<5	<5	45	<10	2.90	0.90	<0.01	4.3	432	<0.05	63	<0.01	1.0	<0.2	0.57	<1	<10	<10
SW-11	165	<5	<5	47	<10	<0.05	1.33	<0.01	2.7	332	<0.05	100	<0.01	2.7	<0.2	<0.05	<1	<10	<10
SW-12	210	<5	<5	48	<10	1.83	<0.05	<0.01	2.8	350	<0.05	97	<0.01	1.7	<0.2	<0.05	<1	<10	<10
SW-13	276	<5	<5	67	<10	0.90	7.40	<0.01	2.2	370	<0.05	91	<0.01	5.3	5.2	<0.05	<1	<10	2800
SW-14	402	<5	<5	52	<10	1.09	2.12	<0.01	3.4	409	<0.05	91	<0.01	6.5	<0.2	<0.05	<1	<10	<10
SW-15	<0.02	<5	<5	55	<10	0.75	1.65	<0.01	3.9	432	<0.05	55	<0.01	8.1	<0.2	0.43	<1	<10	<10
SW-16	840	<5	<5	67	<10	1.37	0.38	<0.01	5.0	700	<0.05	97	<0.01	1.7	<0.2	0.57	<1	<10	<10

Note: A number preceded by "<" indicates the chemical was not detected and the number is the analytical reporting limit from the laboratory. Units are μg/L.

TABLE 6.3

Surface Water Sampling Results for PAHs from the Detention Pond at the Manufactured Gas Plant

Analyte	SW-1	SW-2	SW-3	SW-4	SW-5	SW-6	SW-7	SW-8	SW-9	SW-10	SW-11	SW-12	SW-13	SW-14	SW-15	SW-16	SW-17	SW-18	SW-19	SW-20	SW-21
2-Chloronaphthalene	<5	<5																			
1-Methylnaphthalene	<1	<1	<1	<1	<1	<1	<1	<1	<1	<1	<1	<1	<1	<1	<1						
2-Methylnaphthalene	<0.5	210	<0.5	<0.5	11	<0.5	<0.5	<0.5	<0.5	<0.5	<0.5	<0.5	<0.5	<0.5							
Acenaphthene	<0.2	<0.2	0.37	0.4	0.9	2.6	<0.2	0.25	1.8	160	100	<0.2	<0.2	<0.2	<0.2	<0.2	<0.2	<0.2	<0.2	<0.2	<0.2
Acenaphthylene	10	<0.1	<0.1	<0.1	<0.1	<0.1	<0.1	<0.1	1.2	<0.1	<0.1	8	<0.1	<0.1	<0.1	<0.1	<0.1	<0.1	<0.1	<0.1	<0.1
Anthracene	16	<0.8	<0.8	<0.8	<0.8	<0.8	<0.8	<0.8	10	<0.8	<0.8	<0.8	<0.8	<0.8	<0.8	<0.8	<0.8	<0.8	<0.8	<0.8	<0.8
Benzo[a]anthracene	<0.05	0.09	<0.05	<0.05	<0.05	<0.05	<0.05	<0.05	<0.05	<0.05	<0.05	<0.05	5	<0.05	<0.05	<0.05	<0.05	<0.05	<0.05	<0.05	<0.05
Benzo[a]pyrene	<0.1	0.1	<0.1	<0.1	<0.1	<0.1	<0.1	<0.1	<0.1	<0.1	<0.1	<0.1	5	<0.1	<0.1	<0.1	<0.1	<0.1	<0.1	<0.1	<0.1
Benzo[b] fluoranthene	<0.1	<0.1	<0.1	<0.1	<0.1	<0.1	<0.1	<0.1	<0.1	<0.1	<0.1	<0.1	4.5	<0.1	<0.1	<0.1	<0.1	<0.1	<0.1	<0.1	<0.1
Benzo[g,h,i]perylene	<1	<1	<1	<1	<1	<1	<1	<1	<1	<1	<1	<1	<1	<1	<1	<1	<1	<1	<1	<1	<1
Benzo[k] fluoranthene	<0.1	<0.1	1.3	<0.1	<0.1	<0.1	<0.1	<0.1	<0.1	<0.1	<0.1	<0.1	5	<0.1	<0.1	<0.1	<0.1	<0.1	<0.1	<0.1	<0.1
Chrysene	<0.1	<0.1	<0.1	0.2	<0.1	<0.1	<0.1	<0.1	<0.1	<0.1	<0.1	<0.1	4.8	<0.1	<0.1	<0.1	<0.1	<0.1	<0.1	<0.1	<0.1
Dibenz[a,h] anthracene	<1	<1	<1	<1	<1	<1	<1	<1	<1	<1	<1	<1	<1	<1	<1	<1	<1	<1	<1	<1	<1
Fluoranthene	1.5	<0.07	<0.07	1.8	<0.07	<0.07	3	<0.07	<0.07	<0.07	<0.07	<0.07	6	<0.07	<0.07	0.9	<0.07	<0.07	<0.07	<0.07	<0.07
Fluorene	<0.5	4.4	<0.5	<0.5	<0.5	<0.5	62	<0.5	<0.5	<0.5	<0.5	<0.5	<0.5	<0.5	<0.5	<0.5	<0.5	<0.5	<0.5	<0.5	<0.5
Indeno[1,2,3-c,d] pyrene	<1	<1	<1	<1	<1	<1	<1	<1	<1	<1	5	<1	<1	<1	<1	<1	<1	<1	<1	<1	<1
Naphthalene	<0.2	<0.2	27	<0.2	<0.2	<0.2	8	<0.2	84	<0.2	<0.2	<0.2	<0.2	18	<0.2	<0.2	<0.2	520	<0.2	<0.2	<0.2
Phenanthrene	<0.2	<0.2	5.5	<0.2	<0.2	<0.2	<0.2	<0.2	<0.2	<0.2	<0.2	<0.2	<0.2	<0.2	<0.2	<0.2	<0.2	<0.2	<0.2	<0.2	<0.2
Pyrene	1.7	<0.08	<0.08	3.1	<0.08	<0.08	10	<0.08	<0.08	<0.08	<0.08	2.2	<0.08	<0.08	<0.08	<0.08	1.8	<0.08	<0.08	<0.08	<0.08

Note: A number preceded by "<" indicates the chemical was not detected and the number is the analytical reporting limit from the laboratory. Units are μg/L.

TABLE 6.4

Surface Water Sampling Results for Pesticides from the Detention Pond at the Manufactured Gas Plant

Analyte	SD-1	SD-2	SD-3	SD-4	SD-5	SD-6	SD-7	SD-8	SD-9	SD-10	SD-11	SD-12	SD-13	SD-14	SD-15
Beta-BHC	<0.003	<0.003	<0.003	<0.003	<0.003	<0.003	<0.003	<0.003	<0.003	<0.003	<0.003	<0.003	<0.003	<0.003	<0.003
Delta-BHC	<0.006	<0.006	<0.006	<0.006	<0.006	<0.006	<0.006	<0.006	<0.006	<0.006	<0.006	<0.006	<0.006	<0.006	<0.006
Gamma-BHC (lindane)	<0.005	<0.005	<0.005	<0.005	<0.005	<0.005	<0.005	<0.005	<0.005	<0.005	<0.005	<0.005	<0.005	<0.005	<0.005
Chlordane	<0.014	<0.014	<0.014	<0.014	<0.014	<0.014	<0.014	<0.014	<0.014	<0.014	<0.014	<0.014	<0.014	<0.014	<0.014
4,4'-DDD	<0.011	<0.011	<0.011	<0.011	<0.011	<0.011	<0.011	<0.011	<0.011	<0.011	<0.011	<0.011	<0.011	<0.011	<0.011
4,4'-DDE	<0.004	<0.004	<0.004	<0.004	<0.004	<0.004	<0.004	<0.004	<0.004	<0.004	<0.004	<0.004	<0.004	<0.004	<0.004
4,4'-DDT	<0.012	<0.012	<0.012	<0.012	<0.012	<0.012	<0.012	<0.012	<0.012	<0.012	<0.012	<0.012	<0.012	<0.012	<0.012
Dieldrin	<0.002	<0.002	<0.002	<0.002	<0.002	<0.002	<0.002	<0.002	<0.002	<0.002	<0.002	<0.002	<0.002	<0.002	<0.002
Endosulfan I	<0.007	<0.007	<0.007	<0.007	<0.007	<0.007	<0.007	<0.007	<0.007	<0.007	<0.007	<0.007	<0.007	<0.007	<0.007
Endosulfan II	<0.011	<0.011	<0.011	<0.011	<0.011	<0.011	<0.011	<0.011	<0.011	<0.011	<0.011	<0.011	<0.011	<0.011	<0.011
Endosulfan sulfate	<0.066	<0.066	<0.066	<0.066	<0.066	<0.066	<0.066	<0.066	<0.066	<0.066	<0.066	<0.066	<0.066	<0.066	<0.066
Endrin	<0.015	<0.015	<0.015	<0.015	<0.015	<0.015	<0.015	<0.015	<0.015	<0.015	<0.015	<0.015	<0.015	<0.015	<0.015
Endrin aldehyde	<0.023	<0.023	<0.023	<0.023	<0.023	<0.023	<0.023	<0.023	<0.023	<0.023	<0.023	<0.023	<0.023	<0.023	<0.023
Heptachlor	<0.055	<0.055	<0.055	<0.055	<0.055	<0.055	<0.055	<0.055	<0.055	<0.055	<0.055	<0.055	<0.055	<0.055	<0.055
Heptachlor epoxide	<0.083	<0.083	<0.083	<0.083	<0.083	<0.083	<0.083	<0.083	<0.083	<0.083	<0.083	<0.083	<0.083	<0.083	<0.083
Methoxychlor	<0.1	<0.1	<0.1	<0.1	<0.1	<0.1	<0.1	<0.1	<0.1	<0.1	<0.1	<0.1	<0.1	<0.1	<0.1
Toxaphene	<0.240	<0.240	<0.240	<0.240	<0.240	<0.240	<0.240	<0.240	<0.240	<0.240	<0.240	<0.240	<0.240	<0.240	<0.240

Note: A number preceded by "<" indicates the chemical was not detected and the number is the analytical reporting limit from the laboratory. Units are μg/L.

TABLE 6.5

Surface Water Sampling Results for PCBs and PCDD/Fs from the Detention Pond at the Manufactured Gas Plant

Analyte	SW-1	SD-2	SD-3	SD-4	SD-5	SD-6	SD-7	SD-8	SD-9	SD-10	SD-11	SD-12	SD-13	SD-14
PCBs														
Aroclor-1016	<1	<1	<1	<1	<1	<1	<1	<1	<1	<1	<1	<1	<1	<1
Aroclor-1221	<1	<1	<1	<1	<1	<1	<1	<1	<1	<1	<1	<1	<1	<1
Aroclor-1232	<1	<1	<1	<1	<1	<1	<1	<1	<1	<1	<1	<1	<1	<1
Aroclor-1242	<1	<1	<1	<1	<1	<1	<1	<1	<1	<1	<1	<1	<1	<1
Aroclor-1248	<1	<1	<1	<1	<1	<1	<1	<1	<1	<1	<1	<1	<1	<1
Aroclor-1254	<1	<1	<1	<1	<1	<1	<1	<1	<1	<1	<1	<1	<1	<1
Aroclor-1260	<1	<1	<1	<1	<1	<1	<1	<1	<1	<1	<1	<1	<1	<1
Aroclor-1268	<1	<1	<1	<1	<1	<1	<1	<1	<1	<1	<1	<1	<1	<1
Chlorinated dioxins and furans														
2378-TCDD	4,000	4.9	<2.4	<4.2	12,740	28.5	<3.7	760	<4.6	137	<7.3	<4	38	2.9
12378-PeCDD	14	<1.4	<1.9	<3.4	35.4	<4.1	<3.4	<5.6	<3.1	<5.9	<4.1	<0.5	<1.4	<2
123478-HxCDD	28.3	<1	<1.4	<2.9	<3.5	<2.8	10	<3.7	<8.9	<6.2	1.6	<0.8	<1.2	<4.4
123678-HxCDD	95.3	<0.9	<1.3	<2.7	6.4	<2.6	21.6	<3.5	<8.4	<5.8	<3.3	<0.7	<1.1	<4.5
123789-HxCDD	93.5	<4.1	<6.3	<5.9	<7.7	2.8	6.1	1.8	<6.3	<4.1	<6.3	<4.3	<5.9	<7.7
1234678-HpCDD	7.9	<5.2	31.7	<10.9	3	<3.4	<6.4	5.9	<6.0	12.4	<7.2	<6.8	24.3	<8.3
OCDD	14,890	59.8	30.5	74.7	27,700	726	<44.5	2,990	40.3	240	<19.3	623	<17.1	18.6
TCDF	30.9	<1.3	<2	<3.8	53.1	<3.1	<3.1	<4.3	<2.5	<0.3	<0.7	<1.2	<4	<2.7
12378-PeCDF	36.4	<1.3	<1.7	<3.1	14.1	<2.5	<2.8	<4.8	<2.9	<4.1	<2.8	<0.3	<0.9	<1.3
23478-PeCDF	18.3	<1.3	<1.7	<3.1	28.2	<2.5	<2.8	<4.8	<3	<4.3	<2.9	<0.3	<0.9	<1.3

(continued)

TABLE 6.5 (continued)

Surface Water Sampling Results for PCBs and PCDD/Fs from the Detention Pond at the Manufactured Gas Plant

Analyte	SW-1	SD-2	SD-3	SD-4	SD-5	SD-6	SD-7	SD-8	SD-9	SD-10	SD-11	SD-12	SD-13	SD-14
123478-HxCDF	42.7	<0.7	<1.0	<1.8	2.8	<1.8	16.5	<2.8	<6.4	<4.2	<4.3	<3.1	<3.6	<2.6
123678-HxCDF	34.2	<0.6	<0.8	<1.7	2.8	<1.7	15.5	<2.6	<6	<3.9	1.9	<0.5	<0.8	<3.1
123789-HxCDF	<0.8	<1	<2	6.3	<2.6	2.2	<4.3	<3.2	<7.1	<4.6	<0.2	<0.6	<1	<4
234679-HxCDF	43.6	<0.7	<0.9	<1.8	<2.3	<1.8	16.1	<2.8	<6.3	<4.1	5.9	<0.5	<0.8	<3.3
1234678-HpCDF	580	3.7	2.4	<2.5	29.5	4.3	<3.9	21.6	<6.6	24	<0.7	<1.2	<5.0	<5.3
1234789-HpCDF	50	<1.3	<1.6	<3.3	59.6	<4.3	<3.3	15.5	<11.2	<8.5	6.1	<1	<1.8	<7.6
OCDF	1,180	5.1	6.5	14.5	1,950	56	20.9	300	<6.4	24.7	<16.4	<31.6	<2.0	2.4

Notes: A number preceded by "<" indicates the chemical was not detected and the number is the analytical reporting limit from the laboratory. Units are ng/L. The chemical names for the various dioxin congeners are abbreviated. The full names can be found in Reference 42.

TABLE 6.6

Sediment Sampling Results for Metals from the Detention Pond at the Manufactured Gas Plant

	Al	Sb	As	Ba	Be	Cd	Cr	Co	Cu	Fe	Pb	Mn	Hg	Ni	Se	Ag	Tl	Va	Zn
SD-1	3,450	<0.62	1.1	154	0.25	<0.07	12.1	<3.5	1.39	1400	1230	78.5	0.05	0.64	<0.93	<2.9	<0.43	20.9	2060
SD-2	5,530	<0.62	<0.92	155	0.26	<0.07	8.5	<3.0	0.78	910	278	241	0.54	<0.4	<1	<1.2	<0.63	7.4	1550
SD-3	10,000	<0.98	<0.65	38.6	0.79	<0.07	51.7	2.7	0.51	770	0.46	260	0.66	3.7	<1.3	<0.52	<0.5	5.8	578
SD-4	14,500	<0.7	9.7	97.8	0.65	<0.07	329	2.3	0.91	3000	0.83	155	0.23	2.5	<1.4	0.5	10.4	12.4	235
SD-5	5,850	3.3	11.6	90.8	4.6	<0.07	11.2	5.8	0.51	290	10.2	184	<0.01	0.36	<1.1	9.8	8.9	72.4	1050
SD-6	13,300	2.7	10.9	94	4	<0.07	204	3.6	0.091	69	8.9	214	0.13	1.9	<1.6	9.9	<0.9	45.6	491
SD-7	13,500	37.1	1.8	77.4	1.7	<0.1	215	13.4	2.4	250	52.5	44.1	1.2	1.3	<1.3	<3.1	<0.81	11.1	42.3
SD-8	15,100	<2.6	<10.4	163	1.7	<0.1	176	19.6	1.5	2100	373	219	0.2	0.15	<1.1	<2.7	0.95	8.4	302
SD-9	15,400	3.6	<8.9	133	0.44	<0.1	157	<3.1	1.8	800	54.9	67.6	<0.01	0.73	<1.3	0.55	0.84	6.8	276
SD-10	2,810	<0.72	2	45.1	1	<0.07	4.7	<2.4	2.4	970	511	98.8	<0.01	0.69	<1.3	0.56	1.19	47.3	49.2
SD-11	2,150	<10	2	64.1	0.98	<0.07	4.1	<3.8	0.51	350	11.9	161	<0.01	12.8	<1	<2.2	<1.05	129	86.5
SD-12	6,440	<9	7.8	156	0.79	<0.07	93.9	7.2	1.8	380	587	99	<0.01	92.2	<1.2	<1.8	<0.79	10.6	<36
SD-13	15,100	1.9	6.1	92.5	0.51	<0.039	176	2.3	3.7	530	595	1490	<0.01	0.15	2.1	9.4	0.76	9.5	<43.1
SD-14	14,000	1.8	2.6	163	1.5	0.039	248	<1.4	9.2	60	854	516	<0.01	1.3	1.6	8.2	1.6	10.2	<29.2
SD-15	3,450	0.95	13.7	202	1.4	0.037	147	<1.1	2.1	67	1000	174	<0.01	<5.9	0.94	8.6	1.5	9.8	65
SD-16	5,530	0.93	6	22.6	0.33	0.029	178	3.8	2.4	710	7.7	346	<0.01	1.4	1.1	<1.8	<0.43	7.9	105
SD-17	10,000	1.3	4.5	19.4	0.47	0.021	15.8	1.7	1.5	330	174	81.8	<0.01	1.5	<0.98	8.4	<0.73	9.7	82
SD-18	14,500	2.5	8.3	29.5	1.2	<0.06	6.4	1.9	1.8	1900	350	257	<0.01	<0.039	<0.65	2.7	1.02	19.1	159
SD-19	5,850	<1.1	9.6	30	0.67	<0.06	331	1.5	2.4	1000	719	111	<0.01	<0.039	1.8	10.6	1.04	8.7	191
SD-20	13,300	<0.92	<0.54	22.1	2.9	<0.06	10.3	2.4	2.3	2500	199	149	<0.01	<0.039	<1.2	6.9	0.6	6.3	300
SD-21	13,500	3.3	<0.64	22.3	3.5	<0.1	120	1.4	0.51	4900	411	333	<0.01	<0.039	<0.77	<4.0	0.47	89.3	1930

Note: A number preceded by "<" indicates the chemical was not detected and the number is the analytical reporting limit from the laboratory. Units are mg/kg dry weight.

TABLE 6.7

Sediment Sampling Results for PAHs from the Detention Pond at the Manufactured Gas Plant

Analyte	SD-1	SD-2	SD-3	SD-4	SD-5	SD-6	SD-7	SD-8	SD-9	SD-10	SD-11
2-Chloronaphthalene	570	2400	4600	4200	4800	490	480	2100	9600	990	1900
1-Methylnaphthalene	2300	710	930	180	130	500	120	92	520	950	1500
2-Methylnaphthalene	1900	3800	2400	4700	440	2400	4700	4900	4800	400	3800
Acenaphthene	1900	530	2400	4700	440	2400	4700	4900	4800	480	400
Acenaphthylene	460	430	660	530	410	580	390	940	420	970	570
Anthracene	410	770	490	260	980	1100	79	550	3900	920	2200
Benzo[a]anthracene	2600	1300	3700	3100	380	2600	3900	230	1000	1800	360
Benzo[a]pyrene	400	3400	2400	640	440	380	900	1200	4800	83	400
Benzo[b]fluoranthene	430	3600	2400	750	440	520	950	1500	580	78	400
Benzo[g,h,i]perylene	4800	5000	990	2100	1700	690	1200	4800	4700	3600	6300
Benzo[k]fluoranthene	440	2400	2500	1500	1500	5500	510	130	480	3300	490
Chrysene	1000	1200	2200	4300	3200	2700	1700	4700	19000	3300	5200
Dibenz[a,h]anthracene	1700	130	480	2100	9600	130	1900	2000	2500	3600	400
Fluoranthene	1200	9500	2400	1400	440	900	1600	2500	840	320	46
Fluorene	2200	1800	360	380	2500	9800	2500	320	2500	430	730
Indeno[1,2,3-c,d]pyrene	310	2200	2400	600	440	320	600	950	920	4800	400
Naphthalene	1900	3800	2400	4700	440	2400	4700	4900	4800	410	3800
Phenanthrene	890	7000	2400	580	440	350	820	810	4800	170	400
Pyrene	1000	8100	2400	1200	440	710	1500	2300	620	2100	4100

TABLE 6.7 (continued)
Sediment Sampling Results for PAHs from the Detention Pond at the Manufactured Gas Plant

Analyte	SD-12	SD-13	SD-14	SD-15	SD-16	SD-17	SD-18	SD-19	SD-20	SD-21
2-Chloronaphthalene	2000	2500	3600	400	400	4200	490	490	4600	4,900
1-Methylnaphthalene	580	78	830	4100	9,700	960	1300	9,800	420	430
2-Methylnaphthalene	5800	4100	9700	960	420	4800	4500	1,200	100	830
Acenaphthene	3800	4800	4100	9700	4,800	960	420	4,800	410	4,600
Acenaphthylene	2400	4600	4200	4800	490	480	2100	9,600	62	1,900
Anthracene	1800	360	380	2500	9,800	2500	940	380	750	9,900
Benzo[a]anthracene	62	380	1200	530	250	350	1700	570	2500	280
Benzo[a]pyrene	2600	1200	430	1100	1,300	960	420	810	410	1,000
Benzo[b]fluoranthene	3200	1500	550	1100	3,200	960	55	690	410	1,000
Benzo[g,h,i]perylene	1500	2300	710	930	180	130	500	120	92	100
Benzo[k]fluoranthene	460	340	420	850	59	66	1200	500	1300	3,500
Chrysene	1400	1600	370	260	500	4100	2400	950	440	1,100
Dibenz[a,h]anthracene	400	4200	120	340	920	4900	2500	4,900	500	78
Fluoranthene	5800	3100	740	1900	4,800	960	110	1,200	410	2,300
Fluorene	450	470	940	380	400	9800	420	430	6600	530
Indeno[1,2,3-c,d]pyrene	1700	830	4100	9700	960	1300	420	660	410	830
Naphthalene	9600	4800	420	960	800	2100	950	460	2100	860
Phenanthrene	2900	4100	9700	2200	96,045	510	410	1,100	2700	14,000
Pyrene	1600	3500	960	67	1,400	410	4000	23,000	1400	630

Note: A number preceded by "<" indicates the chemical was not detected and the number is the analytical reporting limit from the laboratory. Units are mg/kg dry weight.

TABLE 6.8

Sediment Sampling Results for PCBs from the Detention Pond at the Manufactured Gas Plant

Analyte	SD-1	SD-2	SD-3	SD-4	SD-5	SD-6	SD-7	SD-8	SD-9	SD-10	SD-11
	Polychlorinated Biphenyls (µg/kg)										
Aroclor-1016	<0.14	<.2	2.5	2.4	2.3	2.3	2.4	6.3	4.7	<5	4.5
Aroclor-1221	<0.29	2.2	2.5	2.4	2.3	2.3	2.4	15.7	11.7	12.6	11.3
Aroclor-1232	<0.14	2.2	2.5	2.4	2.3	2.3	2.4	6.3	4.7	5	4.5
Aroclor-1242	2.5	2.3	2.3	2.3	2.1	2.1	2.2	2.1	2.42	2.1	4.3
Aroclor-1248	2.5	2.4	2.3	2.3	2.4	3.1	2.3	2.5	2.3	2.3	2.3
Aroclor-1254	0.044	2.2	2.5	2.4	2.3	2.3	2.4	3.1	2.3	2.5	2.3
Aroclor-1260	0.032	2.2	2.5	2.4	2.3	2.3	2.4	3.1	2.3	2.5	2.3
Aroclor-1268	2.2	4.8	2.4	2.3	2.3	2.4	6.6	2.3	2.5	2.3	3.3
PCBs (high risk)	5.04	9.1	12.1	11.8	11.4	12.1	11.7	20.3	16.4	11.9	17.9
PCBs (low risk)	2.37	7.2	7.4	7.1	6.9	7	11.4	11.7	9.5	9.8	10.1

TABLE 6.8 (continued)

Sediment Sampling Results for PCBs from the Detention Pond at the Manufactured Gas Plant

Analyte	SD-12	SD-13	SD-14	SD-15	SD-16	SD-17	SD-18	SD-19	SD-20	SD-21
					Polychlorinated Biphenyls (µg/kg)					
Aroclor-1016	4.7	4.6	2.1	4.1	2.2	2.1	2.42	4.2	8.6	<0.035
Aroclor-1221	11.6	11.5	2.1	10.3	2.2	2.1	2.42	10.6	21.5	<0.07
Aroclor-1232	4.7	4.6	2.1	4.1	2.2	2.1	2.42	4.2	8.6	<0.035
Aroclor-1242	<0.035	<0.038	<0.035	<0.035	<0.18	<0.038	<0.038	<0.36	<0.048	2.3
Aroclor-1248	2.1	2.1	2.2	2.1	2.42	2.1	4.3	<0.035	<0.038	<0.035
Aroclor-1254	2.3	2.3	2.1	2.1	2.2	2.1	2.42	2.1	4.3	<0.035
Aroclor-1260	2.3	2.3	2.1	2.1	2.2	<0.1	2.42	2.1	4.3	<0.035
Aroclor-1268	2.3	2.1	2.1	2.2	2.4	2.42	9.1	9.5	<0.035	<0.038
PCBs (high risk)	13.8	13.6	8.5	12.4	9.02	8.4	11.6	10.5	21.5	2.3
PCBs (low risk)	9.3	9	6.3	8.4	6.8	4.62	13.9	15.8	12.9	0.108

Note: A number preceded by "<" indicates the chemical was not detected and the number is the analytical reporting limit from the laboratory. Units are µg/kg dry weight.

TABLE 6.9

Sediment Sampling Results for PCDD/Fs from the Detention Pond at the Manufactured Gas Plant

Analyte	SD-1	SD-2	SD-3	SD-4	SD-5	SD-6	SD-7	SD-8	SD-9	SD-10	SD-11
	Polychlorinated Dioxins and Furans (PCDD/Fs) (µg/kg)										
2378-TCDD	0.12	0.087	0.032	0.404	0.018	0.019	0.048	0.195	0.029	0.161	0.264
12378-PeCDD	<0.007	<0.002	0.0005	0.0032	<0.002	<0.0007	<0.0008	0.002	<0.0005	<0.0009	0.0008
123478-HxCDD	<0.003	<0.001	<0.0003	0.006	<0.0008	<0.0005	<0.0004	0.0032	<0.0006	0.0015	0.0016
123678-HxCDD	0.0076	0.0032	0.0022	0.016	—	0.0025	—	0.0097	0.0009	0.0031	0.004
123789-HxCDD	0.0031	0.0011	0.0022	0.016	<0.0008	0.0015	0.0009	0.0107	0.0017	0.0036	0.0043
1234678-HpCDD	0.215	0.0044	0.0023	0.34	<0.016	<0.045	<0.011	0.190	<0.020	<0.077	<0.089
OCDD	1.30	<0.087	<0.043	<1.9	<0.131	<0.255	<0.077	<1.8	<0.119	<0.36	<0.559
TCDF	0.009	<0.42	<0.26	0.020	0.0013	0.005	0.052	0.019	<0.0004	0.0079	0.007
12378-PeCDF	<0.004	0.0041	0.0039	0.0019	—	—	0.0017	0.0019	0.0006	—	0.0003
23478-PeCDF	<0.003	—	0.0008	0.0024	<0.001	<0.0008	0.0015	0.0026	0.0009	<0.0023	0.0014
123478-HxCDF	—	<0.001	0.0012	0.0051	0.0013	0.0047	0.0022	0.0051	—	—	0.0058
123678-HxCDF	0.0053	—	0.0028	0.0033	—	0.002	0.0019	0.0041	0.0013	—	0.0036
123789-HxCDF	—	—	0.0015	—	—	—	—	—	—	—	—
234679-HxCDF	0.0051	—	—	0.0024	—	0.001	0.001	0.0025	0.0011	—	0.0033
1234678-HpCDF	0.0786	<0.003	0.0012	0.028	<0.001	0.018	<0.0013	0.051	<0.0004	0.015	0.142
1234789-HpCDF	<0.006	<0.003	0.022	<0.0015	<0.001	0.002	<0.0024	<0.0008	<0.0005	0.0023	0.001
OCDF	<0.13	—	0.0013	<0.053	<0.015	<0.023	—	<0.071	<0.009	<0.029	<0.058

TABLE 6.9 (continued)

Sediment Sampling Results for PCDD/Fs from the Detention Pond at the Manufactured Gas Plant

Analyte	SD-12	SD-13	SD-14	SD-15	SD-16	SD-17	SD-18	SD-19	SD-20	SD-21
	Polychlorinated Dioxins and Furans (PCDD/Fs) (μg/kg)									
2378-TCDD	0.019	0.112	0.029	<0.0007	0.469	0.611	7.987	3.534	10.051	7.468
12378-PeCDD	<0.0007	<0.0011	<0.0006	<0.0007	0.0013	0.0019	0.0072	0.0135	0.019	0.0129
123478-HxCDD	<0.0005	0.0013	<0.0003	<0.0006	0.0017	<0.0014	0.0077	0.0235	0.0356	0.0233
123678-HxCDD	—	0.0062	0.0020	—	0.0073	0.0087	0.0422	0.0714	0.089	0.0706
123789-HxCDD	0.0014	0.0053	0.0017	0.0005	0.016	0.0091	0.0422	0.0670	0.0761	0.0844
1234678-HpCDD	<0.01	<0.096	<0.025	<0.0009	<0.159	<0.220	0.709	1.424	1.607	1.787
OCDD	<0.069	<0.878	<0.163	<0.025	<0.942	<1.576	4.568	8.641	9.156	111.541
TCDF	0.0007	0.0061	<0.0005	<0.0006	<0.0009	0.0057	0.025	0.0453	0.047	0.0426
12378-PeCDF	0.0008	0.0022	0.0002	—	0.0013	0.0016	0.0075	0.0094	0.0079	0.0092
23478-PeCDF	0.0007	0.0024	0.0006	<0.0004	0.0021	0.0028	0.0126	0.0115	0.0195	0.0181
123478-HxCDF	0.0014	0.0075	—	—	0.0032	0.0122	0.0336	0.0447	0.0424	—
123678-HxCDF	0.0006	0.0027	0.0014	—	0.0013	0.0045	0.0236	0.0343	0.036	0.0325
123789-HxCDF	—	—	—	—	—	—	—	—	—	—
234679-HxCDF	—	0.0024	0.001	<0.0006	—	0.0029	0.012	0.0213	0.018	0.0221
1234678-HpCDF	0.0046	0.039	0.009	<0.0008	0.022	<0.0031	0.354	0.644	—	<0.0081
1234789-HpCDF	<0.0009	0.0018	<0.0005	—	<0.0012	<0.0034	<0.0053	0.021	0.031	0.0296
OCDF	<0.008	<0.054	<0.011	—	<0.033	<0.088	<0.311	<0.582	—	1.158

Note: A number preceded by "<" indicates the chemical was not detected and the number is the analytical reporting limit from the laboratory. Units are μg/kg dry weight.

TABLE 6.10

Sediment Sampling Results for Pesticides from the Detention Pond at the Manufactured Gas Plant

Analyte	SD-1	SD-2	SD-3	SD-4	SD-5	SD-6	SD-7	SD-8	SD-9	SD-10	SD-11	SD-12	SD-13	SD-14	SD-15	SD-16	SD-17	SD-18	SD-19	SD-20	SD-21
Beta-BHC	<2.4	<2.1	<2.2	<2.5	<2.5	32	<2.4	<2.2	<2.4	2.6	<2.5	<2.0	<1.9	<2.5	<2.6	<2.4	<2.1	<2.2	<2.5	<1.8	<2.0
Delta-BHC	<1.9	<2.0	<2.6	<2.4	<2.5	<2.1	<2.4	<2.5	<2.2	<2.5	<2.0	<2.5	<2.0	<1.9	<2.5	<2.6	<2.4	<2.1	<1.9	<2.3	<2.2
Gamma-BHC (lindane)	2.5	2.5	2.4	2.4	2.4	2.1	2.1	2.1	2.2	2.2	2.2	2.5	2.5	2.5	2.4	2.4	2.4	2.3	2.3	2.3	2.4
Chlordane	<1.9	<2.0	<2.6	<2.5	<2.3	<2.4	18	8.7	7	<.25	<2.0	14	2.7	5.3	2.8	4.1	<2.5	15	<2.1	>2.2	10
4,4'-DDD	5.8	5	4.9	4.8	4.9	4.9	5	5	5	4.8	5	4	7.3	3.5	4	4.6	3.8	3.7	4.9	5	4.9
4,4'-DDE	<3.7	3.8	5	5.2	4.4	4.7	4.8	43	4.8	4.9	3.9	3.8	4.8	31	4.7	4.7	4.8	4.2	4.8	4.6	4
4,4'-DDT	5.1	3.8	5	4.8	4.4	4.7	4.8	6.2	4.8	4.9	3.9	3.8	4.8	4	4.7	4.7	4.8	4.2	4.8	4.6	4
Dieldrin	13	<3.8	<5.0	22	<4.4	13	47	100	23	<4.9	<3.9	<3.8	<4.8	<4.0	<4.7	<4.8	<4.2	41	<4.0	16	7.2
Endosulfan I	<2.4	<2.1	<3.0	<2.1	<4.8	<5.3	<1.9	<2.2	<2.3	<5.5	<2.1	<2.2	<3.5	<2.0	<2.1	<3.9	<4.8	<2.5	<5.0	>2.4	<2.2
Endosulfan II	<3.7	<3.8	<5.0	<4.8	<4.4	<4.7	10	18	<4.8	<4.9	<3.9	<3.8	8.2	88	22	13	<4.8	<4.2	<4.8	<4.0	<4.2
Endosulfan sulfate	<4.	<4.8	<4.2	<4.8	<4.8	<4.6	<4.0	<4.2	<4.8	<4.6	5.4	5.7	<4.3	<4.6	<5.8	<4.7	11	<4.0	<5.0	<4.6	<3.7
Endrin	<3.7	<3.8	<5.0	9.1	<4.4	<4.7	28	<4.9	13	<4.9	<3.9	<3.8	<4.8	<4.0	<4.7	<4.8	<4.6	<4.0	<4.2	21	9.9
Endrin aldehyde	<4.6	<4.4	<4.6	19	<4.6	38	<4.7	3.6	5.4	<4.0	<5.0	<4.6	<4.3	<3.7	<3.6	<4.9	<5.0	<5.8	<4.9	<3.8	<4.0
Heptachlor	2.5	2.5	2.4	2.4	2.4	2.1	2.1	2.1	2.2	2.2	2.2	2.5	2.5	2.5	2.4	2.4	2.4	2.3	2.3	2.3	2.4
Heptachlor epoxide	<4.9	<3.9	<3.8	<4.8	<4.0	<4.7	<4.8	<4.6	<4.0	<4.9	<1.9	<2.0	<2.6	<2.4	<2.5	<1.9	<2.0	<2.6	<2.4	<4.0	<5.0
Methoxychlor	<20	<24	<22	<19	<25	39	<26	<25	<26	<26	<23	<18	<20	<25	39	<26	<25	<26	<24	<22	<19
Toxaphene	<190	<200	<260	<250	<230	<240	<250	<200	<190	<210	<240	<250	<220	<240	<230	<240	<250	<200	<190	<210	<240

Note: A number preceded by "<" indicates the chemical was not detected and the number is the analytical reporting limit from the laboratory. Units are µg/kg dry weight.

TABLE 6.11

Sediment Sampling Results for Metals from the River near the Confluence of the Tributary

	Al	Sb	As	Ba	Be	Cd	Cr	Co	Cu	Fe	Pb	Mn	Hg	Ni	Se	Ag	Tl	Va	Zn
R-SD-1	2967	<0.62	2	9.1	0.19	0.021	13.5	<1.4	1.39	980	3.5	42	0.23	0.5	<1.3	0.5	0.84	9.01	451
R-SD-2	1039	1.8	6	56.1	0.11	<0.06	37.7	2.3	1.39	297	5.6	50	0.66	27.8	<1.2	2.7	1.19	2.16	15
R-SD-3	2465	<0.70	<0.92	15.5	0.05	<0.06	39.7	1.5	1.8	207	0.8	22	<0.01	18.5	<1.2	<1.8	<0.9	4.18	22
R-SD-4	841	0.93	<0.54	1.9	0.17	<0.1	15.6	7.2	1.5	175	23.3	30	<0.01	0.5	<0.65	<1.8	0.6	0.47	71
R-SD-5	5366	3.6	1.1	6.3	0.04	<0.07	16.6	19.6	9.2	287	0.2	567	0.23	0.3	<0.65	9.9	<0.63	2.32	22
R-SD-6	1187	<0.53	<0.54	8.4	0.66	<0.039	71.0	1.5	0.91	117	312.1	23	0.66	30.6	<1.3	<2.2	1.19	2.81	102
R-SD-7	624	1.8	<0.64	11.1	0.06	<0.07	8.8	<3.0	2.3	148	0.1	67	0.66	0.3	<1.2	8.6	<0.81	2.95	13
R-SD-8	482	3.3	13.7	14.1	0.12	<0.06	43.7	1.7	0.091	46	111.6	574	<0.01	0.0	<1.1	<1.2	<0.43	3.79	100
R-SD-9	3471	0.93	2	8.9	0.15	<0.07	4.8	2.3	2.4	79	18.8	52	1.2	30.4	<1	8.6	<0.63	1.65	29
R-SD-10	1203	1.9	<0.65	2.6	0.02	<0.07	30.3	5.8	0.91	31	18.4	32	0.2	0.2	<1.1	<1.2	8.9	3.77	131

Note: A number preceded by "<" indicates the chemical was not detected and the number is the analytical reporting limit from the laboratory. Units are mg/kg dry weight.

TABLE 6.12

Sediment Sampling Results for PCBs and PCDD/Fs from the River near the Confluence of the Tributary

Analyte	R-SD-1	R-SD-2	R-SD-3	R-SD-4	R-SD-5	R-SD-6	R-SD-7	R-SD-8	R-SD-9	R-SD-10
PCBs (µg/kg)										
Aroclor-1016	2.1	<.2	2.2	2.1	<.2	4.2	2.4	8.6	2.1	4.5
Aroclor-1221	2.5	11.7	<0.29	2.4	21.5	2.5	11.7	2.1	11.7	<0.29
Aroclor-1232	2.2	2.2	2.42	4.2	4.6	2.3	4.1	2.2	2.5	2.42
Aroclor-1242	2.3	<0.36	<0.035	2.1	2.2	2.3	2.2	<0.038	<0.038	4.3
Aroclor-1248	2.1	4.3	2.3	2.3	2.3	2.1	2.3	2.3	2.3	<0.035
Aroclor-1254	2.2	2.2	<0.035	4.3	0.044	2.3	2.3	2.3	2.3	2.1
Aroclor-1260	3.1	0.032	2.1	<0.035	2.2	2.3	2.1	2.5	<0.035	2.5
Aroclor-1268	9.5	<0.038	<0.035	6.6	2.1	2.1	2.3	3.3	3.3	9.5
Chlorinated dioxins and furans (µg/kg)										
2378-TCDD	0.0079	0.195	<0.0007	0.029	0.087	0.161	<0.0007	0.123	0.029	0.018
12378-PeCDD	0.0013	0.0072	<0.0007	<0.0007	<0.0008	<0.0007	0.0005	<0.0007	0.002	0.0013
123478-HxCDD	<0.0004	0.0032	0.0356	0.0032	<0.0006	<0.0003	<0.003	0.0077	<0.0006	0.0013
123678-HxCDD	0.0714	0.0062	0.0025	—	0.0022	0.089	0.016	0.002	—	0.016
123789-HxCDD	0.016	0.0036	<0.0008	0.0014	0.0053	0.0422	0.0017	0.0031	0.0017	0.0014

1234678-HpCDD	1.607	0.0023	<0.01	0.0044	<0.089	<0.025	0.215	<0.0009	<0.01	1.424
OCDD	<0.087	<0.878	<0.025	<0.025	<0.043	<0.025	<0.163	<0.163	<0.255	<0.087
TCDF	0.019	0.052	0.0426	<0.0009	0.0061	0.025	<0.0005	0.0057	0.0453	0.052
12378-PeCDF	0.0013	0.0006	<0.004	0.0079	0.0008	0.0006	0.0008	0.0002	0.0022	0.0079
23478-PeCDF	<0.0004	0.0015	0.0126	0.0008	0.0008	0.0115	0.0014	<0.0023	0.0181	0.0006
123478-HxCDF	0.0075	0.0447	0.0424	0.0447	0.0058	0.0336	—	<0.001	0.0032	—
123678-HxCDF	0.0019	0.0236	0.0014	0.0014	0.0013	—	0.0343	0.0045	0.0041	0.002
123789-HxCDF	—	—	—	0.0015	—	—	—	—	—	—
234679-HxCDF	0.012	0.0025	—	0.0025	0.018	—	—	0.0029	0.0025	—
1234678-HpCDF	0.015	0.022	0.039	0.644	0.009	<0.0006	0.022	0.0046	0.644	0.018
1234789-HpCDF	<0.0015	0.0023	0.0296	<0.0012	<0.0053	<0.006	<0.0005	<0.0008	0.0296	<0.0005
OCDF	<0.011	1.158	0.0013	<0.033	0.0013	<0.088	—	—	<0.009	<0.054

Note: A number preceded by "<" indicates the chemical was not detected and the number is the analytical reporting limit from the laboratory. Units are mg/kg dry weight.

TABLE 6.13

Sediment Sampling Results for PAHs from the River near the Confluence of the Tributary

Analyte	SD-1	SD-2	SD-3	SD-4	SD-5	SD-6	SD-7	SD-8	SD-9	SD-10
2-Chloronaphthalene	166	627	57	163	70	229	318	91	58	743
1-Methylnaphthalene	31	122	245	10	18	26	18	28	41	39
2-Methylnaphthalene	83	629	63	351	18	137	71	876	192	104
Acenaphthene	895	46	643	338	673	375	106	630	22	84
Acenaphthylene	113	66	19	28	19	19	1323	8	1184	142
Anthracene	60	26	27	14	105	352	62	217	15	46
Benzo[a]anthracene	14	181	34	20	108	109	176	67	157	308
Benzo[a]pyrene	21	870	91	86	57	24	6	13	444	63
Benzo[b]fluoranthene	77	14	369	166	60	86	51	137	399	6
Benzo[g,h,i]perylene	24	21	34	177	5	12	360	344	550	492
Benzo[k]fluoranthene	16	88	10	708	425	511	61	292	25	148
Chrysene	24	111	52	606	131	111	181	313	187	213
Dibenz[a,h]anthracene	21	434	88	644	29	70	74	328	5	1691
Fluoranthene	262	131	910	10	310	25	3	55	105	55
Fluorene	608	15	157	1681	20	44	106	410	1455	215
Indeno[1,2,3-c,d]pyrene	296	81	80	384	160	200	27	33	133	13
Naphthalene	73	286	394	42	165	18	27	645	485	378
Phenanthrene	43	150	116	141	210	257	23	52	146	1217
Pyrene	175	88	76	79	126	139	4435	961	81	103

Note: A number preceded by "<" indicates the chemical was not detected and the number is the analytical reporting limit from the laboratory. Units are µg/kg dry weight.

TABLE 6.14

Sediment Sampling Results for Pesticides from the River near the Confluence of the Tributary

Analyte	SD-1	SD-2	SD-3	SD-4	SD-5	SD-6	SD-7	SD-8	SD-9	SD-10
Beta-BHC	<2.4	<2.1	<2.2	2.6	<2.4	<2.4	<2.6	<2.2	<2.5	<2.1
Delta-BHC	<2.3	<1.9	<2.4	<2.5	<1.9	<2.5	<2.2	<2.0	<2.4	<2.2
Gamma-BHC (lindane)	2.1	2.4	2.2	2.4	2.5	2.4	2.4	2.3	2.2	2.4
Chlordane	10	<2.5	8.7	<1.9	4.1	8.7	<2.0	<2.5	2.8	<2.6
4,4'-DDD	3.7	3.5	4.9	4.9	4.6	4.9	5	3.8	4.8	4.9
4,4'-DDE	4	3.8	3.8	5.2	4.8	5	4.6	3.8	4.6	4.7
4,4'-DDT	4.2	5	5	5.1	3.9	5	4.2	4.7	4.8	5.1
Dieldrin	<4.4	<3.8	41	<4.7	16	13	<4.2	<4.9	<4.4	13
Endosulfan I	<2.0	<2.1	<5.0	<2.3	<4.8	<2.3	<2.1	<2.2	<5.5	<2.2
Endosulfan II	<4.0	<4.2	22	<5.0	<4.8	<3.9	<4.8	<4.4	18	<4.8
Endosulfan sulfate	11	<4.0	<4.8	<4.8	<5.8	<4.8	<4.2	<4.8	<4.6	<5.0
Endrin	21	<4.4	<4.6	<4.2	<5.0	<3.8	<4.0	<4.4	<4.7	<3.8
Endrin aldehyde	<4.0	<5.0	<4.0	<4.6	<5.0	<4.0	3.6	<4.7	<4.7	<4.9
Heptachlor	2.5	2.4	2.4	2.2	2.4	2.4	2.4	2.1	2.2	2.5
Heptachlor epoxide	<2.6	<4.8	<4.8	<3.9	<4.9	<4.0	<4.0	<2.6	<2.6	<3.9
Methoxychlor	<26	<26	<26	<26	<20	<26	<25	<26	<19	<26
Toxaphene	<200	<250	<240	<190	<240	<250	<240	<190	<240	<210

Note: A number preceded by "<" indicates the chemical was not detected and the number is the analytical reporting limit from the laboratory. Units are µg/kg dry weight.

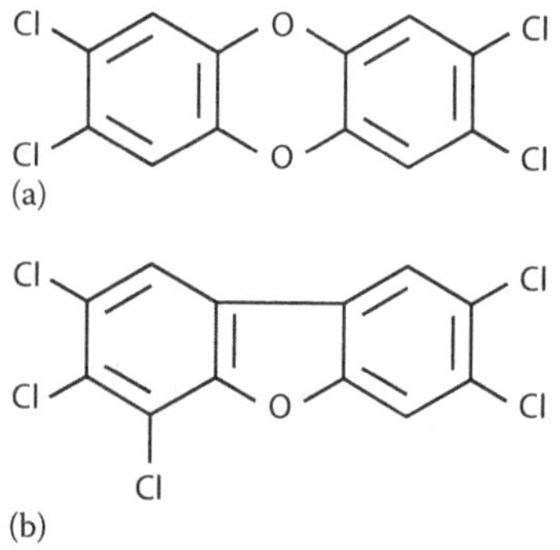

FIGURE 6.3   Structures of two PCDD/F congeners: (a) 2,3,7,8-tetrachlorodibenzodioxin and (b) 2,3,4,7,8-pentachlorodibenzofuran.

weathering processes in the environment. The ecological effects of PAH mixtures can be approximated using estimates of total PAHs. This calculation is most often done by adding up concentrations of each PAH congener in a sample. The surrogate value for non-detect values is calculated as half the reporting limit for the sample.

PCB mixtures contain up to 209 different chemicals, called congeners. There are 209 possible combinations for 1–10 chlorine atoms bound to biphenyl.[44] PCBs produce toxicity by a number of mechanisms.[45] Twelve PCB congeners have dioxin-like properties. Often, these PCB congeners are sampled individually and added to the TEQ calculation discussed in the previous section.[42]

Manufacture of PCBs ceased in the late 1970s. The commercial mixtures of PCBs were used dielectric fluid in transformers and capacitors and were sold as mixtures known as Aroclors. Each Aroclor mixture differed in the weight percent of chlorine present. In the environment, weathering by various biotic and abiotic processes alters the PCB congener composition from the original commercial products. Hence, sampling for Aroclors cannot provide accurate information about the PCB congener mixtures present in environmental media and biota.[46] Notwithstanding, Aroclor analyses are often still used for environmental samples because of cost. Often, TRVs for PCBs are expressed in terms of total PCBs; hence, for some ecological risk assessments, Aroclor analyses can provide a measure of total PCBs and be useful.

Similar to PCBs, the ecological effects of PAH mixtures can be approximated using estimates of total PCBs. This calculation is most often done by adding up concentrations of each PAH congener in a sample. The surrogate value for non-detect values is calculated as half the reporting limit for the sample. The structure of two representative PAHs and a PCB congener is shown in Figure 6.4.

BIOACCUMULATION OF CONTAMINANTS

The most direct and ecologically relevant approach to assessing bioaccumulation is to measure concentrations of contaminants in the tissue of organisms collected or exposed in the field. Sampling of benthic organisms, which provides a much clearer understanding of the bioavailability of contaminants and extent of

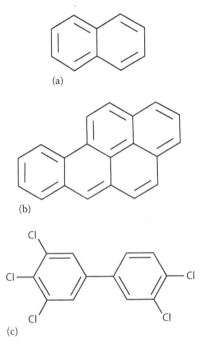

(a)

(b)

(c)

FIGURE 6.4 Structure of two PAHs and one PCB. (a) Naphthalene, (b) benzo-[a]-pyrene, (c) 3,3′,4,4′,5-pentachlorobiphenyl (PCB 126).

contamination, is often limited by the need to obtain a sufficient tissue mass for analysis.[47] Most often, searching the scientific literature can provide values for bioaccumulation/biomagnification factors.

SITE VISIT

Within a mile of the pond, the unnamed stream exiting from the detention pond is not deep enough to support fish—even after a rain. Downstream, the unnamed stream joins a larger stream that flows into the river. The risk assessor had conducted a site visit, and after inspecting the gas plant and detention pond, he had walked the unnamed stream all the way from the pond outlet to the river. Although the hillside on the downstream side of the detention pond was steep, after 200 yd, the gradient became much flatter and the stream made many turns before its confluence with the larger tributary. Near the confluence, the unnamed stream widened to become a slow-moving pool. The risk assessor noted considerable sediment accumulation in the pool. He also noted the presence of a number of aquatic invertebrates, including mayflies, caddis flies, and crayfish. No fish were observed in the unnamed stream or the larger tributary, but when the risk assessor reached the river, he saw several anglers wading about 100 ft offshore. He watched them and noted that one of them was consistently hooking and releasing trout. As he turned to leave, he noticed several mink tracks in the muddy bank. Looking up, he saw a great blue heron rise from the trees 100 ft downstream and take to the air.

Scientific/Management Decision Point #1

The risk assessor prepared a preliminary site visit report for the PM. This report included the conceptual site model shown in Figure 6.5.

In follow-up discussions, the PM noted that the pool in the unnamed stream would likely act as a sediment trap.

"That's exactly right," the risk assessor agreed. "That's what I'm expecting samples to show."

"I had samples taken from the pond and from the river near where the tributary comes in, but none in the stream."

"I doubt there'll be much in the river," said the risk assessor, "but we'll see."

"Samples should be here day after tomorrow. Do the screening for the river first. By the way, I just got this, this morning and printed a copy for you. It's a compilation of dioxin screening levels in sediment.[48] I'll send it by email as well."

"This is pretty complete," said the risk assessor as he scanned the document. "Great!"

When the risk assessor received the sample results, he prepared a set of tables using a spreadsheet so he would have easy access to the data. Tables 6.2 through 6.14 show the results of his work. He reformatted these tables in Microsoft Excel so that they could be easily imported into EPA's ProUCL software to be able to calculate the needed statistics. From these calculations, he prepared screening tables for COPECs for both the detention pond and the river (Tables 6.15 through 6.27). When he showed the tables to the PM, she noted that with the exception of PAHs and dioxins, the concentrations in river sediment were essentially no different than the concentrations in the detention pond.

"Aren't the chemicals all background? No one uses those chlorinated pesticides any more. Haven't they all been banned?" she asked.

"Maybe not," he answered. "The metals in the river sediment are all COPECs in the pond but have lower concentrations. I looked at the sampling locations in the river and the samples were all taken in deposition areas, sediment traps. It's especially hard to say where the metals came from. PCB concentrations in the river and the pond are about the same—of course, spring runoff could have carried the PCBs downstream. The dioxin concentrations are in the river so I suppose they could have been flushed as well. You are right about the legacy pesticides." He showed her the preliminary hazard characterization from steps 1 and 2 conducted with measured concentrations in river sediment (Table 6.28).

"Nothing there but dioxins," she said.

"Depends on the screening value you use. I figure there's more concern for the river so I used a set of 2007 numbers from Oregon. For the detention pond, I used a set of 1993 numbers from the EPA lab in Duluth, Minnesota. That's the difference."[48]

"So we really don't know," said the PM.

"There's another thing to think about," said the risk assessor. "Part of the reason the river is such a good fishery is that it's a tailwater. The water comes out the dam twenty miles upstream. It's a constant 58 degrees Fahrenheit all year—that's

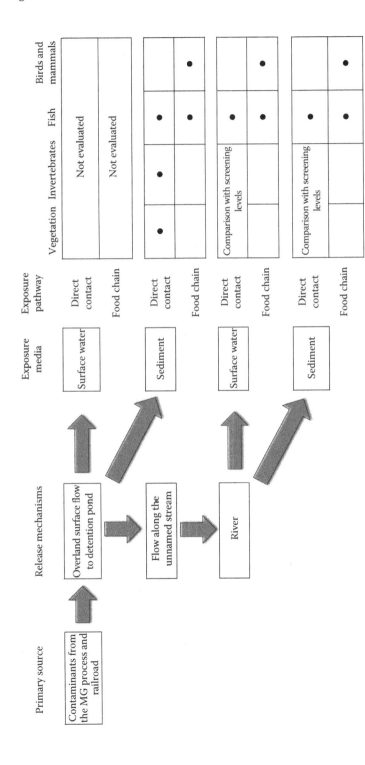

FIGURE 6.5 Conceptual site model for the manufactured gas plant, detention pond, and river.

TABLE 6.15

COPEC Screening for Metals in Surface Water in the Detention Pond

Chemical	N	# Det.	Freq. (%)	Min. ND	Max. ND	Min. Detect	Max. Detect	Arithmetic Mean (ND/2)	NAWQC Acute/Chronic Screening Benchmark	OSWER Ambient Water Quality Criteria	OSWER Tier II Secondary Screening Benchmark	EPA R3 Freshwater Screening Benchmark	Potential for Bioaccum.	COPC (Y/N)	95% UCL
Aluminum	16	13	81	<0.02	<0.02	22	840	152.4	750/87			87		Y	389[a]
Antimony	16	0	0	<5	<5	—	—	—				30		N	
Arsenic	16	0	0	<5	<5	—	—	—	340/150	190		5	Y	N	
Barium	16	16	100	—	—	34	87	45			3.9	4		Y	50.28[b]
Beryllium	16	0	0	<1	<1	—	—	—			5.1	0.66		N	
Cadmium	16	13	81	<0.05	<0.05	0.14	2.9	0.72				0.25	Y	N	
Chromium	16	13	81	<0.05	<0.05	0.27	7.4	1.265	570/74	180		85		N	
Cobalt	16	0	0	<0.01	<0.01	—	—	—			3	23		N	
Copper	16	15	94	<0.05	<0.05	1.6	5	2.802	13/9	11		9	Y	N	

Iron	16	16	100	—	—	220	700	401		1000		300		Y	458.1[c]
Lead	16	2	13	<0.05	<0.05	0.95	27	1.769	65/2.5	2.50		2.50	Y	Y	24.78[d]
Manganese	16	16	100	<0.05	—	42	104	73			80	120	Y	Y	85.59[b]
Mercury	16	0	0	<0.01	<0.01	—	—	—	1.4/0.77	1.3		0.026		N	
Nickel	16	15	94	<1	<1	1	200	21.48	470/52	160		52	Y	Y	97.09[e]
Selenium	16	2	13	<0.2	<0.2	1.1	5.2	0.481		5		1	Y	Y	4.848[d]
Silver	16	7	44	<0.05	<0.05	0.400	0.57	0.227	3.2			3.2	Y	N	
Thallium	16	0	0	<1	<1	—	—	—				0.8		N	
Vanadium	16	0	0	<10	<10	—	—	—			19	20		N	
Zinc	16	2	13	<10	<10	4.5	2800	179.7	120	100		120	Y	Y	2560[d]

Note: Units for all values are µg/L.

[a] 95% KM (Chebyshev).
[b] 95% Approx. gamma.
[c] 95% Student's t.
[d] 99% KM (Chebyshev).
[e] 97.5% KM (Chebyshev).

TABLE 6.16
COPEC Screening for PAHs in Surface Water in the Detention Pond

Chemical	N	# Det.	Freq. (%)	Min. ND	Max. ND	Min. Detect	Max. Detect	Arithmetic Mean (ND/2)	EPA R3/ R5/R6 Freshwater Screening Benchmark	Potential for Bioaccum.	COPEC (Y/N)	95% UCL
2-Chloronaphthalene	2	0	0	<5	<5	—	—	—	54	Y	N	
1-Methylnaphthalene	15	0	0	<1	<1	—	—	—	2.1	Y	N	
2-Methylnaphthalene	14	2	14	<0.5	<0.5	11	210	16	4.7	Y	Y	109.6[a]
Acenaphthene	21	8	38	<0.2	<0.2	0.25	160	12.74	5.8	Y	Y	103.7[b]
Acenaphthylene	21	3	14	<0.1	<0.1	1.2	10	0.96	4840	Y	N	
Anthracene	21	2	10	<0.8	<0.8	10	16	1.6	0.012	Y	Y	10.97[c]
Benzo[a]anthracene	21	2	10	<0.05	<0.05	0.09	5	0.265	34.6	Y	N	3.535[b]
Benzo[a]pyrene	21	2	10	<0.1	<0.1	0.1	5	0.288	0.015	Y	Y	3.537[b]
Benzo[b]fluoranthene	21	1	5	<0.1	<0.1	4.5	4.5	4.5	9.07	Y	N	
Benzo[g,h,i]perylene	21	0	0	<0.1	<0.1	—	—	—	7.64	Y	N	
Benzo[k]fluoranthene	21	2	10	<0.1	<0.1	1.3	5	0.345	NA	Y	N	
Chrysene	21	2	10	<0.1	<0.1	0.2	4.8	0.283	7	Y	N	

											N	
Dibenz[a.h]anthracene	21	0	0	<1	<1	—	—	—	5	Y	N	2.314[d]
Fluoranthene	21	5	24	<0.07	<0.07	0.9	6	0.655	1.9	Y	Y	30.78[e]
Fluorene	21	2	10	<0.5	<0.5	4.4	62	3.388	11	Y	Y	NA[f]
Indeno[1,2,3-c,d]pyrene	21	1	5	<1	<1	5	5	5.00	4.31	Y	Y	83.49[g]
Naphthalene	21	3	14	<0.2	<0.2	8	520	26.51	13	Y	Y	17.58
Phenanthrene	21	3	14	<0.2	<0.2	5.5	84	5.205	0.4	Y	Y	2.937
Pyrene	21	5	24	<0.08	<0.08	1.7	10	0.93	0.025	Y	Y	

Note: Units for all values are µg/L.

[a] 95% KM (Chebyshev).
[b] 99% KM (Chebyshev).
[c] 95% KM (bootstrap).
[d] 95% KM percentile bootstrap.
[e] 97.5% KM (Chebyshev).
[f] Not processed by ProUCL.
[g] 95% KM (t).

TABLE 6.17

COPEC Screening for PCDD/Fs in Surface Water in the Detention Pond

Chemical	Units	N	# Det.	Freq. (%)	Min. ND	Max. ND	Min. Detect	Max. Detect	Arith. Mean of Detections	EPA R4 Chronic Screening Benchmark	EPA R5 ESL Screening Benchmark	EPA R3 Screening Benchmark	Potential for Bioaccum.	COPC (Y/N)	95% UCL
2378-TCDD	pg/L	14	8	57	<2.4	<7.3	2.9	12740	2214						
12378-PeCDD	pg/L	14	2	14	<0.5	<5.9	14	35	24.7						
123478-HxCDD	pg/L	14	3	21	<0.8	<8.9	1.6	28.3	13.3						
123678-HxCDD	pg/L	14	3	21	<0.7	<8.4	6.4	95.30	41.1						
123789-HxCDD	pg/L	14	4	29	<4.1	<7.7	1.8	94	26.05						
1234678-HpCDD	pg/L	14	5	36	<3.4	<24.3	3	31.70	12.18						
OCDD	pg/L	14	11	79	<17.1	<44.5	18.6	27700	4308						
TCDF	pg/L	14	2	14	<0.3	<4.3	30.9	53.10	42						
12378-PeCDF	pg/L	14	1	7	<0.3	<14.1	36.4	36.40	36.4						
23478-PeCDF	pg/L	14	1	7	<0.3	<28.2	18.3	18.3	18.3						
123478-HxCDF	pg/L	14	2	14	<0.7	<16.5	2.8	42.7	22.75						
123678-HxCDF	pg/L	14	3	21	<0.5	<15.5	1.9	34.2	12.97						
123789-HxCDF	pg/L	14	2	14	<0.2	<7.1	2.2	6.3	4.25						
234679-HxCDF	pg/L	14	3	21	<0.5	<6.3	5.9	43.6	21.87						
1234678-HpCDF	pg/L	14	7	50	<0.7	<6.6	2.4	580	95.07						
1234789-HpCDF	pg/L	14	4	29	<1	<11.2	6.1	59.6	32.8						
OCDF	pg/L	14	10	71	<2	<31.6	2.4	1950	356			Screening conducted on TEQs as shown below			
Mammalian TEQ	μg/L	14	14	100			3.12E-06	1.28E-02	1.28E-03	1.00E-05	3.00E-09	3.10E-09	Y	Y	2.836
Avian TEQ	μg/L	14	14	100			3.57E-06	1.29E-02	1.29E-03	1.00E-05	3.00E-09	3.10E-09	Y	Y	3.06
Fish TEQ	μg/L	14	14	100			2.90E-06	1.28E-02	1.28E-03	1.00E-05	3.00E-09	3.10E-09	Y	Y	3.049

Note: All UCL values were calculated used the 95% percentile bootstrap.

TABLE 6.18

COPEC Screening for Metals in Sediment in the Detention Pond

Chemical	N	# Det.	Freq. (%)	Min. ND	Max. ND	Min. Detect	Max. Detect	Arithmetic Mean (ND/2)	EPA R3 Screening Benchmark	EPA R4 Screening Benchmark	EPA R5 Screening Benchmark	EPA R6 Screening Benchmark	COPC (Y/N)	95% UCL
Aluminum	21	21	100			2,150	15,400	9,679					Y	14,280[a]
Antimony	21	11	52	<0.62	<10	0.93	37.1	5.398	2	12		2	Y	14.23[b]
Arsenic	21	15	71	<0.54	<10.4	1.1	13.7	6.513	9.8	7.24	9.79	5.9	Y	6.841[c]
Barium	21	21	100			19.4	202	89.15					Y	119.4[d]
Beryllium	21	21	100			0.25	4.6	1.411					Y	1.977[d]
Cadmium	21	4	19	<0.039	<0.1	0.02	0.04	0.032		1	0.99	0.6	N	
Chromium	21	21	100			4.1	331	119	43.4	52.3	43.4	37.3	Y	196.2[d]
Cobalt	21	21	100			1	19.6	4.19	50		50		N	
Copper	21	21	100			0.091	9.2	2	31.6	18.7	31.6	35.7	Y	2.719[d]
Iron	21	21	100			60	4,900	1,109	20,000			20,000	N	
Lead	21	21	100			46	1,230	353.700	35.8	30.2	35.8	35	Y	696.3[c]
Manganese	21	21	100			44.1	1,490	251	460			460	Y	351.4[d]
Mercury	21	7	33	<0.01	<0.01	0.05	1.20	0.430	0.18	0.13	0.17	0.15	Y	0.352[f]
Nickel	21	15	71	<0.039	<0.64	0.15	92.2	8.439	22.7	15.9	18	20.9	Y	49.82[g]

(continued)

TABLE 6.18 (continued)
COPEC Screening for Metals in Sediment in the Detention Pond

Chemical	N	# Det.	Freq. (%)	Min. ND	Max. ND	Min. Detect	Max. Detect	Arithmetic Mean (ND/2)	EPA R3 Screening Benchmark	EPA R4 Screening Benchmark	EPA R5 Screening Benchmark	EPA R6 Screening Benchmark	COPC (Y/N)	95% UCL
Selenium	21	5	24	<0.65	<1.6	0.94	2.1	1.51	2				Y	1.217[c]
Silver	21	12	57	<0.52	<4	0.500	10.6	6.34	1	2	0.5	1	Y	9.711[b]
Thallium	21	12	57	<0.43	<1.05	0.47	10.4	2.439					Y	2.653[c]
Vanadium	21	21	100			5.8	129	26.10					Y	57.52[a]
Zinc	21	18	86	<29.2	<43.1	42.3	2,060	530.700	121	124	121	150	Y	1,065[h]

Note: Units for all values are mg/kg dry weight.

[a] 95% Chebyshev (mean, SD) UCL.
[b] 97.5% KM (Chebyshev).
[c] 95% KM (t).
[d] 95% Approx. gamma.
[e] 95% Adjusted gamma.
[f] 95% KM (percentile bootstrap).
[g] 99% KM (Chebyshev).
[h] 95% KM (Chebyshev).

TABLE 6.19

COPEC Screening for PAHs in Sediment in the Detention Pond

Chemical	N	# Det.	Freq. (%)	Min. ND	Max. ND	Min. Detect	Max. Detect	Arithmetic Mean (ND/2)	Effect Concentrations	Potential for Bioaccum.	COPEC (Y/N)	95% UCL
2-Chloronaphthalene	21	21	100	—	—	400	9,600	2,653				
1-Methylnaphthalene	21	21	100	—	—	78	9,800	1,720				
2-Methylnaphthalene	21	21	100	—	—	100	9,700	3,174				
Acenaphthene	21	21	100	—	—	400	9,700	3,145				
Acenaphthylene	21	21	100	—	—	62	9,600	1,762				
Anthracene	21	21	100	—	—	79	9,900	1,951		Assessed at total PAHs		
Benzo[a]anthracene	21	21	100	—	—	62	3,900	1,371				
Benzo[a]pyrene	21	21	100	—	—	83	4,800	1,203				
Benzo[b]fluoranthene	21	21	100	—	—	55	3,600	1,158				
Benzo[g,h,i]perylene	21	21	100	—	—	92	6,300	2,021				
Benzo[k]fluoranthene	21	21	100	—	—	59	5,500	1,307				
Chrysene	21	21	100	—	—	260	19,000	2,934				
Dibenz[a,h]anthracene	21	21	100	—	—	78	9,600	2,067				
Fluoranthene	21	21	100	—	—	46	9,500	2,022				
Fluorene	21	21	100	—	—	320	9,800	2,092				
Indeno[1,2,3-c,d]pyrene	21	21	100	—	—	310	9,700	1,660				
Naphthalene	21	21	100	—	—	410	9,600	2,729				
Phenanthrene	21	21	100	—	—	170	96,045	7,254				
Pyrene	21	21	100	—	—	67	23,000	2,926				
									TEC PEC			
Total PAHs	21	21	100	—	—	24,835	1,36,424	45,147	1,610 22,800	Y	Y	53,435[a]

Note: Units for all values are mg/kg dry weight. TEC is the abbreviation for Threshold Effect Concentration and PEC is the abbreviation for Probable Effect Concentration.

[a] 95% approx. gamma UCL.

TABLE 6.20

COPEC Screening for PCBs in Sediment in the Detention Pond

Chemical	N	# Det.	Freq. (%)	Min. ND	Max. ND	Min. Detect	Max. Detect	Arithmetic Mean (ND/2)	Effect Concentrations	Potential for Bioaccum.	COPEC (Y/N)	95% UCL
Aroclor-1016	21	13	62	<0.035	<5	2.1	8.6	3.1	70/7			
Aroclor-1221	21	15	71	<0.07	<0.29	2.1	21.5	6.662				
Aroclor-1232	21	14	67	<0.035	<0.14	2.1	8.6	3.319				
Aroclor-1242	21	6	29	<0.035	<0.36	2.1	4.3	1.401				
Aroclor-1248	21	8	38	<0.035	<0.038	2.1	4.3	2.099	30			
Aroclor-1254	21	9	43	<0.035	<0.035	0.044	4.3	2.204	63/60			
Aroclor-1260	21	9	43	<0.035	<0.1	0.032	4.3	2.106	0.005			
Aroclor-1268	21	11	52	<0.035	<0.038	2.1	9.5	3.001		Assessed as total PCBs		
Total PCBs	21	21	100	2.442		2.442	47.36	24.01	TEC 59.8 PEC 676	Y	N	

Note: Units for all values are μg/kg dry wt.

TABLE 6.21

COPEC Screening for PCDD/Fs in Sediment in the Detention Pond

Chemical	N	# Det.	Freq.	Min. ND	Max. ND	Min. Detect	Max. Detect	Arith. Mean of Detections	Screening Benchmark	Potential for Bioaccum.	COPC (Y/N)	95% UCL
2378-TCDD	21	20	95%	<0.7	<0.7	18	10,050	1,583				
12378-PeCDD	21	10	48%	<0.5	<7	0.5	19	6.23				
123478-HxCDD	21	10	48%	<0.3	<3	1.3	35.6	10.5				
123678-HxCDD	17	17	100%	<0	<0	0.9	89	20.4				
123789-HxCDD	21	20	95%	<0.8	<0.8	0.5	84.4	17.4				
1234678-HpCDD	21	9	43%	<0.9	<220	2.3	1,787	698				
OCDD	21	6	29%	<25	<1,900	1,300	1,11,500	22,830				
TCDF	21	15	71%	<0.4	<420	0.7	52	19.6				
12378-PeCDF	17	16	94%	<4	<4	0.2	9.4	3.41				
23478-PeCDF	20	15	75%	<0.4	<3	0.6	19.5	5.33				
123478-HxCDF	15	14	93%	<1	<1	1.2	44.7	12.2				
123678-HxCDF	17	17	100%	<0	<0	0.6	36	9.48				
123789-HxCDF	1	1	100%	<0	<0	1.5	1.5	1.5				
234679-HxCDF	14	11	79%	<0	<0	1	22.1	6.86				
1234678-HpCDF	20	14	70%	<0.4	<8.1	1.2	644	101				
1234789-HpCDF	21	8	38%	<0.5	<6	1	31	13.8				
OCDF	17	2	12%	<8	<582	1.3	1,158	580				
Mammalian TEQ	21	21	100%			0.923	10,130	1,527	2.5–25	Y	Y	5,719[a]
Avian TEQ	21	21	100%			1.27	10,160	1,551	21–210	Y	Y	5,743[a]
Fish TEQ	21	21	100%			0.978	10,120	1,522	60–100	Y	Y	5,705[a]

Note: Units for all values are ng/kg dry weight.

[a] 97.5% Chebyshev (mean, SD) UCL.

TABLE 6.22

COPEC Screening for Pesticides in Sediment in the Detention Pond

Chemical	N	# Det.	Freq. (%)	Min. ND	Max. ND	Min. Detect	Max. Detect	Arithmetic Mean (ND/2)	EPA R3 Screening Benchmark	EPA R4 Screening Benchmark	EPA R5 Screening Benchmark	EPA R6 Screening Benchmark	COPC (Y/N)	95% UCL
Beta-BHC	21	1	5	<1.8	<2.6	32	32	32	5		5	5	N	
Delta-BHC	21	0	0	<1.9	<2.6				6,400		71,500		N	
Gamma-BHC (lindane)	21	0	0	<2.1	<2.5				2.37	3.3	2.37	0.94	N	
Chlordane	21	10	48	<0.25	<2.6	2.7	18	4.713	3.24	1.7	3.24	4.5	Y	7.655[a]
4,4'-DDD	21	21	100	<3.5		3.5	7.3	4.8	1.22	3.3	4.88	1.22	Y	5.101[b]
4,4'-DDE	21	20	95	<3.7	<3.7	3.8	43	7.512	2.07	3.3	1.42	2.07	Y	17.09[c]
4,4'-DDT	21	1	5	<3.8	<5.1	6.2	6.2	6.2	1.19	3.3	1.19	1.19	N	
Dieldrin	21	9	43	<3.8	<5	7.2	100	14.68	1.9	3.3	295	2.85	Y	26.13[a]
Endosulfan I	21	0	0	<1.9	<5.5				2.9		3.26		N	

Endosulfan II	21	6	29	<3.7	<5	8.2	88	9.148	14		1.94		Y	20.47[a]
Endosulfan sulfate	21	3	14	<3.7	<5.8	5.4	11	2.993	5.4		3.46		Y	6.231[a]
Endrin	21	5	24	<3.7	<5	9.1	28	5.529	2.22	3.3	2.22	2.67	Y	12.75[a]
Endrin aldehyde	21	4	19	<3.6	<5.8	3.6	38	4.964			480		Y	9.477[a]
Heptachlor	21	0	0	<2.1	<2.5				68		0.6		N	
Heptachlor epoxide	21	0	0	<1.9	<4.9				2.47		2.47	0.6	N	
Methoxychlor	21	2	10	<18	<26	39	39	39	18.7		13.6		Y	39[d]
Toxaphene	21	0	0	<190	<260				0.1		0.077		N	

Note: Units for all values are µg/kg dry weight.

[a] 95% KM (t) UCL.

[b] 95% Student's t UCL.

[c] 95% KM (Chebyshev) UCL.

[d] Maximum value.

TABLE 6.23

COPEC Screening for Metals in Sediment in the River

Chemical	N	# Det.	Freq. (%)	Min. ND	Max. ND	Min. Detect	Max. Detect	Arithmetic Mean (ND/2)	EPA R3 Screening Benchmark	EPA R4 Screening Benchmark	EPA R5 Screening Benchmark	EPA R6 Screening Benchmark	COPC (Y/N)	95% UCL
Aluminum	10	10	100				482	5366					N	
Antimony	10	7	70	<0.53	<0.7	0.93	3.6	1.519	2	12		2	Y	2.305[a]
Arsenic	10	5	50	<0.54	<0.92	1.1	13.7	2.645	9.8	7.24	9.79	5.9	Y	5.516[a]
Barium	10	10	100			1.87	56.1	13.41					N	
Beryllium	10	10	100			0.0166	0.66	0.157					N	
Cadmium	10	1	10	<0.039	<0.07	0.21	0.21	0.21		1	0.99	0.6	N	
Chromium	10	10	100			4.76	71.0	28.17	43.4	52.3	43.4	37.3	Y	39.97[b]
Cobalt	10	1	10	<1.4	<3	1.5	19.6	4.41	50		50		N	
Copper	10	10	100			0.091	9.2	2.189	31.6	18.7	31.6	35.7	N	
Iron	10	10	100			31.4	980	236.7	20,000			20,000	N	

Lead	10	10	100			0.123	312.1	49.44	35.8	30.2	35.8	35	Y	235.4[c]
Manganese	10	10	100			22.0	573.6	145.9	460			460	Y	455[d]
Mercury	10	7	70	<0.01	<0.01	0.2	1.2	0.386	0.18	0.13	0.17	0.15	Y	0.646[a]
Nickel	10	10	100			0.022	30.6	10.9	22.7	15.9	18	20.9	N	
Selenium	10	0	0	<0.65	<1.3				2				N	
Silver	10	5	50	<1.2	<2.2	0.5	9.9	3.44	1	2	0.5	1	Y	5.77[a]
Thallium	10	5	50	<0.43	<0.9	0.6	8.9	1.442					N	
Vanadium	10	10	100			0.466	9.02	3.312					N	
Zinc	10	10	100			12.67	450.7	95.5	121	124	121	150	N	

Note: Units for all values are mg/kg dry weight.

[a] 95% KM (t) UCL.
[b] 95% Student's t UCL.
[c] 95% Adj. gamma UCL.
[d] 95% Chebyshev (mean, SD) UCL.

TABLE 6.24

COPEC Screening for PAHs in Sediment in the River

Chemical	N	# Det.	Freq. (%)	Min. ND	Max. ND	Min. Detect	Max. Detect	Arithmetic Mean (ND/2)	Effect Concentrations	Potential for Bioaccum.	COPEC (Y/N)	95% UCL
2-Chloronaphthalene	10	10	100	—	—	0.057	0.473	0.252				
1-Methylnaphthalene	10	10	100	—	—	0.01	0.245	0.058				
2-Methylnaphthalene	10	10	100	—	—	0.018	0.876	0.252				
Acenaphthene	10	10	100	—	—	0.022	0.895	0.381				
Acenaphthylene	10	10	100	—	—	0.008	1.323	0.292				
Anthracene	10	10	100	—	—	0.014	0.352	0.092				
Benzo[a]anthracene	10	10	100	—	—	0.014	0.308	0.117				
Benzo[a]pyrene	10	10	100	—	—	0.006	0.870	0.168				
Benzo[b]fluoranthene	10	10	100	—	—	0.006	0.399	0.137				
Benzo[g,h,i]perylene	10	10	100	—	—	0.005	0.550	0.202				
Benzo[k]fluoranthene	10	10	100	—	—	0.01	0.708	0.228				
Chrysene	10	10	100	—	—	0.024	0.606	0.193				
Dibenz[a,h]anthracene	10	10	100	—	—	0.005	1.691	0.338				
Fluoranthene	10	10	100	—	—	0.003	0.910	0.187				
Fluorene	10	10	100	—	—	0.015	1.681	0.471				
Indeno[1,2,3-c,d]pyrene	10	10	100	—	—	0.013	0.384	0.141				
Naphthalene	10	10	100	—	—	0.018	0.645	0.251				
Phenanthrene	10	10	100	—	—	0.023	1.217	0.236				
Pyrene	10	10	100	—	—	0.076	4.435	0.626		Assessed as total PAHs		
									TEC	PEC		
Total PAHs	10	10	100			2.71	7.43	4.62	1.610	22.8		5.58[a]

Note: Units for all values are mg/kg dry weight.

[a] 95% Students-t UCL.

TABLE 6.25
COPEC Screening for PCBs in Sediment in the River

Chemical	N	# Det.	Freq. (%)	Min. ND	Max. ND	Min. Detect	Max. Detect	Arithmetic Mean (ND/2)	Effect Concentrations	Potential for Bioaccum.	COPEC (Y/N)	95% UCL
Aroclor-1016	10	9	90	<0.2	<0.2	0.2	8.6	2.85				
Aroclor-1221	10	8	80	<0.29	<0.29	2.1	21.5	6.639				
Aroclor-1232	10	10	100			2.2	4.6	2.914				
Aroclor-1242	10	6	60	<0.035	<0.36	2.1	4.3	1.564				
Aroclor-1248	10	9	90	<0.035	<0.035	2.1	4.3	2.232		Assessed as total PCBs		
Aroclor-1254	10	9	90	<0.035	<0.035	0.044	4.3	2.237				
Aroclor-1260	10	8	80	<0.035	<0.035	0.032	3.1	1.687				
Aroclor-1268	10	8	80	<0.035	<0.038	2.1	9.5	3.874				
									TEC PEC			
Total PCBs	10	10				9.22	35.04	23.77	59.8 676			27.63[a]

Note: Units for all values are µg/kg dry weight.

[a] 95% Student's t UCL.

TABLE 6.26
COPEC Screening for PCDD/Fs in Sediment in the River

Chemical	N	# Det.	Freq. (%)	Min. ND	Max. ND	Min. Detect	Max. Detect	Arith. Mean of Detections	Screening Benchmark Range	Potential for Bioaccum.	COPC (Y/N)	95% UCL
2378-TCDD	10	8	80	<0.7	<0.7	7.9	195	86				
12378-PeCDD	10	5	50	<0.7	<0.8	0.5	7.2	2.46				
123478-HxCDD	10	5	50	<0.3	<3	1.3	35.6	10.2				
123678-HxCDD	8	8	100	<0	<0	2	89	25.7				
123789-HxCDD	10	9	90	<0.8	<0.8	1.4	42.2	8.49				
1234678-HpCDD	10	6	60	<0.9	<89	2.3	1607	544				
OCDD	10	0	0	<10	<10							
TCDF	10	8	80	<0.5	<0.9	5.7	52	31				
12378-PeCDF	10	9	90	<4	<4	0.2	7.9	2.48				
23478-PeCDF	10	8	80	<0.4	<2.3	0.6	18.1	5.91				
123478-HxCDF	8	7	88	<1	<1	3.2	44.7	26				
123678-HxCDF	9	9	100	<0	<0	1.3	34.3	8.28				
123789-HxCDF	1	1	100	<0	<0	1.5	1.5	1.5				
234679-HxCDF	6	6	100	<0	<0	2.5	18	6.73				
1234678-HpCDF	10	9	90	<0.6	<0.6	4.6	644	158				
1234789-HpCDF	10	3	30	<0.5	<6	2.3	29.6	20.5				
OCDF	8	3	38	<9	<88	1.3	1158	387				
Mammalian TEQ	10	10				9.34	217	82.6	0.052–1.4	Y	Y	124[a]
Avian TEQ	10	10				6.86	264	102	0.07–3.5	Y	Y	149[a]
Fish TEQ	10	10				6.35	215	78	0.56	Y	Y	120[a]

Note: Units for all values are ng/kg dry weight.
[a] 95% Student's t UCL.

TABLE 6.27
COPEC Screening for Pesticides in Sediment in the River

Chemical	N	# Det.	Freq. (%)	Min. ND	Max. ND	Min. Detect	Max. Detect	Arithmetic Mean (ND/2)	EPA R3 Screening Benchmark	EPA R4 Screening Benchmark	EPA R5 Screening Benchmark	EPA R6 Screening Benchmark	COPC (Y/N)	95% UCL
Beta-BHC	10	0		<2.1	<2.6				5		5	5	N	
Delta-BHC	10	0		<1.9	<2.5				6,400		71,500		N	
Gamma-BHC (lindane)	10	10				2.1	2.5	2.33	2.37	3.3	2.37	0.94	N	
Chlordane	10	5	50	<1.9	<2.6	2.8	10	4.005	3.24	1.7	3.24	4.5	N	
4,4'-DDD	10	10				3.5	5	4.5	1.22	3.3	4.88	1.22	Y	4.841[a]
4,4'-DDE	10	10				3.8	5.2	4.43	2.07	3.3	1.42	2.07	Y	4.739[a]
4,4'-DDT	10	10				3.9	5.1	4.7	1.19	3.3	1.19	1.19	Y	4.978[b]
Dieldrin	10	4	40	<3.8	<4.9	13	41	9.62	1.9	3.3	295	2.85	Y	
Endosulfan I	10	0		<2	<5.5				2.9		3.26		N	
Endosulfan II	10	2	20	<3.9	<5	18	22	5.795	14		1.94		Y	19.38[c]
Endosulfan sulfate	10	1	10	<4	<5	11	11		5.4		3.46		N	
Endrin	10	1	10	<3.8	<5	21	21		2.22	3.3	2.22	2.67	N	
Endrin aldehyde	10	0		<3.6	<5						480		N	
Heptachlor	10	10				2.1	2.5	2.35	68		0.6	0.6	N	
Heptachlor epoxide	10	0		<2.6	<4.9				2.47		2.47		N	
Methoxychlor	10	0		<19	<26				18.7		13.6		N	
Toxaphene	10	0		<190	<250				0.1		0.077		N	

Note: Units for all values are µg/kg dry weight.

[a] 95% Student's t UCL.

[b] 95% Approx. gamma UCL.

[c] 95% KM (t) UCL.

TABLE 6.28
Preliminary Range of HQs for COPECs in Sediment in the River

Chemical	Units	Average Conc.	95% UCL	Max. Detect	Threshold Effect Level[a]	Probable Effect Level[b]	Range of HQs
Inorganic substances							
Antimony	mg/kg	1.52	2.305	3.6	2	25	0.06–2
Arsenic	mg/kg	2.65	5.52	13.7	9.79	33	0.08–1
Chromium	mg/kg	28.2	39.97	71	43.4	111	0.25–2
Lead	mg/kg	49.4	235.4	312	35.8	128	0.4–9
Manganese	mg/kg	146	574	455	460		0.1–0.7
Mercury	mg/kg	0.386	0.646	1.2	0.18	1.06	0.4–7
Silver	mg/kg	3.44	5.77	3.44	1.000	4	0.9–6
PAHs							
Total PAHs	mg/kg	4.62	5.58	7.43	1.61	23	0.2–5
PCBs							
Total PCBs	µg/kg	23.77	27.63	35.04	59.8	676	0.04–0.6
PCDD/Fs							
PCDD/Fs (TEQ-mammalian)	ng/kg	82.6	124	217	0.52	1	60–400
PCDD/Fs (TEQ-avian)	ng/kg	102	149	264	0.7	4	30–400
PCDD/Fs (TEQ-fish)	ng/kg	78	120	215	0.56	NA	140–500

[a] For metals, PAHs, and PCBs, the TEL was the consensus TEC; for dioxin-like chemicals (PCDDs/Fs), the threshold effect level was the lower end of the range from Oregon DEQ (2007) reported in Klamath.

[b] For metals, PAHs, and PCBs, the PEL was the consensus PEC; for dioxin-like chemicals (PCDDs/Fs), the threshold effect level was the upper end of the range from Oregon DEQ (2007) reported in Klamath.

why the fishing is so good. Much of what's in the river sediment could have come from the lake sediment above the dam."

"You know there used to be a paper mill above the lake," she said. "That could be where the dioxin came from." Both of them were aware of the reputation of paper mills. "Do you agree that it makes no sense to try to clean up the pond?" asked the PM. "Don't we just need to show that it's not likely to impact the river?

"I'll do a food chain model for birds and mammals. That should allay any doubt that the river is impacted. It's easy enough to do a mink and a great blue heron. I even saw a heron on the site visit. I don't know how I'll do the fish, but I'll figure something out."

"You can do it," she told him. "I know what a smart guy you are."

"Can we collect fish and get 'em analyzed?" he asked.

"Sure. But no more than ten. You know how expensive the analytical gets."

TABLE 6.29
Sampling Results from Trout Obtained
from the River (ng/kg Wet Weight)

Congener	#1	#2	#3	#4	#5	#6	#7	#8	#9	#10
TCDD	1.74	8.41	6.74	4.03	6.69	3.67	3.87	11.10	10.53	10.59
12378-PeCDD	2.24	0.60	2.31	4.37	1.97	2.88	2.33	1.13	1.72	0.81
123478-HxCDD	1.69	0.37	1.89	4.00	0.44	1.23	1.72	1.10	1.07	0.21
123678-HxCDD	0.69	1.91	1.27	3.83	5.01	2.33	3.28	1.68	0.84	10.26
123789-HxCDD	1.05	0.51	0.04	0.64	1.67	1.01	0.64	0.25	1.35	0.72
1234678-HpCDD	1.27	0.98	14.59	24.51	3.67	11.99	0.76	1.20	9.62	5.12
OCDD	16.18	23.61	59.01	179.06	19.86	42.00	31.46	10.87	25.33	6.95
TCDF	0.94	0.34	0.35	0.40	0.14	1.23	0.71	0.41	1.19	0.25
12378-PeCDF	1.20	1.08	1.65	1.64	0.68	1.27	1.15	2.08	2.96	2.99
23478-PeCDF	9.47	15.86	7.81	8.26	15.14	35.21	9.97	18.63	13.43	28.04
123478-HxCDF	0.33	0.72	1.02	1.48	2.01	0.74	0.25	0.42	0.34	3.49
123678-HxCDF	0.19	0.52	0.26	0.57	0.17	0.11	0.11	0.43	0.26	0.26
123789-HxCDF	2.79	1.23	1.15	3.15	0.56	1.65	1.83	0.84	1.17	0.43
234678-HxCDF	2.64	1.22	1.38	1.24	0.24	0.43	2.88	1.16	1.38	0.46
1234678-HpCDF	2.20	1.21	0.89	0.90	0.20	0.42	1.47	0.71	0.74	0.24
1234789-HpCDF	4.12	2.18	1.76	1.27	0.33	0.47	2.04	1.01	0.94	0.43
OCDF	17.60	40.05	56.19	123.93	21.07	36.21	28.42	8.52	27.58	11.77

"I'll collect 'em myself. I got a fishin' buddy in the analytical lab we use. I'll take him fishing and maybe we can get a deal."

Once back at his desk, the risk assessor planned a fishing trip with his friend. He figured between them that they could get 10 trout to analyze. He called the state DNR and got a permit for fish collection.

The risk assessor had a successful fishing trip and the analytical results are shown in Table 6.29. Summary statistics and ProUCL outputs are also provided in Table 6.30.

ASSESSMENT AND MEASUREMENT ENDPOINTS

From his conversation with the PM, the risk assessor identified assessment endpoints as toxicity to growth and reproduction to fish and high-trophic-level avian and mammalian piscivores. These classifications are often known as "guilds" and are characterized by feeding strategy, food source, and trophic level. The great blue heron (*Ardea herodias*) and mink (*Mustela vison*) were chosen to represent these two guilds. The most sensitive processes for the toxicity of PCDD/Fs in these guilds are reproduction and embryonic development.

TABLE 6.30
Summary Statistics and UCL Calculation for PCDD/F Congeners in Trout

Chemical	Units	N	# Det.	Freq. (%)	Min. ND	Max. ND	Min. Detect	Max. Detect	Arith. Mean of Detections	95% UCL	Method
2378-TCDD	ng/kg	10	10	100	—	—	1.74	11.1	6.737	8.679	95% Student's t
12378-PeCDD	ng/kg	10	10	100	—	—	0.6	4.37	2.036	2.673	95% Student's t
123478-HxCDD	ng/kg	10	10	100	—	—	0.21	4.0	1.372	2.007	95% Student's t
123678-HxCDD	ng/kg	10	10	100	—	—	0.69	10.26	3.11	5.29	95% Approx. gamma
123789-HxCDD	ng/kg	10	10	100	—	—	0.04	1.67	0.788	1.074	95% Student's t
1234678-HpCDD	ng/kg	10	10	100	—	—	0.76	24.51	7.371	15.55	95% Approx. gamma
OCDD	ng/kg	10	10	100	—	—	6.95	179.1	41.43	76.92	95% Approx. gamma
TCDF	ng/kg	10	10	100	—	—	0.14	1.23	0.596	0.826	95% Student's t
12378-PeCDF	ng/kg	10	10	100	—	—	0.68	2.99	1.67	2.125	95% Student's t
23478-PeCDF	ng/kg	10	10	100	—	—	7.81	35.21	16.18	21.42	95% Student's t
123478-HxCDF	ng/kg	10	10	100	—	—	0.25	3.49	1.08	1.885	95% Approx. gamma
123678-HxCDF	ng/kg	10	10	100	—	—	0.11	0.57	0.288	0.383	95% Student's t
123789-HxCDF	ng/kg	10	10	100	—	—	0.43	3.15	1.48	2.001	95% Student's t
234679-HxCDF	ng/kg	10	10	100	—	—	0.24	2.88	1.303	1.811	95% Student's t
1234678-HpCDF	ng/kg	10	10	100	—	—	0.2	2.2	0.898	1.251	95% Student's t
1234789-HpCDF	ng/kg	10	10	100	—	—	0.33	4.12	1.455	2.12	95% Student's t
OCDF	ng/kg	10	10	100	—	—	8.52	123.9	37.13	61.31	95% Approx. gamma
Mammalian TEQ	ng/kg	10	10	100	—	—	9.886	27.24	18.05	21.32	95% Student's t
Fish TEQ	ng/kg	10	10	100	—	—	10.35	26.27	18.16	21.23	95% Student's t

Note: Concentration is ng/kg wet weight. Avian TEQ is not calculated here because these will be applied to the modeled concentrations in the eggs of great blue heron.

Hence, the assessment endpoints would be the following:

- Assessment of toxicity to fish
- Assessment of development risk to avian piscivore embryos
- Assessment of development risk to mammalian piscivore embryos

The corresponding measurement endpoints would be the following:

- Measured fish tissue concentrations
- Modeled concentrations in the eggs of avian piscivores
- Modeled PCDD/F intakes in mammalian piscivores

Generally, ecological risk assessments compare risks or concentrations in a reference area with the area under consideration. The risk assessor knew he needed a reference area that would be relatively similar to the river but would have background levels of PCDD/Fs. It proved to be impossible to find a suitable reference area; he provided a description of his efforts in a memo to the PM.

Toxicity Reference Values

Prior to the fishing trip, the risk assessor had planned to use bioaccumulation factors to estimate fish tissue concentrations based on those in river sediment and had even found a publication on the energy requirements of trout that would enable him to calculate feeding rates.[49] However, the planned fish collection would make that part of the food chain modeling unnecessary. Then his real work began—collection of appropriate TRVs and development of the exposure factors used in food chain modeling.

For TRVs in fish, the U.S. Army Corps of Engineers had published a straightforward method for deriving fish tissue benchmarks from published data.[50] He used the methods in that paper and the data on trout. Combining the data from this chapter for brook trout, lake trout, and rainbow trout, the GM value for the no-effect level in tissue was 0.06 µg/kg and the lowest effect level in tissue was 0.104 µg/kg (Table 6.31).

TRVs derived from both laboratory- and field-based studies were used to assess the modeled concentrations in great blue heron eggs. EPA previously developed an egg-based TRV by taking the GM of the effect-concentrations (NOAEC/LOAEC) in three double-crested cormorant (*Phalacrocorax auritus*) egg-injection studies.[51] In these studies, PCDD/Fs were injected into cormorant eggs that were artificially incubated until hatching. Based on embryo mortality, the resulting no observed adverse effect concentration (NOAEC) and lowest observed adverse effect concentration (LOAEC) in eggs were 3,670 and 11,090 ng total avian TEQ/kg wet weight, respectively. In addition, field studies of great blue heron that included egg collection provided field-based NOAEC and LOAEC values. In one study, a mean concentration of 220 ng TEQ/kg wet weight in eggs was not associated with reduction in the number of successful nests and number of fledglings per nest.[52] However, in a second study, great blue heron eggs with a mean concentration of 360 ng TEQ/kg wet weight in eggs revealed health effects in hatchlings produced by incubation.[53]

TABLE 6.31
Derivation of No-Effect and Low-Effect Tissue
Levels in Trout

Species	No-Effect Tissue Conc.	Lowest-Effect Tissue Conc.
Brook trout	0.084	0.156
Brook trout	0.135	0.185
Lake trout	0.035	
Lake trout	0.023	0.05
Lake trout	0.034	0.04
Lake trout	0.044	0.055
Lake trout	0.034	0.055
Lake trout	0.033	0.044
Rainbow trout		0.279
Rainbow trout	0.194	0.291
Rainbow trout	0.176	0.244
Geometric mean	0.06	0.104

Note: Units are ng/g or μg/kg wet weight.

The predominant environmental exposure pathway in mink is through ingestion.[54] A study on the mink collected from near the Saginaw River in Michigan was considered most appropriate—the study employed multiple doses and used environmental sources of PCDD/Fs that had undergone many years of weathering and were thus likely similar to the PCDD/Fs found in the river considered in this risk assessment.[55] Concentrations of PCBs, PCDDs, and PCDFs were measured in the diet. This study tested doses of 2.1 (control group), 22.4, 36.5, and 56.6 ng TEQ/kg (wet wt. in food) over a period of approximately 120 days. Eight of the kits per dose group were also maintained on treatment dosing until they were 27 weeks old. No adverse effects on reproductive or developmental endpoints were observed at any of the doses tested. No adverse effects on reproductive or developmental endpoints, including breeding success, whelping success, gestation length, litter size, or offspring survival, were observed at any of the doses tested. Hence, the largest dietary dose in this study—56.6 ng TEQ/kg, wet wt—is the NOAEL.

RISK ASSESSMENT RESULTS

Risks to all receptors were assessed using an HQ approach. This involved calculation of the ratio between the measurement endpoint and the TRV expressed in the same units. This process is very similar to the HQ approach in human health where the measure of risk is the ratio between an estimated intake dose and an RfD.

Development of Exposure Concentrations

Exposure concentrations for trout, great blue heron, and mink were calculated using a TEF approach and measured concentrations in fish, modeled concentrations in heron eggs, and modeled ingestion doses to mink. The calculations and exposure concentrations are shown in Table 6.32.

Risk to Trout

The wet weight concentrations in whole fish of 23.95 ng TEQ/kg wet weight (Table 6.32) were compared to the NOAEL and LOAEL TRVs in trout (Table 6.31). Please note that the units are different and the TRVs will need to be multiplied by 1000 to express them in similar units. Dividing the TRV values by the tissue concentrations gives a NOAEL-based HI of 0.4 and a LOAEL-based HI of 0.2. Hence, the risk of PCDD/Fs to trout in the river would be considered very low.

Risk to Great Blue Heron

The risk to great blue heron was assessed as the ratio of the modeled wet weight concentrations in eggs to the NOAEL and LOAEL values of 220 and 360 ng TEQ/kg wet weight, discussed previously. The modeled TEQ in great blue heron eggs was 948.3 ng TEQ/kg wet weight (Table 6.32). The NOAEL-based HI would be 4 and the LOAEL-based HI would be 3. Hence, the predicted risk to great blue heron using the field-based TRVs would warrant further investigation—possibly, the collection of bird eggs for PCDD/F analysis. However, using the laboratory-based TRVs (NOAEL = 3,670 ng/kg; LOAEL = 11,090 ng/kg), the NOAEL-based HI would be 0.3 and the LOAEL-based HI would be 0.09.

Risk to Mink

The exposure factors for the mink are shown in Table 6.33. Again, care with unit conversions is warranted. A mink weighing 550 g would consume 121 g of food per day. For this assessment, trout were assumed to compose 100% of the diet. Hence, using the intake concentration for mink from Table 6.32, the daily intake of TEQ would be 2.89 ng/day and the daily dose would be 5.26 ng/kg BW/day. This value is over 10-fold lower than the NOAEL of 56.6 ng/kg/day and the HI would be 0.1.

The risk assessor also prepared a summary table so that all the information would be in one place. This table is not shown; instead, one of the exercises at the end of this chapter is to prepare such a summary table.

Scientific/Management Decision Point #2

"So the only HIs above one are the heron, and that's only for the field-based TRVs, right?" asked the PM, perusing the tables the risk assessor had prepared. "Highest risk is 4—that's not much."

"I agree," said the risk assessor. "You mentioned a paper mill. Do you know any more about that? If any sediment from the lake got flushed into the river, that could account for these dioxin levels. I really don't think this came from the pond—otherwise, I think we'd see some PCBs and the PAHs would likely be higher as well."

TABLE 6.32

Exposure Concentrations of Toxic Equivalents of PCDD/Fs in Fish, Modeled Toxic Equivalents in Great Blue Heron Eggs Based on Fish Tissue/Bird Egg Biomagnification Factors, and Ingestion Doses for Mink

Congener/Group	95% UCL Concentration in Fish (ng/kg Wet Wt.)	Fish TEF	TEQ in Fish	Selected BMF (Fish Tissue/ Bird Egg)	Bird TEF	Modeled TEQ in GBH Eggs (ng/kg Wet Wt.)	Mammalian TEF	Fish Intake TEQ for Mink (ng/kg Wet Wt.)
TCDD	8.679	1	8.68	21	1	182.26	1	8.68
12378-PeCDD	2.673	1	2.67	9.7	1	25.93	1	2.67
123478-HxCDD	2.007	0.5	1.00	125	0.05	125.44	0.1	0.20
123678-HxCDD	5.29	0.01	0.05	154	0.01	8.15	0.1	0.53
123789-HxCDD	1.074	0.01	0.01	125	0.1	1.34	0.1	0.11
1234678-HpCDD	15.55	0.001	0.02	154	0.001	2.39	0.01	0.16
OCDD	76.92	0.0001	0.01	174	0.0001	1.34	0.0001	0.01
TCDF	0.826	0.05	0.04	0.42	1	0.02	0.1	0.08
12378-PeCDF	2.125	0.05	0.11	32	0.1	3.40	0.05	0.11
23478-PeCDF	21.42	0.5	10.71	32	1	342.72	0.5	10.71
123478-HxCDF	1.885	0.1	0.19	15	0.1	2.83	0.1	0.19
123678-HxCDF	0.383	0.1	0.04	15	0.1	0.57	0.1	0.04
123789-HxCDF	2.001	0.1	0.20	15	0.1	3.00	0.1	0.20
234678-HxCDF	1.811	0.1	0.18	15	0.1	2.72	0.1	0.18
1234678-HpCDF	1.251	0.01	0.01	15	0.01	0.19	0.01	0.01
1234789-HpCDF	2.12	0.01	0.02	15	0.01	0.32	0.01	0.02
OCDF	61.31	0.0001	0.01	5.7	0.0001	0.03	0.0001	0.01
Exposure concentrations in fish tissue, bird eggs, and intakes for mink (TEQ ng/kg wet wt.)			23.95			948.3		23.90

TABLE 6.33
Exposure Factors for the Mink

Functional Group	Default Primary Indicator Species	Body Weight, BW (kg)	Food Ingestion Rate (g Wet Wt./g-d)	Dietary Composition	Water Ingestion Rate (g/g-d)	Home Range, HR (ha)
Freshwater mammalian piscivore	Mink (*M. vison*)	0.55	0.22	Trout 78% Forage fish 9% Arthropods 13% Assumed all trout	0.11	20.4

"There's no way to clean up," she said. "I can't make a case for dredging an internationally famous trout fishery and then hoping the fishing will recover."

"Agreed. Here's what I'll do—write up the results and show the range of HQs for each receptor. I think you can make a case that there's no real risk to the birds—and if there's still some concern, I can set up to collect eggs and have them analyzed."

"I hate to do things like that," she said. "Collecting new data, well, you don't know where it'll lead."

"We really need those egg results to put this to bed," said the risk assessor. "I can do it myself, but I need a team of two other guys. One has to be a tree climber, spikes and all. I know where to hire them. The eggs are in the nests about the end of May and hatch from July 1 on. So, we've got a little time."

"I'll get back to you on that," she said. "Anyway, now you have fish concentrations, there's one other thing I need—human health. If there's no risk to people, it'll be easier to communicate no risk to critters."

"Sorry," said the risk assessor. "Can't help you there. Joe, my esteemed colleague who sits in the next office down from me, does human health. I'll tell him to come and see you."

"You can't do it?"

"Nope. I don't do human health. But then, Joe doesn't fish nor does he know how to collect bird eggs."

"Geez! And I thought we were silo-ed!"

EXERCISES FOR THOUGHT AND DISCUSSION

Summary Table

Prepare a summary table of risks from the ecological risk assessment shown here. Realize that the details such as TEQ, the many calculations, may not be of great interest to the PM. What is the most relevant information? What is the most uncertain information? How can the risks and uncertainties be best communicated in a single table?

Toxicity Equivalence Factors for Wildlife

Please use the Internet to obtain a free copy of Van den Berg et al.[42] Also, download EPA's Framework for Application of the Toxicity Equivalence Methodology for Polychlorinated Dioxins, Furans, and Biphenyls in Ecological Risk Assessment from http://www.epa.gov/raf/tefframework/. After reading both documents, can you think of any ways to improve the risk assessment presented in this chapter?

Ecological Risk Assessment of the Former Gold Mine

Using the data provided in Chapter 5 on the former gold mine, prepare a SLERA. Assume the gold mine is in Alaska and try to find state-specific guidance that will direct your efforts. This guidance is available on the Internet.

REFERENCES

1. Council on Environmental Quality. *A Citizen's Guide to NEPA: Having Your Voice Heard*. Washington, DC, December, 2007. http://energy.gov/nepa/downloads/citizens-guide-nepa-having-your-voice-heard
2. Suter GW. *Ecological Risk Assessment*. Boca Raton, FL: Taylor & Francis Group, 2007.
3. United States Environmental Protection Agency (USEPA). *Summary Report on Issues in Ecological Risk Assessment*. EPA/625/3-91/018. Washington, DC, February, 1991. http://www.epa.gov/raf/publications/summary-report-issues-eco-ra.htm
4. United States Environmental Protection Agency (USEPA). *Framework for Ecological Risk Assessment*. EPA/630/R-92/001. Risk Assessment Forum. Washington, DC, February, 1992. http://www.epa.gov/raf/publications/framework-eco-risk-assessment.htm
5. United States Environmental Protection Agency (USEPA). *Risk Assessment Guidance for Superfund: Volume II: Environmental Evaluation Manual. Interim Final*. EPA/540/1-89/001. Washington, DC, March, 1989. http://rais.ornl.gov/documents/RASUPEV.pdf
6. United States Environmental Protection Agency (USEPA). *DRAFT Proposed Guidelines for Ecological Risk Assessment*. Washington, DC, 1995. http://www.ntis.gov/search/product.aspx?ABBR=PB96146907&starDB=GRAHIST
7. United States Environmental Protection Agency (USEPA). *Ecological Risk Assessment Guidance for Superfund: Process for Designing and Conducting Ecological Risk Assessments—(Interim Final)*. EPA 540-R-97-006. OSWER 9285.7-25. PB97-963211. Washington, DC, June, 1997. http://www.epa.gov/oswer/riskassessment/ecorisk/ecorisk.htm
8. United States Environmental Protection Agency (USEPA). *Guidelines for Ecological Risk Assessment*. Risk Assessment Forum. Washington, DC, May 14, 1998. http://www.epa.gov/raf/publications/guidelines-ecological-risk-assessment.htm
9. United States Environmental Protection Agency (USEPA). *Endangered and Threatened Species Effects Determinations*. Office of Pesticide Programs. Washington, DC, January 23, 2004. http://www.epa.gov/scipoly/sap/meetings/2008/october/overview_doc_2004.pdf
10. United States Environmental Protection Agency (USEPA). *Safer Rodenticide Products | Pesticides | US EPA*, March, 2013. http://www.epa.gov/pesticides/mice-and-rats/ (accessed April 12, 2013).

11. Jones DS, Suter II GW, and Hull RN. *Toxicological Benchmarks for Screening Contaminants of Potential Concern for Effects on Sediment-Associated Biota: 1997 Revision.* ES/ER/TM-95/R4. Oak Ridge, TN, November, 1997. http://rais.ornl.gov/guidance/tm.html

12. Efroymson RA, Will ME, and Suter II GW. *Toxicological Benchmarks for Contaminants of Potential Concern for Effects on Soil and Litter Invertebrates and Heterotrophic Processes: 1997 Revision.* ES/ER/TM-126/R2. Oak Ridge, TN, November, 1997. http://rais.ornl.gov/guidance/tm.html

13. Efroymson RA, Will ME, Suter II GW, and Wooten AC. *Toxicological Benchmarks for Screening Contaminants of Potential Concern for Effects on Terrestrial Plants: 1997 Revision.* ES/ER/TM-85/R3. Oak Ridge, TN, November, 1997. http://rais.ornl.gov/guidance/tm.html

14. Sample BE, Opresko DM, and Suter II GW. *Toxicological Benchmarks for Wildlife: 1996 Revision.* ES/ER/TM-86/R3. Oak Ridge, TN, June, 1996. http://rais.ornl.gov/guidance/tm.html

15. Suter II GW and Tsao CL. *Toxicological Benchmarks for Screening of Potential Contaminants of Concern for Effects on Aquatic Biota on Oak Ridge Reservation: 1996 Revision.* ES/ER/TM-96/R2. Oak Ridge, TN, June, 1996. http://rais.ornl.gov/guidance/tm.html

16. Sample BE, Suter II GW, Efroymson RA, and Jones DS. *A Guide to the ORNL Ecotoxicological Screening Benchmarks: Background, Development and Application.* ORNL/TM-13615. Oak Ridge, TN, May 1998. http://www.esd.ornl.gov/programs/ecorisk/documents/whtppr21.pdf

17. Swartjes FA. Risk-based assessment of soil and groundwater quality in The Netherlands: Standards and remediation urgency. *Risk Anal.* 1999, December;19(6):1235–1249.

18. Friday G. *Ecological Screening Values for Surface Water, Sediment, and Soil.* WSRC-TR-98-00110. 1998. http://www.osti.gov/scitech/servlets/purl/4764

19. United States Environmental Protection Agency (USEPA). *Guidance for Developing Ecological Soil Screening Levels.* OSWER Directive 9285.7-55. Washington, DC, February, 2005. http://www.epa.gov/ecotox/ecossl/SOPs.htm

20. Wang F, Goulet RR, and Chapman PM. Testing sediment biological effects with the freshwater amphipod *Hyalella azteca*: The gap between laboratory and nature. *Chemosphere.* 2004, December;57(11):1713–1724.

21. Persaud D, Jaagumagi R, and Hayton A. *Guidelines for the Protection and Management of Aquatic Sediment Quality in Ontario.* ISBN 0-7778-9248-7. August, 1993. http://www.itrcweb.org/contseds-bioavailability/References/guide_aquatic_sed93.pdf

22. Di Toro DM, Zarba CS, Hansen DJ, Berry WJ, Swartz RC, Cowan CE et al. Technical basis for establishing sediment quality criteria for nonionic organic chemicals using equilibrium partitioning. *Environ Toxicol Chem.* 1991, December;10(12):1541–1583.

23. Ankley GT, Di Toro DM, Hansen DJ, and Berry WJ. Assessing the ecological risk of metals in sediments. *Environ Toxicol Chem.* 1996, December;15(12):2053–2055.

24. Berry WJ, Hansen DJ, Boothman WS, Mahony JD, Robson DL, Di Toro DM et al. Predicting the toxicity of metal-spiked laboratory sediments using acid-volatile sulfide and interstitial water normalizations. *Environ Toxicol Chem.* 1996, December;15(12):2067–2079.

25. Di Toro DM, McGrath JA, Hansen DJ, Berry WJ, Paquin PR, Mathew R et al. Predicting sediment metal toxicity using a sediment biotic ligand model: Methodology and initial application. *Environ Toxicol Chem.* 2005;24(10):2410.

26. Long ER and MacDonald DD. Recommended uses of empirically derived, sediment quality guidelines for marine and estuarine ecosystems. *Hum Ecol Risk Assess.* 1998;4(5):1019–1039.
27. MacDonald DD, Ingersoll CG, and Berger TA. Development and evaluation of consensus-based sediment quality guidelines for freshwater ecosystems. *Arch Environ Contam Toxicol.* 2000, July;39(1):20–31.
28. Ingersoll CG, MacDonald DD, Wang N, Crane JL, Field LJ, Haverland PS et al. Predictions of sediment toxicity using consensus-based freshwater sediment quality guidelines. *Arch Environ Contam Toxicol.* 2001, July;41(1):8–21.
29. MacDonald DD, Carr RS, Eckenrod D, Greening H, Grabe S, Ingersoll CG et al. Development, evaluation, and application of sediment quality targets for assessing and managing contaminated sediments in Tampa Bay, Florida. *Arch Environ Contam Toxicol.* 2004, February;46(2):147–161.
30. United States Environmental Protection Agency (USEPA). *Sources of Screening Values for Sediment and Soil Based Entirely or Partially on Ecological Risk* by James Chapman. Chicago, IL, September 29, 2004. http://www.epa.gov/R5Super/ecology/benchmemo.htm (accessed October 3, 2013).
31. United States Environmental Protection Agency (USEPA). Memorandum from Ted Simon to Jon D. Johnston and Earl Bozeman. *Amended Guidance on Ecological Risk Assessment at Military Bases: Process Considerations, Timing of Activities, and Inclusion of Stakeholders.* June 23, 2000. http://rais.ornl.gov/documents/ecoproc2.pdf
32. United States Environmental Protection Agency (USEPA). *Guidance for Comparing Background and Chemical Concentrations in Soil for CERCLA Sites.* EPA 540-R-01-003. OSWER 9285.7-41. Washington, DC, September, 2002. http://www.epa.gov/oswer/riskassessment/pdf/background.pdf (accessed October 3, 2013).
33. United States Environmental Protection Agency (USEPA). *Guidance on Systematic Planning using the Data Quality Objectives Process (QA/G-4).* EPA/240/B-06/001. Washington, DC, February, 2006. http://www.epa.gov/quality/qa_docs.html
34. United States Environmental Protection Agency (USEPA). *Guidance on Choosing a Sampling Design for Environmental Data Collection (QA/G-5S).* EPA/240/R-02/005. Washington, DC, December, 2002. http://www.epa.gov/quality/qa_docs.html
35. United States Environmental Protection Agency (USEPA). Systematic Planning: A Case Study for Hazardous Waste Site Investigations (QA/CS-1). EPA/240/B-06/004. Washington, DC, February, 2006. http://www.epa.gov/quality/qa_docs.html
36. United States Environmental Protection Agency (USEPA). *Guidance for Quality Assurance Project Plans (QA/G-5).* EPA/240/R-02/009. Washington, DC, December, 2002. http://www.epa.gov/quality/qa_docs.html
37. United States Environmental Protection Agency (USEPA). *Data Quality Assessment: A Reviewer's Guide (QA/G-9R).* EPA/240/B-06/002. Washington, DC, February, 2006. http://www.epa.gov/quality/qa_docs.html (accessed October 3, 2013).
38. United States Environmental Protection Agency (USEPA). *Data Quality Assessment: Statistical Tools for Practitioners (QA/G-9S).* EPA/240/B-06/003. Washington, DC, February, 2006. http://www.epa.gov/quality/qa_docs.html (accessed October 3, 2013).
39. MacKnight SD and Mudroch A. *Handbook of Techniques for Aquatic Sediments Sampling [...] XA-GB.* 2nd edn., illustrated, revised. Boca Raton, FL: CRC Press, 1994.
40. Hill AB. The environment and disease: Association or causation? *Proc R Soc Med.* 1965, May;58:295–300.
41. Van den Berg M, Birnbaum LS, Denison M, De Vito M, Farland W, Feeley M et al. The 2005 World Health Organization reevaluation of human and Mammalian toxic equivalency factors for dioxins and dioxin-like compounds. *Toxicol Sci.* 2006, October;93(2):223–241.

42. Van den Berg M, Birnbaum L, Bosveld AT, Brunström B, Cook P, Feeley M et al. Toxic equivalency factors (TEFs) for PCBs, PCDDs, PCDFs for humans and wildlife. *Environ Health Perspect.* 1998, December;106(12):775–792.

43. Haws LC, Su SH, Harris M, Devito MJ, Walker NJ, Farland WH et al. Development of a refined database of mammalian relative potency estimates for dioxin-like compounds. *Toxicol Sci.* 2006, January;89(1):4–30.

44. Frame GM, Wagner R, Carnaham JC, Brown JF, May RJ, Smullen LA, and Bedard DL. Comprehensive, quantitative, congener-specific analyses of eight Aroclors and complete PCB congener assignments on Db-I capillary Gc columns. *Chemosphere.* 1996;33(4):603–623.

45. Simon T, Britt JK, and James RC. Development of a neurotoxic equivalence scheme of relative potency for assessing the risk of PCB mixtures. *Regul Toxicol Pharmacol.* 2007, July;48(2):148–170.

46. Cleverly D. Memorandum: *Response to Ecological Risk Assessment Forum Request for Information on the Benefits of PCB Congener-Specific Analyses.* Washington, DC: U.S. Environmental Protection Agency, 2005.

47. Mackay D and Fraser A. Kenneth Mellanby Review Award. Bioaccumulation of persistent organic chemicals: Mechanisms and models. *Environ Pollut.* 2000, December;110(3):375–391.

48. United States Environmental Protection Agency (USEPA). Memorandum from Brian Ross and Erika Hoffman. *Compilation and discussion of sediment quality values for dioxin, and their relevance to potential removal of dams on the Klamath River.* San Francisco, CA, January 13, 2010. http://klamathrestoration.gov/sites/klamathrestoration.gov/files/EPA%20Klamath%20dioxin%20memo%201-13-10%20final.pdf (accessed October 3, 2013).

49. Mambrini M, Médale F, Sanchez MP, Recalde B, Chevassus B, Labbé L et al. Selection for growth in brown trout increases feed intake capacity without affecting maintenance and growth requirements. *J Anim Sci.* 2004, October;82(10):2865–2875.

50. Steevens JA, Reiss MR, and Pawlisz AV. A methodology for deriving tissue residue benchmarks for aquatic biota: A case study for fish exposed to 2, 3, 7, 8-tetrachlorodibenzo-p-dioxin and equivalents. *Integr Environ Assess Manage.* 2005;1(2):142–151.

51. United States Environmental Protection Agency (USEPA). *Analyses of Laboratory and Field Studies of Reproductive Toxicity in Birds Exposed to Dioxin-Like Compounds for Use in Ecological Risk Assessment.* EPA/600/R-03/114F. Cincinnati, OH, April, 2003. http://cfpub.epa.gov/ncea/cfm/recordisplay.cfm?deid=56937#Download

52. Elliott JE, Harris ML, Wilson LK, Whitehead PE, and Norstrom RJ. Monitoring temporal and spatial trends in polychlorinated dibenzo-p-dioxins (PCDDs) and dibenzofurans (PCDFs) in eggs of great blue heron (*Ardea herodias*) on the coast of British Columbia, Canada, 1983–1998. *Ambio.* 2001, November;30(7):416–428.

53. Sanderson JT, Elliott JE, Norstrom RJ, Whitehead PE, Hart LE, Cheng KM, and Bellward GD. Monitoring biological effects of polychlorinated dibenzo-p-dioxins, dibenzofurans, and biphenyls in great blue heron chicks (*Ardea herodias*) in British Columbia. *J Toxicol Environ Health.* 1994, April;41(4):435–450.

54. United States Environmental Protection Agency (USEPA). *Wildlife Exposure Factors Handbook.* EPA/600/R-93/187. Washington, DC, December, 1993. http://cfpub.epa.gov/ncea/cfm/recordisplay.cfm?deid=2799#Download

55. Bursian SJ, Beckett KJ, Yamini B, Martin PA, Kannan K, Shields KL, and Mohr FC. Assessment of effects in mink caused by consumption of carp collected from the Saginaw River, Michigan, USA. *Arch Environ Contam Toxicol.* 2006, May;50(4):614–623.

7 Future of Risk Assessment

I believe that man will not merely endure: he will prevail. He is immortal, not because he alone among creatures has an inexhaustible voice, but because he has a soul, a spirit capable of compassion and sacrifice and endurance.

William Faulkner
Nobel Banquet Speech, 1950

The present geologic epoch has come to be called the "Anthropocene" in the popular press. The name signifies the increasing effect of humans in changing the global environment. In 2012, the NRC pointed out that the problems faced by humans in the Anthropocene epoch are "wicked problems."[1] Wicked problems have multiple causes, are resistant to solutions, difficult to define, and socially and politically complex with multiple stakeholders holding differing views on both the desired outcome and how it should be achieved. Furthermore, wicked problems span the understanding of several scientific disciplines, each with a set of "deep" uncertainties surrounding the problem.[1]

Wicked problems occur on very large scales with temporal, scientific, and social aspects. Arriving at a conclusion regarding the appropriate response to a wicked problem becomes complicated as new data and information are constantly becoming available or changing. In addition, the understanding of the dimensions of the problem is incomplete due to unknown interactions and feedback loops. Nonetheless, because society has embraced the science as a means of understanding and the scientific method as a means of finding answers, scientists (including risk assessors) must be prepared to offer a theory supported by evidence that includes choices for an appropriate response to the problem. This set of choices includes doing nothing, and all the choices are informed or shaped by our best understanding of the problem based on the evidence. Doing nothing may, at times, be the best choice because acting on incomplete or faulty information has the potential to exacerbate the situation.[2]

Many scientists and decision makers have come to appreciate the "unknown unknowns" and the potential liability of acting on incomplete information. The idea of the "unknown unknowns" was made popular in a 2002 speech by the Secretary of Defense Donald Rumsfeld. Shortly before Rumsfeld's speech, Nassim Nicholas Taleb, author of *The Black Swan: The Impact of the Highly Improbable*, gave a speech at the Department of Defense that likely spurred Rumsfeld's thinking.[3]

In their book, *Resilience: Why Things Bounce Back*, Andrew Zolli and Ann Marie Healy chronicle how an increased investment in domestic corn production for ethanol following Hurricane Katrina led to food riots in Mexico 2 years later. They provide other examples of how the volatility and interconnectedness of today's world has produced a sense of disruption and vulnerability.[4] A growing number of thinkers in the disciplines of social science, economics, cognitive science, and philosophy share the view that the instability of today's world and resulting unease will require an adaptive shift toward resilience—the ability to bounce back from unforeseen shocks and surprises. Taleb has characterized this ability as anti-fragility—the idea embodied in the famous quote by Friedrich Nietzsche from *The Twilight of the Idols*—"What does not kill me, makes me stronger."[5,6]

It might seem that risk assessment, as a predictive activity, is increasingly out of place in an unpredictable world. Frankly, in an unpredictable world where resilience and anti-fragility are the hallmarks of survivors, the expectation that predictions of experienced scientists based on up-to-date observation, experiments, and analyses can provide a sufficient basis for decision making, risk management or other interventions may well be an expression of hubris—so why bother? The body of scientific knowledge, whether weak or compelling, seems irrelevant in the face of the unknown unknowns.

VALUE OF THE PRECAUTIONARY PRINCIPLE

There are instances in which the use of a precautionary approach in decision making is justified. The major justification for the precautionary principle is that scientific knowledge is uncertain. Part of the sense of disruption and vulnerability discussed earlier is that humankind's scientific knowledge is uncertain and not just because of inconclusive or absent data but because all science is uncertain. Acceptance of a particular level of uncertainty for scientific knowledge supporting a decision is not a question of science but a question of values and interests.

Science remains the best way for humankind to understand the universe. However, science will forever be incomplete, and, as noted by Sir Austin Bradford Hill, the recognition that so-called scientific truth is liable to modification with the advancement of knowledge does not confer the license to ignore a suspected or known hazard or to postpone addressing this hazard.[7] This is the essence of the precautionary principle.

Another way to think about the precautionary principle is that it involves a shift from an expectation that science is objective and certain to the recognition of the uncertainty of science. The precautionary principle recognizes that while science may provide information for decisions, this information is uncertain and societal decision making involves values and interests as well as scientific knowledge.[8] The sovereign authority envisioned by the philosopher John Locke in *Two Treatises of Government* is embodied in the regulations that prescribe risk assessment as the basis for environmental decision making.[9] Increasingly, the sovereign has had to address scientific uncertainty and data gaps. The precautionary principle provides the legal—not scientific—means of addressing uncertainty with a bias toward safety. Hence, the role of science in society forms much of the basis of the precautionary principle; defining this role, and thus, the role of the sovereign in filling these data gaps involves issues of scientific communication, the problems and heuristics of uncertainty itself, and ignorance about and trust/mistrust of regulatory risk policies by the lay public.[10]

Differing views on the meaning, value, and application of the precautionary principle have led to a misunderstanding by the public. As expressed by the EU in 2000, the proportionality of response is part of the precautionary principle and means "tailoring measures to the chosen level of protection." Hence, the use of a total ban may not be advisable in all cases.[11]

Embodied in the precautionary principle is the motto that captures the thoughts of John Locke in his 1689 work, *An Essay Concerning Human Understanding*, and recently popularized by a number of thinkers: "Absence of evidence is not evidence of absence." John Locke characterized this as the argument from ignorance.[12] However, scientific evidence is of varying quality, both high and low—study design, statistical power, risk of bias, and other aspects of evidentiary quality must be taken into account when basing decisions on scientific information.[13]

However, in contrast to the appropriate application of the precautionary principle, what the argument from ignorance proposes for decision making in the face of uncertainty is that acting on the basis of an absence of evidence is equally valid as acting on the basis of evidence. The sentiment behind the argument from ignorance is that the absence of evidence is equal in probative value to evidence, and thus, both evidence and absence of evidence provide sufficient proof to support taking action. This position is emphatically not that of the precautionary principle; rather, the principle gives greater weight to consideration of uncertainty as a reason for taking action.

For example, in Chapter 2, the Belladonna Blues Band and their song "Bad Blood Blues" about the occurrence of many environmental chemicals in human blood were mentioned. Logically, the absence of any observable evidence of harm from these chemicals does not mean the chemicals in our blood and bodies are not producing harm. This is exactly the type of scientific uncertainty addressed by the precautionary principle. The measured opinion of the NRC in their 2006 report on human biomonitoring is an example of the appropriate use of precautionary thinking. The NRC states:

> ... absence of evidence effects is not identical with evidence of absence effects-a distinction that must be clear to constituents. Otherwise there is a large practical communication and ethical risk attached to simply saying that the presence of chemicals in human tissue does not imply health effects.[14]

The presence of measurable albeit vanishingly tiny concentrations of environmental chemicals in our bodies seems to pale when compared with the increase in the human lifespan in the developed world that occurred simultaneously with the increase of the number of chemicals in commerce during the twentieth century.

However, from a purely evidentiary point of view, the position of the NRC is exactly correct—the presence of chemicals and simultaneous observation of no easily observable effects do not imply that effects are absent altogether.

The fact that the precautionary principle is a legitimate means of addressing scientific uncertainty for the purposes of regulation implies that some evidence is needed in order to arrive at a decision to take regulatory action. The most appropriate application of the precautionary principle occurs when the regulatory decision is based upon the following three factors: (1) the uncertainty associated with the scientific basis of a decision, (2) the consequences of the proposed regulatory action, and (3) the consequences of taking no action.

BENEFITS OF A PRECAUTIONARY APPROACH

Societal benefits may indeed accrue from applying the precautionary principle as appropriate. There seems to be a general public perception that dietary supplements are safe and do not need to be tested. However, there are a growing number of cases suggesting that in some cases, dietary herbal supplements or traditional Chinese medicine (TCM) may produce significant adverse health effects when combined with western pharmaceuticals.[15] Efforts have begun to explore these interactions and provide a rational approach for combining health-care products from different cultures.[16]

A ban on the use of any TCM or herbal products would not be a proportional response. However, dissemination and education of the potential for adverse health effects in these situations would be an appropriate expression of a risk management action consistent with the precautionary principle.

PROPORTIONALITY: HOW MUCH PRECAUTION IS WARRANTED?

In California, the Safe Drinking Water and Toxic Enforcement Act of 1986 was approved by voters as a means of addressing concerns about exposure to toxic chemicals. That initiative is known as Proposition 65. To comply, the state publishes a list of chemicals "known to cause cancer or birth defects." The list is updated yearly and now includes about 800 chemicals. Under Proposition 65, businesses are required to provide notification about significant amounts of chemicals in their products.

The Office of Environmental Health Hazard Assessment (OEHHA) of the California Environmental Protection Agency (Cal/EPA) administers Proposition 65 and evaluates all currently available scientific information on substances considered for placement on the Proposition 65 list. Part of this evaluation is the development of maximum allowable dose levels (MADLs).

The passage of Proposition 65 by voters reflects the risk-averse view of many individuals. Go into any Starbucks in California, and look for the warning sign that the carcinogen acrylamide is present in the coffee and pastries one would otherwise be enjoying (Figure 7.1).

The basis for the Proposition 65 warning for coffee and pastries can be found at the website of California's OEHHA at http://oehha.ca.gov/prop65/CRNR_notices/pdf_zip/MADL022610.pdf.

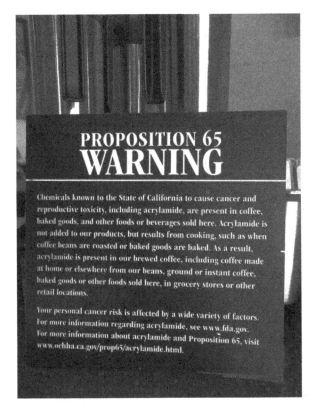

FIGURE 7.1   Proposition 65 warning about acrylamide in coffee and pastries, seen in a Starbucks coffee shop in California.

In this document, OEHHA scientists present a derivation of an MADL of acrylamide. You may wish to consider some of the aspects of the MADL for acrylamide such as the critical endpoint, human relevance of this endpoint, and the calculation of the MADL value. In addition, you might wish to discover the amount of acrylamide in a cup of coffee or slice of pumpkin loaf. With all this information in hand, decide for yourself if this Proposition 65 warning is warranted for acrylamide in coffee.

FEAR AND POSSIBILISTIC THINKING

Risk assessment is about probabilities. However, the sense of disruption and vulnerability has created a type of cultural pessimism that has given rise to what can be called "possibilistic" risk assessment. This type of risk assessment is based on the expectation of the worst possible outcomes and a fatalistic view of the future.[17] Possibilistic thinking attempts to legitimize this cultural pessimism by calling it science. However, possibilistic thinking has a cavalier attitude toward the evidentiary nature of science.

The Delaney Clause that banned the presence of carcinogens in food was discussed in Chapter 1. Think how likely one could find any food item entirely free of chemicals that were associated with cancer. In Box 4.3 in Chapter 4, a dose–response assessment of the carcinogenicity of *Ginkgo biloba* in rats was presented. Under the Delaney Clause, *G. biloba* would need to be banned, and even the US Congress was guilty of improper possibilistic thinking when it passed the Delaney Clause.

In an exercise at the end of Chapter 1, you were asked to watch a video called "The Story of Cosmetics." This campaign is funded by the Tides Foundation that supports many worthy efforts in philanthropy and social justice (http://www.tides.org/). However, this project seems to be based on fear mongering, the devaluation of knowledge, and the institutionalization of ignorance. The video plays on the discomfort many people feel about engaging and comprehending uncertainty.

The news media often take advantage of this discomfort. The attacks on science by Peter Waldman of the *Wall Street Journal* and Mark Obmascik of *The Denver Post* are discussed in, respectively, Chapters 1 and 3 in this book and are examples of such fear mongering. The overall result is a mood of confusion, pessimism, vulnerability, powerlessness, and the institutionalization of insecurity that has enabled the acceptance of possibilistic risk assessment. Those who use this cultural pessimism for their own ends are "fear entrepreneurs." Their efforts often have harmful effects on society because they diminish the role of science in shaping policy.

Risk Assessment Is about Probabilities Not Possibilities

The fear entrepreneurs have rejected a scientific approach, evidence-based decision making, and seemingly, the whole idea of risks as probabilities. Given that their message is to view innovation and "thinking outside the box" with dread and loathing, these merchants of fear reject as irresponsible and dangerous any sort of risk evaluation that honestly considers probabilities.

The pioneers of risk estimation were gamblers—Pascal, Galileo, Bernoulli—they possessed a gambler's hope and they longed for a challenge.[18] The best practice of risk assessment determines the most accurate probabilities of various outcomes to inform decisions. The numbers that were calculated in Chapter 5 of the likelihood of cancer were probabilities. In fact, probabilistic risk assessment has been fully endorsed in the risk assessment community (Appendix A).[19–21]

The argument made by the advocates of possibilistic thinking is that because the threats faced by humanity are unknown, there is insufficient information upon which to base any realistic estimate of probabilities. In contrast, probabilistic thinking offers a concrete and evidence-based approach to problem solving, and, consistent with the precautionary principle, a respect for the individual expression of values associated with scientific uncertainty.

This chapter began with a quote from William Faulkner's Nobel Prize acceptance speech. It is a message of hope and speaks of the infinite capabilities of humankind to better its fortunes. The foreword to this book discussed the personal value of entering a field that needs new blood and quick young minds capable of engaging in the details of a field of exponentially growing complexity. An environmental risk practitioner who can see the practical side of an issue but who has scientific ability as well as the ethics and confidence to challenge the norms will be in constant demand.

Science has enabled humankind to know and understand the world; if you believe this, then you also believe that a science-based risk assessment, conducted in good faith and including an appropriate characterization of uncertainty, provides a valuable and necessary decision tool, indeed, the best decision tool. This viewpoint is exactly where those on the leading edge of environmental risk assessment in the twenty-first century have landed.

TWENTY-FIRST CENTURY RISK ASSESSMENT: NEW DATA SOURCES AND NEW METHODS

Indeed, the twenty-first century is an exciting time to be entering the field. There are so many new methods and types of data to be considered and potentially used for informing risk assessment. Several of these will be discussed here. It seems doubtful they will completely supplant traditional risk assessment methodology that was presented in the first six chapters of this book. However, scientists are just beginning to learn how to use the new methods and data, and in such a situation, a good idea is an opportunity—just like the old advice about building a better mousetrap.

However, there are philosophical and conceptual roadblocks to the implementation of these new methods, and after presentation of the new data and tools, these obstacles will be considered.

TOXICITY TESTING IN THE TWENTY-FIRST CENTURY

Risk assessment and regulation of chemicals are both undergoing a massive transformation in terms of data sources, the generation of these data, and the manner in which the data are translated into evidence to support risk management. The reasons are (1) the growing number of substances in commerce for which there exist little or no toxicity information, (2) animal welfare concerns, and (3) the realization that the results of high-dose in vivo animal testing do not represent responses in a heterogeneous human population exposed to much lower doses.

Recent advances in toxicogenomics, bioinformatics, systems biology, and computational toxicology are both remarkable and humbling. The goal of twenty-first century toxicology is to change the nature of toxicity evaluations and transform hazard evaluation and risk assessment from a system that uses high-dose in vivo animal bioassays to one based primarily on computational profiling and in vitro methods. If this transformation is to be successful, a clear focus is needed for identifying relevant human health risks with a defined degree of confidence from the in vitro results. Initially, these results would be used for prioritization and screening of chemicals; with time and experience, hazard prediction and even quantitative risk assessment may be possible.[22–24]

To implement this in vitro testing, EPA has initiated a number of activities in an effort to incorporate the use of HT/HC in vitro assays into risk assessment. The most visible of these is EPA's ToxCast™ program, consisting of a battery of both commercial and publically developed HT/HC assays. Initially, the ToxCast approach has been designed to utilize the vast array of commercially available HT/HC assays to screen substances of interest to EPA. ToxCast is part of EPA's contribution to a collaboration that includes the FDA and the National Institutes of Health (NIH).

This approach has both advantages and disadvantages, and the two primary areas of discussion are validation of the assays and the use of prediction models.[24] Many of these commercial methods used in ToxCast are proprietary, so details about assay development, replicability, sensitivity, and specificity are not necessarily available for independent evaluation and scientific peer review—in many ways, these proprietary assays are "black boxes." However, one distinct and obvious advantage is that these assays can be robotically automated to generate data very quickly.

The majority of HT and HC assays being applied in ToxCast™ were first developed for pharmaceutical purposes and were later adapted by EPA for screening of commodity and environmental chemicals. In the pharmaceutical industry, the assays are used to search for likely drug candidates with potent biological activities already predicted by computational methods such as QSAR. In contrast, commodity chemicals are selected based on their physicochemical properties with the goal of improving the specific performance of a product—commodity chemicals typically possess much lower biological activity than do drug candidate molecules.

This difference between commodity chemicals and drug candidates begs the question of whether these in vitro assays are even capable of producing meaningful information in a risk assessment context. First, exposure must be considered, and testing commodity chemicals at artificially high concentrations in HT/HC assay systems simply to elicit measurable responses will obviously have little or no real-world significance. Second, each assay or collection of assays must be anchored in biological knowledge—without this anchor, neither evaluation of the biological context within one or more modes of action nor an understanding of the meaning of the assay doses in terms of real-world exposures is possible.

Details of ToxCast™

Phase I of ToxCast was completed in 2009 and profiled in more than 300 well-studied chemicals, primarily pesticide active ingredients.[25] The entire phase I ToxCast dataset is available at http://epa.gov/ncct/toxcast/data.html. ACToR stands for the Aggregated Computational Toxicology Resource.[26] ACToR is an online database that is EPA's online warehouse of publicly available chemical toxicity data, including ToxCast. Table 7.1 shows assay results for 17β-estradiol and the pesticide methoxychlor in two ToxCast assays from Attagene that measure gene transactivation by the estrogen receptor; Figure 7.2 shows plots and Hill model fits to these data. The Hill model is one of the continuous models in

TABLE 7.1

ToxCast™ Data for Estradiol and Methoxychlor

Source Name SID	CASRN	Chemical Name	Assay Component	Concentration (μM)	Value	Value Type
PC01		Estradiol	ATG_ERa_TRANS	0.000101611	1.158695025	Fold_change
PC01		Estradiol	ATG_ERa_TRANS	0.000304832	1.692233664	Fold_change
PC01		Estradiol	ATG_ERa_TRANS	0.000914495	3.793282648	Fold_change
PC01		Estradiol	ATG_ERa_TRANS	0.002743484	6.273I2097	Fold_change
PC01		Estradiol	ATG_ERa_TRANS	0.008230453	8.936194784	Fold_change
PC01		Estradiol	ATG_ERa_TRANS	0.024691358	23.26166875	Fold_change
PC01		Estradiol	ATG_ERa_TRANS	0.074074074	22.6788567	Fold_change
PC01		Estradiol	ATG_ERa_TRANS	0.222222222	21.63025695	Fold_change
PC01		Estradiol	ATG_ERa_TRANS	0.666666667	21.28919257	Fold_change
PC01		Estradiol	ATG_ERa_TRANS	2	18.52064286	Fold_change
PC01		Estradiol	ATG_ERE_CIS	0.000101611	3.617676324	Fold_change
PC01		Estradiol	ATG_ERE_CIS	0.000304832	5.957038338	Fold_change
PC01		Estradiol	ATG_ERE_CIS	0.000914495	8.570793093	Fold_change
PC01		Estradiol	ATG_ERE_CIS	0.002743484	8.881709102	Fold_change
PC01		Estradiol	ATG_ERE_CIS	0.008230453	9.27690957	Fold_change
PC01		Estradiol	ATG_ERE_CIS	0.024691358	9.297512438	Fold_change
PC01		Estradiol	ATG_ERE_CIS	0.074074074	8.320749195	Fold_change
PC01		Estradiol	ATG_ERE_CIS	0.222222222	9.338718174	Fold_change
PC01		Estradiol	ATG_ERE_CIS	0.666666667	8.110974539	Fold_change
PC01		Estradiol	ATG_ERE_CIS	2	7.766344747	Fold_change
DSSTOX_40522	72-43-5	Methoxychlor	ATG_ERa_TRANS	0.030786013	0.906776181	Fold_change
DSSTOX_40522	72-43-5	Methoxychlor	ATG_ERa_TRANS	0.09329095	0.922053388	Fold_change

(continued)

TABLE 7.1 (continued)
ToxCast™ Data for Estradiol and Methoxychlor

Source Name SID	CASRN	Chemical Name	Assay Component	Concentration (μM)	Value	Value Type
DSSTOX_40522	72-43-5	Methoxychlor	ATG_ERa_TRANS	0.282699848	0.93338809	Fold_change
DSSTOX_40522	72-43-5	Methoxychlor	ATG_ERa_TRANS	0.856666206	1.365585216	Fold_change
DSSTOX_40522	72-43-5	Methoxychlor	ATG_ERa_TRANS	2.5959582	4.597453799	Fold_change
DSSTOX_40522	72-43-5	Methoxychlor	ATG_ERa_TRANS	7.86654	10.58464066	Fold_change
DSSTOX_40522	72-43-5	Methoxychlor	ATG_ERa_TRANS	23.838	9.274828257	Fold_change
DSSTOX_40522	72-43-5	Methoxychlor	ATG_ERE_CIS	0.030786013	0.997885196	Fold_change
DSSTOX_40522	72-43-5	Methoxychlor	ATG_ERE_CIS	0.09329095	0.977945619	Fold_change
DSSTOX_40522	72-43-5	Methoxychlor	ATG_ERE_CIS	0.282699848	0.912688822	Fold_change
DSSTOX_40522	72-43-5	Methoxychlor	ATG_ERE_CIS	0.856666206	1.249848943	Fold_change
DSSTOX_40522	72-43-5	Methoxychlor	ATG_ERE_CIS	2.5959582	1.975830816	Fold_change
DSSTOX_40522	72-43-5	Methoxychlor	ATG_ERE_CIS	7.86654	3.812084592	Fold_change
DSSTOX_40522	72-43-5	Methoxychlor	ATG_ERE_CIS	23.838	4.409195062	Fold_change

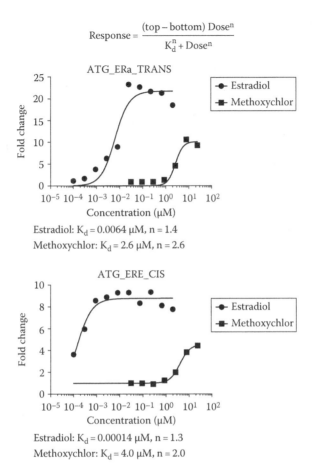

$$Response = \frac{(top - bottom)\, Dose^n}{K_d^n + Dose^n}$$

ATG_ERa_TRANS

Estradiol: $K_d = 0.0064\ \mu M$, $n = 1.4$
Methoxychlor: $K_d = 2.6\ \mu M$, $n = 2.6$

ATG_ERE_CIS

Estradiol: $K_d = 0.00014\ \mu M$, $n = 1.3$
Methoxychlor: $K_d = 4.0\ \mu M$, $n = 2.0$

FIGURE 7.2 Hill model fits of 17β-estradiol and methoxychlor from two of Attagene transcription assays in the ToxCast™. The equation for the Hill model is shown above the two plots.

EPA's BMDS discussed in Chapter 4. For completeness, the Hill model equation is shown in Figure 7.2. As an exercise, look up the EC_{50} values for methoxychlor in these two assays in the ACToR database at http://actor.epa.gov/actor/faces/ACToRHome.jsp. From the plots, what else can you say about the potency and intrinsic activity of methoxychlor relative to estradiol?

In conjunction with ToxCast phase I, EPA assembled a large database of mammalian in vivo toxicity data for these substances from pesticide registration studies submitted to the agency. This database is known as ToxRefDB. This resource was developed to help facilitate investigations of the correlations between ToxCast results and in vivo effects.

ToxCast is based on the premise that toxicity was driven by interactions between chemicals and biomolecular targets. Since these targets, and indeed the MOA for most chemicals, have yet to be fully understood, the focus of ToxCast was a multiple target matrix with multiple data domains ordered in terms of increasing biological relevance and increasing cost.[27]

Phase I of ToxCast was meant to be a proof-of-concept phase using approximately 300 substances for which extensive animal testing results were available. These chemicals were chosen to afford the opportunity to determine the correspondence between the resulting ToxCast data and in vivo results.[28]

Activity profiles of the phase I ToxCast chemicals have to date revealed both expected and unexpected results for chemicals in signaling and metabolic pathways.[25] Phase II of ToxCast, which commenced in 2010, is currently screening almost 1000 chemicals from industrial and consumer products, food additives, and failed pharmaceuticals using the assays as in phase I and some additional assays.[29]

ToxCast activities are ongoing, and information on progress is continually emerging. As mid-2013, not all the data and information are currently unavailable, and this creates an unfortunate impression of the lack of transparency. Sharing and publicizing details about the assays and the prediction models fosters scientific debate and leads to a demonstration of the true level of rigor of the methods.

Knowledge of Mode of Action Is Necessary to Understand and Use ToxCast™ Results

MOA is the main focus of EPA's *Guidelines for Carcinogen Risk Assessment*[30] and of this textbook. MOA should be the main focus of the way in which to use the results of in vitro assays and HT/HC advanced screening approaches. One of the uses of MOA is to address the question of human relevance of responses observed in animals, and MOA should also be used to determine the human relevance of responses from in vitro assays and HT/HC approaches.[22,24,31]

Knowledge of MOA should also be used to inform the prediction models used for translating assay results to real-world exposures. Therefore, these predictive models must be built upon a firm understanding of the interaction of a substance with all relevant biological processes. Knowledge of toxicokinetics, toxicodynamics, and dose–response relationships of key events is needed for a full exploration of the MOA underlying an adverse outcome. Ideally, individual assays or assay suites will be associated with biomarkers or measured biological responses. If the quantitative relationships (1) between the assay results and the biomarkers associated with the assays and (2) between the biomarkers and the occurrence of adverse effects both become known, then the assays can be used for hazard prediction, and the vision expressed by NRC in *Toxicity Testing in the 21st Century* will be realized.[32]

A weight-of-evidence approach will be needed to examine the evidentiary support (or lack of it) between assay results, biomarkers, and disease.[33–37] Toxicity pathways are in fact biological response pathways, and, in general, they are either linked or identical to normal cellular response pathways that maintain homeostasis and normal function in the face of both internal and external stressors.[32] Toxicity can be thought of as failures of homeostasis and resulting maladaptive and biologically inappropriate activity in these response pathways.

The measurable events further along the pathway that lead to and are experimentally or toxicologically associated with the adverse outcome may be observable by biomarkers. If necessary to the occurrence of the adverse outcome, an event is then truly a key event in the MOA.[38]

Early Prediction Models

Early in the history of ToxCast, hazard prediction was both the primary goal and primary challenge.[27] However, the use of biochemical or genomic assays to build predictive models of toxicity was limited because of incomplete knowledge of the underlying biology.[39] Some studies used regression/correlation approaches to examine the relationship between in vivo lower exposure limit (LEL) values and in vitro half-maximal activity concentration (AC50) values from ToxRefDB and ToxCast DB, respectively.[28]

One particular thorny aspect of the prediction models was dosimetry or pharmacokinetics. An early attempt in incorporating dosimetry used a simple one-compartment toxicokinetic model that incorporated metabolism and renal excretion to predict oral human equivalent doses in mg/kg/day from AC50 or lowest effective concentration (LEC) values from ToxCast. These were compared to oral human exposure values estimated from NHANES.[40,41]

Recently, ToxCast phase I data were evaluated comprehensively by attempting to predict 60 in vivo endpoints using 84 different statistical classification models with the data from more than 600 in vitro assays.[42] In addition, the predictive power of these statistical models was compared with that of QSAR and chemical descriptors. The predictive power of the assays was not any better than that of the chemical descriptors, and the conclusion was that at best, the assays could be used to identify "risk factors" for particular chemicals that could conceivably be useful in screening.

In summary, both the diversity and complexity of the prediction models used with ToxCast data are increasing. This is hardly surprising given the relative "newness" of the data and prediction models. The happy consequence of this diversity of approaches is that the field of in vitro-to-in vivo prediction models is rapidly maturing.

If the new approaches are to be successful, they must be strongly focused on identifying human health risks with a defined degree of confidence relative to that of existing animal testing methods. Also, discriminating those assay responses that are relevant to human health risk from those that are irrelevant will require anchoring the assay to one or more key events within a MOA and confidence in the translation of assay results to environmentally relevant exposure levels.

How Many Toxicity Pathways? Are They All Covered by ToxCast™?

ToxCast assays were selected on the basis of convenience, and this has led to another area of uncertainty: does the suite of ToxCast assays cover the entire range of toxicity pathways in humans? The large number of assays within ToxCast (currently over 1000) suggests that the range of toxicity pathways and modes of action may be largely covered.[28,43] For understanding these results within the context of MOA, the distinction between an adverse effect and an adaptive effect needs to be made more clear—as noted in Chapter 4, changes in enzyme levels are considered adverse in some IRIS assessments.[44,45]

As discussed in Chapter 2, the question of coverage by ToxCast of the entire domain of toxicity pathways remains unknown, and the question of whether a sufficient number of pathways are represented remains unanswered.[46,47] The set of pathways that are important for toxicity are not yet known, and, for the ones that are identified with toxicity, the magnitude of perturbations of the pathway that result in toxicity also remains to be discovered. ToxCast assays were chosen for convenience and availability. What is likely is that ToxCast will end up being a hypothesis generator and may help identify new pathways or parts of pathways that are important for toxicity.

SCIENCE AND DECISIONS: ADVANCING RISK ASSESSMENT

In 2009, in response to a request from EPA, the NRC published this report that immediately became controversial because of the recommended approach to dose–response. Due to the burgeoning amount of scientific data, risk analyses of increasingly difficult issues were being sought; these issues included multiple chemical exposures, variation in susceptibility, cumulative risk assessment, life cycle impacts, cost–benefit considerations, and risk–risk tradeoffs. In order to meet these demands, the report focused on recommendations to improve both the quality of the technical analysis in risk assessments and the utility of risk assessments for decision making.

Improvements in Problem Formulation

One area of concern to the authors of the report was problem formulation. This aspect of risk assessment was discussed at length in Chapter 2. *Science and Decisions* urged a greater focus on the upfront stages of risk assessment—planning, scoping, and problem formulation.[48]

The report also recommended consideration of uncertainty and variability in all phases of risk assessment. In 2001, EPA's Superfund Program released *Risk Assessment Guidance for Superfund (RAGS) Volume III—Part A: Process for Conducting Probabilistic Risk Assessment.*[20] This document actively discouraged the application of probabilistic methods to dose–response assessment. Of course, including variation in exposure only in a probabilistic risk assessment gives only part of the picture.[49] Hence, *Science and Decisions* encouraged inclusion of quantitative estimates of uncertainty and variability at all key computational steps in a risk assessment.[48,50]

Replacing Defaults with Data

Back in 1983, the Red Book recommended the use of uniform inference guidelines, as discussed in Chapter 1. This recommendation resulted in a number of advances such as the cancer guidelines; the recommendation also resulted in the proliferation of default values for many widely used quantitative factors in risk assessment. Over time, these default values became "set in stone," and the original thinking and scientific basis of these numbers were more often than not forgotten. In many risk assessments, the ascendancy of the defaults often trumped measured values that were directly applicable to the problem at hand.[48,50]

For example, considering the example in Chapter 5 of the former gold mine, using EPA default value of 60% for arsenic bioavailability or even the earlier regulatory practice of not accepting any value other than 100% for bioavailability would have significantly changed the outcome of the risk assessment.

The NRC committee recommended using alternative assumptions or values in lieu of the default if the alternative can be demonstrated to be superior. The committee also indicated that many defaults without any seeming basis had become ingrained in risk assessment practice. Often, regulators are hesitant to abandon defaults, especially when data indicate that the default is overly conservative/protective.[51]

Silver Book Recommendations for Dose–Response

For a number of years, EPA has been attempting (or saying so) to "harmonize" noncancer and cancer risk assessment. In 1997, a colloquium organized by EPA's RAF concluded that the agency needed to "push the envelope" in terms of consideration of MOA, and such consideration was likely the means of harmonization.[52] One of the goals of benchmark dose modeling is harmonization of the cancer and noncancer approaches.[53,54]

The NRC committee also wanted to address the scientific limitations with current approaches to dose–response assessment, as presented in Chapter 4. The committee reasoned that because of background exposures and ongoing disease processes, variations in susceptibility would exist within the human population. These variations could contribute to the appearance of a linear dose–response at the population level.[55] For example, early studies indicated the possibility of low-dose linearity in mortality associated with exposure to PM.[56,57] However, recent studies that examined a range of potential causes of mortality suggest that weather may play a greater role than exposure to PM and that the dose-response may actually show hormesis.[58–62] A no-effect level cannot be determined for the dose–response between blood lead concentration and reduction in IQ, although the exact shape of the dose-response curve is far from certain.[63–65]

EPA's current dose–response paradigm includes an immediate and artificial separation between cancer and noncancer outcomes. When one uses the slope factor methodology for cancer risk assessment that assumes low-dose linearity, quantitative estimates of risk are available on both an individual and population bases. Hence, these estimates, correct or not, can be used in cost–benefit analyses and consideration of risk–risk tradeoffs.

In contrast, a hazard index from a noncancer assessment provides no such quantitative information. How can one quantify risk either above or below the reference dose in a way that enables quantitative risk comparisons? The regulatory cancer risk range of 10^{-6} to 10^{-4} can represent a regulatory target at the population level, but for noncancer hazard, no similar quantity can provide insight into the magnitude of population risk that might be considered acceptable—the "bright line" of an HI value of unity provides no insight at all as to the likelihood of adversity.

The uncertainty factors used for low-dose extrapolation and interspecies extrapolation represent a mixture of uncertainty and variability, and thus their use is contrary to EPA's 1997 *Policy for the Use of Probabilistic Analysis in Risk Assessment*.[19] At present, the utility of noncancer assessment is limited and hinders the ability to conduct analyses of risk–risk tradeoffs, to weigh costs and benefits, or to provide transparency in decisions.

The Silver Book recommended a unified approach to dose–response, using MOA to determine the shape of the dose–response curve in the low-dose region. The exploration of a possible default value for interindividual variation in cancer susceptibility was also recommended. The examples provided in the Silver Book did not explicitly demonstrate that linear low-dose extrapolation would be used for both cancer and noncancer dose–response. The NRC report suffered from being less than clear about the specifics of low-dose extrapolation, and this lack of clarity engendered a great deal of comment within the risk assessment community.[66–72]

The Silver Book described three conceptual models for dose–response to be chosen based on whether the low dose–response was linear on an individual or population level. As an exercise at the end of this chapter, these three models will be explored.[48] This exercise should help you decide for yourself whether the Silver Book did indeed recommend the assumption of a linear dose–response in the low-dose region as appropriate for all adverse effects. Notwithstanding this controversy, there remains great benefit in providing a dose–response assessment resulting in quantitative estimates of toxicity appropriate for use in a decision-analytic approach that can determine the societal value of a range of risk management options.

Cumulative Risk Assessment

In 2003, EPA's Risk Assessment Forum released the *Framework for Cumulative Risk Assessment*.[73] Before and after that, the idea of cumulative risk assessment had been the subject of many publications.[74–76] The idea was likely popularized at EPA by Dr. Gershon Bergeisen who served as the special assistant to the director of Superfund in the 1990s. Dr. Bergeisen now operates a family medical practice in Hawaii. Dr. Bergeisen's influence may have been the start of the wish at EPA to move away from single-chemical single-pathway assessments and toward more holistic considerations of multiple exposures in real-world community contexts. This more holistic approach provides both challenges and opportunities—challenges when the stressor more likely related to the health outcome of concern is beyond EPA's regulatory mandate, for example, smoking, and opportunities to utilize the newer in vitro approaches and knowledge of MOA for assessment of multiple chemicals with multiple dimensions of exposure.[51]

Toxicology by itself cannot provide much information about psychosocial and lifestyle factors that may influence the occurrence of adverse outcomes; nonetheless, epidemiology may be able to address these factors but is limited in its ability to demonstrate causal connections with multifactorial health outcomes. Indeed, the interface between toxicology and epidemiology is an area of growing interest among risk assessors.[77] Creativity in applying methods and results from a variety of disciplines to develop insights into cumulative risk methods and techniques will be a growing field. Diet and nutrition are certainly factors that affect susceptibility to chemical exposures.[78,79] The interaction of smoking, radon exposure, and other factors that influence lung cancer is being explored and will likely provide insight into how to account for the effects of multiple stressors on a common endpoint.[80–82] A number of computational techniques for cumulative risk assessment are emerging.[83–86] The use of biomarkers and community-based assessments will also provide another point of view.[87–90]

For smoking, poor nutrition, lack of exercise, poverty, and other factors that may produce significant exacerbation of the risk of health outcomes related to environmental chemical exposure, regulators will be in a difficult position—the decision must be made whether to regulate industrial facilities with pollution control measures based on the attributable fraction of risk from each of those facilities or rather to lobby other government agencies such as the CDC to increase their efforts toward smoking cessation/health education. In addition, the presentation of comparative risks that includes consideration of lifestyle factors may alienate many of the stakeholders.

Psychosocial stress and socioeconomic status (SES) are often mentioned as contributors to risk. Much of this psychosocial stress stems from poverty.[91–95] Poverty is obviously not a problem that can be addressed within the current rubric of environmental regulation, and there is a larger issue—does poverty stem from accident of birth or is it the result of an individual's life decisions? This question is well beyond the scope of this book. However, if SES is indeed considered as a factor in a risk assessment, what would be done with this information? How could regulatory activities address any risk attributable to SES? What, if any, actions could be taken in a capitalist democratic society that would constitute an appropriate response to SES as a risk factor?

Stressors other than chemical exposure may contribute a large portion of the disease burden in a community. In such a case, the most appropriate question is whether reduction of chemical exposures would have benefits that exceeded the cost. Experienced risk communicators are well aware that such risk comparisons often serve only to drive a wedge between risk assessors and the community they are trying to protect.[96] For consideration of the risk of nonchemical stressors, such as psychosocial stress or SES, boundaries and limits need to be established upfront that both define the purpose of including nonchemical stressors in the risk assessment and also make clear the limited extent of a government's ability to regulate these stressors.

The Silver Book points out that in any process in which the government, communities, and other stakeholders come together for a decision, there will be inevitable differences in the ability of different stakeholders to influence the process. These differences stem from imbalance in political and economic power. Risk assessors should be aware of these issues, and addressing them should be discussed during the problem formulation stage of any cumulative risk assessment.

Alliance for Risk Assessment

Because of the general negative response to the dose–response assessment presented in the Silver Book, Dr. Michael Dourson of Toxicology Excellence in Risk Assessment assembled the Alliance for Risk Assessment (ARA), a broad-based nonprofit, government, and NGO coalition. ARA recruited well-known scientists with expertise in risk assessment to serve on a science panel. Others presented case studies that explored various aspects of the Silver Book methodology to the science panel. The science panel and interested workshop participants developed an interactive framework for organizing case study methods, and the panel used the framework to identify additional case studies that address important gaps in methodology. ARA has been holding workshops for about 2 years, and the series is transitioning to an "evergreen" approach including a standing panel that reviews methods and issues on a semiannual basis, leading to updating of the framework. Readers are encouraged to explore the ARA website at http://www.allianceforrisk.org/.

EXPOSURE: SHOULD WE EVEN CARE ABOUT TOXICITY?

One consequence of the transformation in toxicity testing is that exposure characterization of humans using biomarkers in blood and urine as currently conducted by the NHANES of the CDC can be combined with relatively simple pharmacokinetics and used for screening.

ToxCast data were used in one such evaluation. ToxCast data are reported as the AC50, which is the concentration producing 50% of the maximum activity in the assay, or LEC, which is the lowest effective concentration observed in the assay. The values are reported in units of μM. Reliance on the reported assay results could potentially misrepresent in vivo effects of chemicals—what is lacking is any information on toxicokinetics that would determine actual internal exposure. To address this data gap, hepatic metabolism and plasma protein binding were experimentally measured for 239 ToxCast phase I chemicals, and these data were used in a population-level toxicokinetic model that performed the necessary in vitro-to-in vivo extrapolation (IVIVE). For each chemical, the model produced an estimate of the distribution of the daily human oral dose that would result in steady-state in vivo blood concentrations equivalent to in vitro AC50 or LEC. The estimated steady-state oral equivalent doses associated with the in vitro assays were compared with chronic aggregate human oral exposure estimates to assess whether in vitro bioactivity would be expected at the dose-equivalent level of human exposure. For 90% of the 239 chemicals, the 95th percentile value of exposure estimated from human urinary concentrations was below the range of oral equivalent doses estimated from ToxCast data, often by several orders of magnitude. The remaining 10% of exposure estimates overlapped the lower end of the range of oral equivalent doses.[97]

What this study shows is the ability to use exposure data along with in vitro results to determine if a chemical poses a risk at current human exposure levels. In essence, this is an exposure-based risk assessment that uses biological activity as a measure and asks whether current human exposures could result in biological activity. This clever approach does away with the need for both HI and dose–response assessment. Instead of estimating exposure with the uncertain models presented in Chapter 3, this method uses the plethora of NHANES biomarkers now being measured routinely.

The exposome has been defined as all environmental exposures from conception onwards, including those related to diet and lifestyle.[98] The exposome also includes endogenous exposures to chemicals, such as formaldehyde, that occur naturally within the human body.[99] The endogenous processes include inflammation, lipid peroxidation, oxidative stress, infections, and activity of gut flora. Because the exposome represents these combined exposures in their entirety, it provides an unbiased agnostic assessment for evaluating the causes of disease, environmental or otherwise.[100] Biomarkers, as discussed in Chapter 4, also represent the exposome.

In this regard, the exposome, as assessed by sampling and analysis of body fluids, represents a top-down approach to exposure, whereas measurements of chemicals in soil, air, water, and food would represent a bottom-up approach to exposure.

The exposome is an intriguing concept, but the necessary details can never be completely measured. One would need to measure an ever-increasing number of factors in the body at ever-smaller time intervals even to hope of getting a handle on the variation of an individual's exposome over time. Currently, a snapshot in time of an individual's exposome would likely consist of the list of environmental toxicants currently in the NHANES suite; blood lipid and enzyme profiles; screening tests for a variety of diseases; DNA and hemoglobin adducts; alterations in lipids; genomic, epigenomic, and proteomic changes; alterations in the microbiome; and likely others. As currently envisioned, the exposome is a dream come true for proponents of big data.

In 2012, the NRC released a report titled *Exposure Science in the 21st Century: A Vision and a Strategy*.[101] The report recommended that information on biomarkers be combined with data derived from remote sensing, GPS, satellite imaging, and other sources using an informatics approach. The development of the informatics necessary to combine these diverse data in a meaningful way was emphasized in the document. EPA's ExpoCast program was recently developed as a complement to ToxCast; the details have yet to be made public.

What some envision is the advent of environment-wide association studies (EWAS), comparable to the genome-wide association studies (GWAS) in genetic epidemiology. Recently, an EWAS revealed a relationship between the metabolic products of the gut flora and cardiovascular disease.[102] However, experience with GWAS indicates that careful examination of the results is warranted before using the results for any sort of prognostication.[103]

The exposome concept is continuing to evolve.[104] What will likely come about is an integrated approach that uses both internal and external exposure assessment and combines these into a top-to-bottom approach. In sum, the picture of exposure science in the twenty-first century is still emerging.

There is an additional general point to be made about the importance of exposure assessment as a screening tool. Currently, the estimate of the number of untested chemicals in commerce today ranges up to 100,000. Animal testing for these chemicals is clearly not possible in a timely fashion and would be extremely costly. Addressing these chemicals is one of the goals of EPA's ToxCast program—although the ToxCast assays measure only biological activity, not toxicity. However, if some measure of exposure to a particular chemical could be determined—through an estimate developed from the use scenario or even from a biomarker—then, as discussed earlier, this exposure estimate could be compared with an estimate of biological activity from an in vitro assay. If the exposure estimate were a hundred fold less than the toxicity estimate, this might indicate a lack of concern was justified, and the resulting margin of exposure estimate would likely be sufficient for risk management purposes.

MODE OF ACTION

EPA's cancer guidelines were not the first document to highlight MOA. A very recent paper by the ARA science panel written in response to the recommendations in *Science and Decisions*[48] advanced the concept of a fit-for-purpose dose–response assessment.[105] The idea of fit-for-purpose is that the level of complexity and effort of the dose–response assessment be no greater than that needed to select between risk management alternatives—consistent with the increased focus on problem formulation.[48]

ARA expressed dismay that MOA was used only rarely in the dose–response assessment. In fact, the leadership of EPA's IRIS program seems intent on ignoring MOA. ARA indicated that the integration of MOA information into the dose–response assessment should occur as early as possible. Ignoring MOA until later in the dose–response assessment limits the ability to use this information for low-dose and interspecies extrapolation, identification of susceptible populations, and population-level estimates of the range of variability in response. The consideration of MOA should fit the purpose of the risk assessment and conform in complexity and scope to the risk management decision.

Early in the history of risk assessment, the notion was advanced that background disease processes and exposures could linearize the dose–response.[106] The statement was originally made well before the idea of MOA even existed, and this same idea made its way into the Silver Book—"effects of exposures that add to background processes and background endogenous and exogenous exposures can lack a threshold if a baseline level of dysfunction occurs without the toxicant and the toxicant adds to or augments the background process."[48]

A recent risk assessment for the pesticide chlorpyrifos that acts by inhibiting AChE demonstrated that human variability in background cholinesterase activity, differing levels of enzymatic activity for degradation of the bioactive moiety, and lifestyle factors would not appreciably add to the predicted level of AChE inhibition at current exposure levels.[107–111] This demonstration required a great deal of data on MOA as well as toxicokinetic and toxicodynamic data.

The adoption of the LNT by the Safe Drinking Water Committee of the NAS (Chapter 4) was partly based on the idea that DNA is pristine. In fact, DNA is far from pristine. Every cell has a steady-state background of at least 50,000 endogenous DNA lesions. Under conditions of oxidative stress, this number is expected to increase. If DNA replication occurs before these lesions are repaired, the result can be mutations. Recent work employing formaldehyde labeled with the heavy isotopes carbon-13 and deuterium demonstrated the ability to separate endogenous DNA adducts from those produced by exposure to double-labeled formaldehyde. In EPA's 2010 formaldehyde risk assessment developed by the IRIS program, an upper-bound inhalation unit risk value of 0.081 ppm was derived. Hence, at 1 ppm, the risk of nasopharyngeal cancer, Hodgkin's lymphoma, or leukemia would be 8.1%. The corresponding risk-specific dose at a risk of 1 in 10,000 would be 1.2 parts per billion (ppb). When data on exogenous adducts were used as the basis of the risk assessment, the risks were up to 20,000 times lower than estimated by the EPA.[112] Just to put this risk-specific dose in context, the range of formaldehyde concentrations occurring in normal human breath may be up to 50 ppb. Higher levels may occur as the result of disease and are being explored as clinical biomarkers of disease.[99,113,114]

One aspect of consideration of MOA is that it seems daunting for some—especially those not schooled in biology. In this regard, individuals come to the field of risk assessment from a variety of disciplines. Many from engineering or fields such as operational research, while highly knowledgeable in areas such as statistics or mathematical modeling, are frankly uncomfortable dealing with biology. This discomfort is a possible reason that MOA has been given short shrift within the IRIS program. What would help greatly is a catalog of potential modes of action. Thomas Hartung, director of the CAAT at Johns Hopkins, suggests there cannot be all that many toxicity pathways or modes of action for adverse effects.[46] One of the projects in CAAT is to map the human toxome. To provide guidance on selecting MOAs consistent with current biological knowledge, the International QSAR Foundation, the American Society of Cellular and Computational Toxicology, and others are developing collections of known MOAs that can be assembled into a Wiki.[115] Read-across and other methods that develop a limited number of modes of action from chemical structure will also likely be informative.[22,24,31]

ADVENT OF EVIDENCE-BASED TOXICOLOGY

An evidence-based approach to toxicology is an area of intense interest. The model for this approach is that of evidence-based medicine (EBM).[116] The term "evidence-based toxicology" (EBT) was first used in 2005.[37] Dr. Hartung is a pioneer in the application of evidence-based approaches in toxicology.[36]

EBM is based on assessing the totality of evidence regarding a particular medical intervention. EBT attempts to apply similar approaches to the assessment of the totality of evidence regarding the toxicity of substances. This evidence includes studies in animals, humans, cells or tissues in vitro, computational toxicology, and predictive methods such as QSAR. The practices of systematic reviews of evidence,[117] transparency in decisions,[118] open data disclosure,[117,119] synthesis of different types of evidence,[77] and assessment of bias/credibility[120,121] are just beginning to be applied in toxicology.

OBSTACLES TO THE ADVANCEMENT OF RISK ASSESSMENT

When science is used to support societal decisions, the quality of the science itself, the relevance of the information to the decision at hand, and the acceptability of that decision based on nonscientific factors must all be determined. Societal decisions may be taken for a variety of reasons—science-based or otherwise. What is most important is transparency in the decision and the communication of the reasons a particular choice was made.

Decision makers who wish to "adjust" the risk assessment to support a particular decision made for reasons other than risk should instead provide an honest account of the basis for the decision. For example, at the former gold mine example in Chapter 5, let us assume the project manager wished to remove the tailings pile as the risk management outcome—she/he may have considered it an eyesore and a potential detriment to future business at the park. She/he might have instructed the risk assessor to use a value of 100% for arsenic bioavailability and not to consider any lower value. This action would have been inappropriate and unethical on her part and contrary to the idea of explicit separation of risk assessment and risk management.

What is absolutely necessary for both risk assessors and risk managers is that an honest statement of the basis for the decision and the evidence for and against this decision need to be provided. At times, some decision makers may wish to "tweak" or "adjust" the scientific evidence to lend support to the preferred decision alternative. Honest scientists who strive to be objective and as free from bias as possible provide the best support to decision makers, even if the evidence might not support the wishes of these decision makers.

In this section, two obstacles to progress in risk assessment will be discussed. These are COI and bias, and resistance to change.

CONFLICT OF INTEREST AND BIAS: THEY ARE EVERYWHERE!

Scientists at any academic, industry, government, or nongovernment institution may have conflicts of interest, both financial and nonfinancial. One can define COI as follows: an individual is in conflict if he/she owes a duty of loyalty or responsibility to two distinct parties, both of which are likely to be affected in different ways by the activity in which the individual is engaged. COI policies generally address financial but not nonfinancial conflicts of interest. Policies for handling financial COI issues generally rely on disclosure as a mechanism to manage COI.[122]

COIs or competing interests may arise because of financial interests, organizational affiliations, or personal biases. Scientists working to support the chemical industry are assumed by environmental activists to have competing interests because of financial sponsorship—other sources of bias are ignored by the activists.[123] The unfortunate misdeeds of some in the tobacco industry have produced much skepticism about the scientific credibility of all industries.[124–127] The Society of Toxicology provides useful definitions of COI, bias, and advocacy.[128]

Academic scientists are not immune from COI. Unfortunately, the publication bias against negative results may influence some academics. The publish-or-perish climate of academia and the cutthroat competition for a decreasing pool of government grant money have negative influences on science.[129] Journals and journal editors have a bias toward publishing positive results.[130,131] Negative results just do not seem as interesting to editors whose primary goal may be to drive up the impact factor of their journal. Thus, many academics have become prisoners of publication bias.

Notwithstanding the almost universal practice of disclosure of financial support, more often than not individuals in conflict have trouble recognizing their own conflicts—even individuals of high character and morals. An aspect of the human condition may be a perceptual blind spot that hinders honest and critical self-examination. If so, disclosure as a policy for addressing COI is worthless.[122]

Bias, but not COI, is inherent in all scientific endeavors. Hypothesis generation and testing involves expectation and creates a bias in the author of the hypothesis.[132] All scientists have opinions about their work and that of others in their field—they cannot help having bias. Money aside, scientists are often unconsciously biased toward results that confirm their preconceptions.[133,134]

There exists a growing literature purporting to find that research funded by private industry produces results that favor the industry funding the research more often than research on the same topic relying on other funding sources. Bias in pharmaceutical and medical device studies has been considered in a Cochrane review, and the evidence of bias in drug studies was "convincing and consistent."[135,136]

There appear to be a number of sources of bias in toxicological studies.[137] Good laboratory practice (GLP) is one means of reducing bias. Studies carried out under GLP are subject to external inspection that includes an audit of raw data, data completeness, and data accuracy. With GLP, the results of experiments that do not meet the experimenter's preconceptions cannot be trashed because an accounting of supplies and animals are part of GLP.[138] Hence, one can have confidence in both positive and negative results in studies conducted with GLP.

For about the last decade, there have been calls by environmental activists to dismiss the results of any study funded by private industry without even a cursory examination of the scientific merits of the study. For example, in November 2012, EPA held a public stakeholder meeting on the IRIS program at its offices in Crystal City, Virginia. One of the invited speakers was Dr. Richard Denison of the Environmental Defense Fund. Dr. Denison indicated in his comments that industry-funded toxicology studies should never be used in IRIS assessments—implying that such studies were too biased to use in regulatory decisions.

Industry-funded studies are conducted with GLP, and thus, there is a great deal of accountability. Often, industry-funded toxicological investigations use an external peer review committee that reviews study design, results, and prepublication drafts—for example, the recent MOA study of hexavalent chromium occurring in 2011 and 2012.[139]

When it comes to bias, funding source may actually be a result of a scientist's bias and not a cause of it.[140] Scientific interpretations will always reflect the biases of the authors. Even scientists who appear apolitical and disinterested may possess a worldview by virtue of their education or disciplinary orientation that others might view as bias. Some scientists have suggested that information about funding sources should not be included in scientific publications. This would force the readers to judge the quality of the science itself.[132,141,142]

More recently, various sets of considerations/criteria have been advanced as a basis for judging the quality of science with a view to better supporting regulatory determinations.[120,121,143] The most recent includes ten specific criteria, including disclosure of the funding source, independence of the investigator from undue influence, and reproducibility of the finding by other investigators.[120,123,144,145] These criteria are shown in Box 7.1.

BOX 7.1   CRITERIA FOR ASSESSING THE CREDIBILITY OF SCIENTIFIC WORK[120]

1. Whether the principal investigator (PI) has fully disclosed sources of funding and other "competing interests"
2. Whether the PI is legally guaranteed the right to (a) publish the results of the study without prior sponsor approval, (b) analyze and interpret the resulting data, and (c), where appropriate, control the study design
3. Whether the investigator or sponsor has publicly released the research data or test method to allow others to review them and seek to replicate the analysis
4. Whether the investigator conducted research that was designed objectively and reported factually, so that, according to accepted principles of scientific inquiry, the research design adequately tests an appropriately phrased hypothesis
5. Whether the work was peer reviewed
6. Whether, before its commencement, the study was included on a public registry of research intended for use in policy making
7. Whether the PI's remuneration was geared to a particular experimental outcome
8. Whether a sponsor or a PI participated in an arrangement by which the sponsor would pay the PI to lend his or her name to a presentation or article actually drafted by someone else
9. Whether a PI working under the auspices of a contract research organization (CRO) or other entity has maintained clarity between that entity and the PI's academic or other affiliation
10. Whether the sponsoring organization employs systematic external review of research and testing programs to promote a culture of scientific integrity

The article presenting these criteria can be obtained at no charge at http:// www.ncbi.nlm.nih.gov/pmc/articles/PMC3114808/?report=classic. The article also discusses additional considerations such as the use of a qualitative or quantitative weighting scheme and the next steps for implementing a formal consensus process.[120]

Bias, Scientific Misconduct, and the Food Quality Protection Act of 1996 (FQPA): A Cautionary Tale

The latest buzzword in emerging environmental threats is endocrine disruption. This idea may have advanced first by Lou Guillette, a professor at the University of Florida, who investigated the effects of a pesticide spill on wildlife at Lake Apopka in central Florida. In 1941, a levee was built in order to develop the shallow wetlands on the north side of Lake Apopka into a muck farm, the rich soil of which supported commercial growing of vegetables. In 1980, pesticides manufactured by the Tower Chemical Company were spilled in Lake Apopka. Following the spill, Guillette found reproductive abnormalities in alligators in Lake Apopka.[146–148] Guillette went on to publish more than 70 papers on development of the reproductive system in alligators. Two of Guillette's early collaborators were John McLachlan and Steven F. Arnold of Tulane University.

In 1996, Arnold, Guillette, McLachlan, and others published a paper in *Science* claiming that pesticides and hydroxylated PCBs, although weakly estrogenic when acting alone, were synergistic when tested in combination, and together produced an effect 1000 fold more potent in mimicking estrogen.[149] In 1997, McLachlan, the senior author, withdrew the article from *Science*, indicating that neither his laboratory personnel nor other researchers could replicate the results.[150,151] In 1999, John LaRosa, then chancellor of Tulane, wrote in *Science* that McLachlan did not commit or have knowledge of scientific misconduct; however, an independent review of Arnold's data did not support the conclusions of the paper.[152] Arnold resigned from Tulane in 1997, found work at the Roswell Park Cancer Institute in Buffalo, New York, and is reported to have subsequently left science to attend business school.[153] In 2001, the Department of Health and Human Services found Steven Arnold guilty of scientific misconduct "by providing falsified and fabricated materials to investigating officials at Tulane University in response to a request for original data to support the research results and conclusions reported in the *Science* paper."[154]

About 2 months before the publication of the original Arnold paper that claimed 1000-fold synergism, Theo Colborn, Dianne Dumanoski, and John Peterson Myers published a popular science book called *Our Stolen Future*.[155] This hugely popular book touts Guillette, Arnold, and McLachlan as heroic pioneers and claimed that even tiny amounts of pesticides and other manufactured chemicals could act as endocrine disruptors with eloquently written and compelling personal anecdotes as well as carefully selected scientific references. Both this book and the falsified results from Tulane increased public awareness and general anxiety about pesticides and other chemicals in food. Legislative committees accelerated their activities to be the first to enact provisions for endocrine disruption into law.

The Tulane results likely had an effect on pushing the FQPA into law.[156] The FQPA specifically required EPA to consider aggregate exposure assessment to many substances and also children's exposure. As part of FQPA, the House Commerce Committee, chaired by Rep. Tom Bliley (R-Va.), wrote into the FQPA an additional 10-fold margin of safety for the pesticide chemical residues to be applied for infants and children.[157] The Commerce Committee relied heavily on a 1993 NRC report, *Pesticides in the Diet of Infants and Children,* that indicated children may have both higher exposure and higher susceptibility to pesticides.[158]

Neither the House Commerce Committee nor EPA seemed troubled by the withdrawal of Arnold's synergism paper and the suggestion that part of the FQPA was based on inaccurate science. Daniel Byrd, a consulting toxicologist practicing in Washington, DC, opined on the effect of the FQPA:

> The far more certain health risks are more expensive food, diminished food availability and poorer food quality. The middle class will not starve or even significantly reduce its consumption of fresh fruits and vegetables when the food supply gets expensive, but disadvantaged groups will suffer. Thus, the ironically named FQPA may well protect us some of us from quality food.[156]

Whether Arnold, McLachlan, Guillette, or their collaborators were biased by reading *Our Stolen Future* is not known. What is possible is that Arnold, consciously or not, fudged his experimental data to match his preconceptions. The sad ending to this chain of events is that Byrd's predictions regarding the quality of food available to the poor have come true. The decreased availability of fresh produce in the so-called food deserts may be a factor in the occurrence of obesity, type 2 diabetes, and related health problems.[159–161] The price of food is determined by many factors, and whether, as predicted by Byrd, any part of the health problems of the poor can be attributed to the hasty passage of the FQPA remains an unanswered question.

Shielding Oneself from Bias

All of us are biased. Our preconceptions result from our entire past life experience. Good scientists strive for objectivity in spite of bias. Karl Popper, the twentieth century philosopher and exponent of the scientific method, opined that all good theories are falsifiable and that it is vital to be able to test a theory by virtue of experiment or observation. The evidence provided by these experiments or observations would either support or falsify the theory.[162]

By considering science as evidence and the scientific methods as the testing of theories and hypotheses, a hypothetico-deductive approach using the specifics of the evidence has formed the basis of current thinking about MOA.[163,164] Deductive reasoning and strict adherence to the scientific method can remove some bias. Good scientists most often try to recreate the mindset and reasoning of their critics to see if their conclusions hold up—and thus try to examine their own biases.

The stridency with which some suggest that use of industry-funded science in developing toxicity criteria is inappropriate provides an example of such bias. With good reason, the pharmaceutical industry, the medical industry, and the tobacco industry are replete with examples of how science can be corrupted by the exigencies of decision making.[124–127,135,136,165,166] Today, most scientific journals require that the funding source of the work be acknowledged explicitly, making it all too easy for readers, displeased with the results because of their own biases, to dismiss any inconvenient evidence based solely on funding source. Some scientists have likened this criticism to an assault on science driven by postmodern thinking; this type of thinking questions the very existence of a real world. Postmodern thinking claims that scientists have just as much right to believe in the big bang theory or human evolution as nonscientists have a right to believe the moon is made of green cheese or that flying spaghetti monsters once ruled the skies—with the consequence that scientific knowledge becomes just another of many cultural realities.[167] While everyone has an opinion, not all opinions are created equal—the acceptance of certain ideas or theories and rejection of others should be based on the weight of the evidence.

The work of many activist groups has been important in raising the consciousness of environmental issues—those who work for such groups are motivated by the honest desire to make the world a better place. Everyone has biases, one of which being the source of financial support.[168,169] Earlier in this chapter, the Tides Foundation that provided support for "The Story of Cosmetics" was discussed. The Beldon Fund (http://www.beldon.org/) provides funding to states and environmental activist groups for a variety of activities. The explicit aims of this funding include the following:

- To expose and prevent the destructive influence of corporate interests on scientific research, publications, and science-based policy
- To ensure that the science used by government in developing regulations of toxic chemicals is free from manipulation by industry groups
- To combat industry influence on the science and on the defense of precautionary toxic chemical policy through coalitions with state government officials
- To ensure that state and federal regulatory agencies can rely upon unbiased and reliable scientific information and advice in implementing health, safety, and environmental laws

All of this information is available at the Beldon Fund's website in its annual reports at http://www.beldon.org/programs-impact.html.

MISINFORMATION AND THE LACK OF SCIENTIFIC LITERACY

The legitimization of alternative worldviews and diminishing of science as a means of understanding has its roots in education (or lack thereof), the role of the media, and the hesitation of scientists to engage in science communication.[170–173] The misinterpretation of the precautionary principle by the public may be due to these factors, and the failure to recognize and communicate the risks of various management options (including maintaining the status quo) must be balanced against each other.[174]

A glaring example of "not knowing what you don't know" is provided by Professor Rena Steinzor of the University of Maryland School of Law during a meeting of the Subcommittee on Oversight of the House Science, Space and Technology Committee. Professor Steinzor also serves as president of the Center for Progressive Reform that receives support from the Beldon Fund. This subcommittee hearing was titled "EPA's IRIS Program: Evaluating the Science and Process Behind Chemical Risk Assessment." At about 2 h and 14 min into the hearing, Professor Steinzor said:

> My son, who's twenty, is sitting behind me. One of the most distressing things I've heard today is that he has formaldehyde in his body and he exhales it at levels much higher than the reference dose set by EPA's database. That didn't happen because he's walking through a natural paradise on the Chesapeake Bay. It's because the air is polluted. We live in a non-attainment area that is awash in toxics ...

The archived recording of the subcommittee hearing can be found at http://science. house.gov/hearing/investigations-and-oversight-hearing-epas-iris-program.[175]

What Professor Steinzor apparently didn't know were two facts—one, that formaldehyde occurs naturally in human breath and, two, that formaldehyde is not one of the six common air pollutants included in the NAAQS that provide the basis of a non-attainment designation by the EPA.[99,113,114,176] Professor Steinzor's testimony was no doubt well intended—to elevate concern about chemical exposures using formaldehyde as an example. However, by her conflating two unrelated and incorrect facts, the example failed to make her point. This cautionary tale suggests that all of us—risk assessors and others as well—should refrain from making substantive statements about disciplines other than those they have studied in depth.

ALL THE UNCERTAINTY YOU COULD WANT

Wicked problems, as discussed at the start of this chapter, are those with deep uncertainties. Experts disagree about the outcomes of various policy alternatives. Without sufficient objective evidence to support rational decision making and conflict resolution, the decision outcomes are often dictated by passions and unwarranted convictions.[177] Such wicked problems include the effects of climate change, cyberterrorism, and the threat of weapons of mass destruction.[17]

In any risk assessment activity, one strives to reduce uncertainty. Reducing uncertainty has become almost a slogan and, in some cases, means increased complexity and almost certainly, less transparency. The degree of uncertainty in a risk assessment is always relative. For example, an extrapolation factor for development of a reference dose for a chemical is proposed, and the basis is a quantitative comparison of key events in humans and animals; the question that must be addressed is whether the use of this extrapolation factor will reduce the overall uncertainty in the risk assessment relative to the use of the default uncertainty factor, generally having a value of 10.

Usually, one compares the uncertainty of proposed changes in risk methodology relative to commonly used defaults, but this is not always the case. The use of biomarkers in lieu of estimating exposure based on assumptions of human behavior is a way to reduce uncertainty in exposure relative to the exposure estimation based on human behavior in specific scenarios as presented in Chapter 3 that may use scenario-specific exposure factors rather than defaults.

The important question to ask oneself when considering a change in a risk assessment is whether the change will reduce uncertainty compared to not changing the assessment. With wicked problems, one may have no clue. In addition, there will be political and social pressure to do something. When will sufficient information be available to support a credible decision? Value-of-information methods, discussed earlier, may be a means of getting an answer. However, this question will be increasingly difficult to answer—especially for wicked problems—but those who choose the field of risk assessment as a career will need to provide answers many times over.

CONCLUSIONS

During the 1970s, environmental activists were instrumental in raising the public consciousness of the need for environmental protection and did the world a great service by their efforts. However, in the twenty-first century, the problems highlighted by activists have no easy fixes—unlike the 1970s, now the problems are wicked.

Of course, the future is difficult to predict—especially if the world is as unpredictable as some think. Maybe society should abandon risk assessment and put these resources toward building a culture of individual and societal resilience and anti-fragility. Is such a transition even possible? For now, risk assessment provides the "best of the worst" means of determining societal responses to complex threats and problems. Environmental risk assessment has many strengths and weaknesses. Changes will not come from experienced risk assessors, but from the set of new minds that take up this challenge in the twenty-first century. While the problems and challenges may not have changed that much, the amount of scientific information that can be brought to bear has increased hugely. It will take fresh young minds and new ways of seeing problems for progress to occur. This situation should not be viewed as a burden—rather it is a challenge to be met and a great opportunity for those with the imagination and drive to make a change.

EXERCISES FOR THOUGHT AND DISCUSSION

These following exercises have no correct answers. If you have difficulty with these, realize that so does the rest of the risk assessment community.

EXPLORING THE THREE CONCEPTUAL MODELS
FOR DOSE–RESPONSE FROM THE SILVER BOOK

Chapter 5 of Ref. [48] describes three conceptual models for dose–response assessment of chemicals. The choice of the model is dependent on background processes and exposures, the biological effects of the chemical being considered, the nature of human variability, and other possible factors. The three models are the following:

1. Nonlinear individual response, low-dose linear population response with background dependence
2. Low-dose nonlinear individual and population response, low dose–response independent of background (i.e., a threshold response for which a reference dose is most appropriate)
3. Low-dose linear individual and population dose–response (i.e., a non-threshold response for which a slope factor is most appropriate)

Discuss what sorts of data could be collected to inform one about the variability in the population response based on background exposure and ongoing disease processes.

CONFLICT OF INTEREST: YOUR OWN INVESTIGATION.

Obtain from your university library a copy of a 2007 paper by Hardell et al. (*Am J Ind Med* 50:227–233).[178] This paper engendered a number of letters to the editor that are printed in the same issue. Read the paper and these letters, and discuss them with a view to characterizing Hardell's view of COI.

CUMULATIVE RISK ASSESSMENT

Find the following papers in your university library:

- Love et al. (2010) *Am J Epidemiol* 172:127–134
- Geronimus et al. (2006) *Am J Public Health* 95:826–833
- Chakraborty et al. (2011) *Am J Public Health* 101, Suppl. 1:S27–S36

Discuss these papers, and determine the feasibility of conducting a cumulative risk assessment that involves psychosocial stress.

REFERENCES

1. National Research Council NRC. *Science for Environmental Protection: The Road Ahead.* Washington, DC: National Academy Press, 2012.
2. Kriebel D, Tickner J, Epstein P, Lemons J, Levins R, Loechler EL et al. The precautionary principle in environmental science. *Environ Health Perspect.* 2001, Sep;109(9):871–876.
3. Rumsfeld D, Myers R. DoD News Briefing—Secretary Rumsfeld and Gen. Myers. U.S. Department of Defense. Office of the Assistant Secretary of Defense (Public Affairs). News Transcript. http://www.defense.gov/Transcripts/Transcript. aspx?TranscriptID=2636 (accessed October 22, 2013).

4. Zolli A and Healy AM. *Resilience: Why Things Bounce Back.* New York: Free Press, 2012.
5. Taleb NN. *Antifragile: Things that Gain from Disorder.* Random House, New York. ISBN 978-0-679-64527-6. 2012.
6. Nietzsche FW. *Twilight of the Idols and The Anti-Christ.* Harmondsworth, London: Penguin, 1990.
7. Hill AB. The environment and disease: Association or causation? *Proc R Soc Med.* 1965, May;58:295–300.
8. Tallacchini M. Before and beyond the precautionary principle: Epistemology of uncertainty in science and law. *Toxicol Appl Pharmacol.* 2005, Sep 9;207(2 Suppl):645–651.
9. Locke J. *Two Treatises of Government.* New Haven, CT: Yale University Press, 2003.
10. Christoforou T. The precautionary principle and democratizing expertise: A European legal perspective. *Sci Publ Policy.* 2003;30(3):205–211.
11. European Commission (EC). *Communication from the Commission on the Precautionary Principle: COM.* Brussels, Belgium, February 2, 2000. http://ec.europa. eu/dgs/health_consumer/library/pub/pub07_en.pdf (accessed October 4, 2013).
12. Locke J and Nidditch PH. *An Essay Concerning Human Understanding.* Oxford, U.K.: Clarendon Press, 1979.
13. Altman DG and Bland JM. Absence of evidence is not evidence of absence. *Br Med J.* 1995, Aug 8;311(7003):485.
14. National Research Council NRC. *Human Biomonitoring for Environmental Chemicals.* Washington, DC: National Academies Press, 2006.
15. Abdel-Rahman A, Anyangwe N, Carlacci L, Casper S, Danam RP, Enongene E et al. The safety and regulation of natural products used as foods and food ingredients. *Toxicol Sci.* 2011, Oct;123(2):333–348.
16. Chan E, Tan M, Xin J, Sudarsanam S, and Johnson DE. Interactions between traditional Chinese medicines and Western. *Curr Opin Drug Discov Dev.* 2010;13(1):50–65.
17. Furedi F. Precautionary culture and the rise of possibilistic risk assessment. *Erasmus L. Rev.* 2009; 2(2):197–220. http://papers.ssrn.com/sol3/papers. cfm?abstract_id=1498432 (accessed May 2, 2013).
18. Bernstein PL. *Against the Gods: The Remarkable Story of Risk.* New York: John Wiley & Sons, Inc., 1998.
19. United States Environmental Protection Agency (USEPA). *Policy for the Use of Probabilistic Analysis In Risk Assessment.* Office of the Science Advisor. May 15, 1997. http://www.epa.gov/stpc/2probana.htm (accessed September 1, 2012).
20. United States Environmental Protection Agency (USEPA). *Risk Assessment Guidance for Superfund (RAGS) Volume III—Part A: Process for Conducting Probabilistic Risk Assessment.* EPA 540-R-02-002. OSWER 9285.7-45. PB2002 963302. Washington, DC, December, 2001. http://www.epa.gov/oswer/riskassessment/rags3adt/ (accessed September 1, 2012).
21. Bogen KT, Cullen AC, Frey HC, and Price PS. Probabilistic exposure analysis for chemical risk characterization. *Toxicol Sci.* 2009, May;109(1):4–17.
22. Stephens ML, Andersen M, Becker RA, Betts K, Boekelheide K, Carney E et al. Evidence-based toxicology for the 21st century: Opportunities and challenges. *ALTEX.* 2013;30(1):74–103.
23. Judson R, Kavlock R, Martin M, Reif D, Houck K, Knudsen T et al. Perspectives on validation of high-throughput assays supporting 21st century toxicity testing. *ALTEX.* 2013;30(1):51–56.
24. Patlewicz G, Simon T, Goyak K, Phillips RD, Rowlands JC, Seidel SD, and Becker RA. Use and validation of HT/HC assays to support 21st century toxicity evaluations. *Regul Toxicol Pharmacol.* 2013, Mar;65(2):259–268.

25. Knudsen TB, Houck KA, Sipes NS, Singh AV, Judson RS, Martin MT et al. Activity profiles of 309 ToxCast™ chemicals evaluated across 292 biochemical targets. *Toxicology.* 2011, Mar 3;282(1–2):1–15.

26. Judson R, Richard A, Dix D, Houck K, Elloumi F, Martin M et al. ACToR— Aggregated computational toxicology resource. *Toxicol Appl Pharmacol.* 2008, Nov 11;233(1):7–13.

27. Dix DJ, Houck KA, Martin MT, Richard AM, Setzer RW, and Kavlock RJ. The ToxCast program for prioritizing toxicity testing of environmental chemicals. *Toxicol Sci.* 2007, Jan;95(1):5–12.

28. Judson RS, Houck KA, Kavlock RJ, Knudsen TB, Martin MT, Mortensen HM et al. In vitro screening of environmental chemicals for targeted testing prioritization: The ToxCast project. *Environ Health Perspect.* 2010, Apr;118(4):485–492.

29. Sipes NS, Martin MT, Kothiya P, Reif DM, Judson RS, Richard AM et al. Profiling 976 ToxCast chemicals across 331 enzymatic and receptor signaling assays. *Chem Res Toxicol.* 2013, May 5;26(6):878–895.

30. United States Environmental Protection Agency (USEPA) USEPA. Risk Assessment Forum. *Guidelines for Carcinogen Risk Assessment.* EPA/630/P-03/001F. March, 2005. http://www.epa.gov/raf/publications/pdfs/CANCER_GUIDELINES_FINAL_3-25-05.PDF

31. Patlewicz GY and Lander DR. A step change towards risk assessment in the 21st century. *Front Biosci (Elite Ed).* 2013;5:418–434.

32. National Research Council (NRC). *Toxicity Testing in the 21st Century: A Vision and a Strategy.* Washington, DC: The National Academies Press, 2007. http://www.nap.edu/catalog.php?record_id=11970

33. National Research Council (NRC). *Applications of Toxicogenomic Technologies to Predictive Toxicology and Risk Assessment.* Washington, DC: National Academies Press, 2007. http://www.nap.edu/catalog.php?record_id=12037

34. Hoffmann S and Hartung T. Toward an evidence-based toxicology. *Hum Exp Toxicol.* 2006, Sep;25(9):497–513.

35. Hartung T. Food for thought… on evidence-based toxicology. *ALTEX.* 2009;26(2):75–82.

36. Hartung T. Evidence-based toxicology—The toolbox of validation for the 21st century? *ALTEX.* 2010;27(4):253–263.

37. Guzelian PS, Victoroff MS, Halmes NC, James RC, and Guzelian CP. Evidence-based toxicology: A comprehensive framework for causation. *Hum Exp Toxicol.* 2005, Apr;24(4):161–201.

38. Julien E, Boobis AR, Olin SS, and Ilsi Research Foundation Threshold Working Group. The key events dose-response framework: A cross-disciplinary mode-of-action based approach to examining dose-response and thresholds. *Crit Rev Food Sci Nutr.* 2009, Sep;49(8):682–689.

39. Martin MT, Judson RS, Reif DM, Kavlock RJ, and Dix DJ. Profiling chemicals based on chronic toxicity results from the U.S. EPA ToxRef database. *Environ Health Perspect.* 2009, Mar;117(3):392–399.

40. Rotroff DM, Wetmore BA, Dix DJ, Ferguson SS, Clewell HJ, Houck KA et al. Incorporating human dosimetry and exposure into high-throughput in vitro toxicity screening. *Toxicol Sci.* 2010, Oct;117(2):348–358.

41. Sipes NS, Martin MT, Reif DM, Kleinstreuer NC, Judson RS, Singh AV et al. Predictive models of prenatal developmental toxicity from ToxCast high-throughput screening data. *Toxicol Sci.* 2011, Nov;124(1):109–127.

42. Thomas RS, Black MB, Li L, Healy E, Chu T-M, Bao W et al. A comprehensive statistical analysis of predicting in vivo hazard using high-throughput in vitro screening. *Toxicol Sci.* 2012, Aug;128(2):398–417.

43. Boekelheide K and Campion SN. Toxicity testing in the 21st century: Using the new toxicity testing paradigm to create a taxonomy of adverse effects. *Toxicol Sci.* 2010, Mar;114(1):20–24.

44. Boekelheide K and Andersen ME. A mechanistic redefinition of adverse effects - A key step in the toxicity testing paradigm shift. *ALTEX.* 2010;27(4):243–252.

45. Keller DA, Juberg DR, Catlin N, Farland WH, Hess FG, Wolf DC, and Doerrer NG. Identification and characterization of adverse effects in 21st century toxicology. *Toxicol Sci.* 2012, Apr;126(2):291–297.

46. Hartung T and McBride M. Food for thought… on mapping the human toxome. *ALTEX.* 2011;28(2):83–93.

47. Mayr E. *The Growth of Biological Thought: Diversity, Evolution and Inheritance.* Cambridge, MA: Harvard University Press, 1982.

48. National Research Council (NRC). *Science and Decisions: Advancing Risk Assessment.* Washington, DC: The National Academies Press, 2009. http://www.nap.edu/catalog.php?record_id=12209

49. Simon T. Just who is at risk? The ethics of environmental regulation. *Hum Exp Toxicol.* 2011, Aug;30(8):795–819.

50. Abt E, Rodricks JV, Levy JI, Zeise L, and Burke TA. Science and decisions: Advancing risk assessment. *Risk Anal.* 2010, Jul;30(7):1028–1036.

51. Rodricks JV and Levy JI. Science and decisions: Advancing toxicology to advance risk assessment. *Toxicol Sci.* 2013, Jan;131(1):1–8.

52. United States Environmental Protection Agency (USEPA). *Summary of the US EPA Colloquium on a framework for human health risk assessment. Colloquium #1.* Assessment Forum. Washington, DC, November 24, 1997. http://www.epa.gov/raf/publications/sum-framework-human-ra-vol-1.htm (accessed October 4, 2013).

53. United States Environmental Protection Agency (USEPA). Risk Assessment Forum. *Benchmark Dose Technical Guidance.* EPA/100/R-12/001. Washington, DC, June, 2012. http://www.epa.gov/raf/publications/pdfs/benchmark_dose_guidance.pdf

54. United States Environmental Protection Agency (USEPA). Risk Assessment Forum. *Benchmark Dose Technical Guidance Document (External Review Draft).* EPA/630/R-00/001. Washington, DC, October, 2000. http://www.epa.gov/raf/publications/pdfs/BMD-EXTERNAL_10_13_2000.PDF

55. Lutz WK, Gaylor DW, Conolly RB, and Lutz RW. Nonlinearity and thresholds in dose-response relationships for carcinogenicity due to sampling variation, logarithmic dose scaling, or small differences in individual susceptibility. *Toxicol Appl Pharmacol.* 2005, Sep 9;207(2 Suppl):565–569.

56. Daniels MJ, Dominici F, Samet JM, and Zeger SL. Estimating particulate matter-mortality dose-response curves and threshold levels: An analysis of daily time-series for the 20 largest US cities. *Am J Epidemiol.* 2000, Sep 9;152(5):397–406.

57. Schwartz J, Laden F, and Zanobetti A. The concentration–response relation between PM2.5 and daily deaths. *Environ. Health Perspect.* 2002;110(10):1025–1029.

58. Cox LA. Miscommunicating risk, uncertainty, and causation: fine particulate air pollution and mortality risk as an example. *Risk Anal.* 2012, May;32(5):765–767; author reply 768–770.

59. Cox LA, Popken DA, and Berman DW. Causal versus spurious spatial exposure-response associations in health risk analysis. *Crit Rev Toxicol.* 2013;43 Suppl 126–138.

60. Cox LA, Popken DA, and Ricci PF. Warmer is healthier: effects on mortality rates of changes in average fine particulate matter (PM2.5) concentrations and temperatures in 100 U.S. cities. *Regul Toxicol Pharmacol.* 2013, Aug;66(3):336–346.

61. Cox LA. Caveats for causal interpretations of linear regression coefficients for fine particulate (PM2.5) air pollution health effects. *Risk Anal.* 2013, Jun;n/a. DOI: 10.1111/risa.12084

62. Cox LA. Hormesis for fine particulate matter (PM 2.5). *Dose Response.* 2012;10(2):209–218.

63. Lucchini RG, Zoni S, Guazzetti S, Bontempi E, Micheletti S, Broberg K et al. Inverse association of intellectual function with very low blood lead but not with manganese exposure in Italian adolescents. *Environ Res.* 2012, Oct;118:65–71.

64. Strayhorn JC and Strayhorn JM. Lead exposure and the 2010 achievement test scores of children in New York counties. *Child Adolesc Psychiatry Ment Health.* 2012;6(1):4.

65. Budtz-J.rgensen E, Bellinger D, Lanphear B, Grandjean P, and International Pooled Lead Study Investigators. An international pooled analysis for obtaining a benchmark dose for environmental lead exposure in children. *Risk Anal.* 2013, Mar;33(3):450–461.

66. White RH, Cote I, Zeise L, Fox M, Dominici F, Burke TA et al. State-of-the-science workshop report: Issues and approaches in low-dose-response extrapolation for environmental health risk assessment. *Environ Health Perspect.* 2009, Feb;117(2):283–287.

67. Rhomberg LR. Practical risk assessment and management issues arising were we to adopt low-dose linearity for all endpoints. *Dose Response.* 2011;9(2):144–157.

68. Rhomberg LR, Goodman JE, Haber LT, Dourson M, Andersen ME, Klaunig JE et al. Linear low-dose extrapolation for noncancer heath effects is the exception, not the rule. *Crit Rev Toxicol.* 2011, Jan;41(1):1–19.

69. Rhomberg LR. Linear low-dose extrapolation for noncancer responses is not generally appropriate. *Environ Health Perspect.* 2009, Apr;117(4):A141–A142; author reply A142–A143.

70. Crump KS, Chiu WA, and Subramaniam RP. Issues in using human variability distributions to estimate low-dose risk. *Environ Health Perspect.* 2010, Mar;118(3):387–393.

71. Crump KS, Chen C, Chiu WA, Louis TA, Portier CJ, Subramaniam RP, and White PD. What role for biologically based dose-response models in estimating low-dose risk? *Environ Health Perspect.* 2010, May;118(5):585–588.

72. Crump KS. Use of threshold and mode of action in risk assessment. *Crit Rev Toxicol.* 2011, Sep;41(8):637–650.

73. United States Environmental Protection Agency (USEPA). *Framework for Cumulative Risk Assessment.* EPA/630/P-02/001F. Risk Assessment Forum. Washington, DC, May 2003. http://www.epa.gov/raf/publications/framework-cra.htm

74. Putzrath RM. Reducing uncertainty of risk estimates for mixtures of chemicals within regulatory constraints. *Regul Toxicol Pharmacol.* 2000, Feb;31(1):44–52.

75. Wilkinson CF, Christoph GR, Julien E, Kelley JM, Kronenberg J, McCarthy J, and Reiss R. Assessing the risks of exposures to multiple chemicals with a common mechanism of toxicity: How to cumulate? *Regul Toxicol Pharmacol.* 2000, Feb;31(1):30–43.

76. Mileson BE, Chambers JE, Chen WL, Dettbarn W, Ehrich M, Eldefrawi AT et al. Common mechanism of toxicity: A case study of organophosphorus pesticides. *Toxicol Sci.* 1998, Jan;41(1):8–20.

77. Adami H-O, Berry SCL, Breckenridge CB, Smith LL, Swenberg JA, Trichopoulos D et al. Toxicology and epidemiology: Improving the science with a framework for combining toxicological and epidemiological evidence to establish causal inference. *Toxicol Sci.* 2011, Aug;122(2):223–234.

78. Hennig B, Ormsbee L, McClain CJ, Watkins BA, Blumberg B, Bachas LG et al. Nutrition can modulate the toxicity of environmental pollutants: Implications in risk assessment and human health. *Environ Health Perspect.* 2012, Jun;120(6):771–774.

79. Oates L and Cohen M. Assessing diet as a modifiable risk factor for pesticide exposure. *Int J Environ Res Public Health.* 2011, Jun;8(6):1792–1804.

80. Chahine T, Schultz BD, Zartarian VG, Xue J, Subramanian SV, and Levy JI. Modeling joint exposures and health outcomes for cumulative risk assessment: The case of radon and smoking. *Int J Environ Res Public Health.* 2011, Sep;8(9):3688–3711.

81. Tse LA, Yu IT-S, Qiu H, Au JSK, and Wang X-R. A case-referent study of lung cancer and incense smoke, smoking, and residential radon in Chinese men. *Environ Health Perspect.* 2011, Nov;119(11):1641–1646.

82. Lubin JH and Steindorf K. Cigarette use and the estimation of lung cancer attributable to radon in the United States. *Radiat Res.* 1995, Jan;141(1):79–85.

83. Wason SC, Smith TJ, Perry MJ, and Levy JI. Using physiologically-based pharmacokinetic models to incorporate chemical and non-chemical stressors into cumulative risk assessment: A case study of pesticide exposures. *Int J Environ Res Public Health.* 2012, May;9(5):1971–1983.

84. Yu QJ, Cao Q, and Connell DW. An overall risk probability-based method for quantification of synergistic and antagonistic effects in health risk assessment for mixtures: Theoretical concepts. *Environ Sci Pollut Res Int.* 2011, Aug;19(7):2627–2633.

85. Price PS and Han X. Maximum cumulative ratio (MCR) as a tool for assessing the value of performing a cumulative risk assessment. *Int J Environ Res Public Health.* 2011, Jun;8(6):2212–2225.

86. Tan Y-M, Clewell H, Campbell J, and Andersen M. Evaluating pharmacokinetic and pharmacodynamic interactions with computational models in supporting cumulative risk assessment. *Int J Environ Res Public Health.* 2011, May;8(5):1613–1630.

87. Williams PRD, Dotson GS, and Maier A. Cumulative risk assessment (CRA): Transforming the way we assess health risks. *Environ Sci Technol.* 2012, Oct 10;46(20):10868–10874.

88. Silins I and Högberg J. Combined toxic exposures and human health: Biomarkers of exposure and effect. *Int J Environ Res Public Health.* 2011, Mar;8(3):629–647.

89. Sexton K and Linder SH. Cumulative risk assessment for combined health effects from chemical and nonchemical stressors. *Am J Public Health.* 2011, Dec;101 Suppl 1:S81–S88.

90. Sexton K. Cumulative risk assessment: An overview of methodological approaches for evaluating combined health effects from exposure to multiple environmental stressors. *Int J Environ Res Public Health.* 2012, Feb;9(2):370–390.

91. Love C, David RJ, Rankin KM, and Collins JW. Exploring weathering: Effects of lifelong economic environment and maternal age on low birth weight, small for gestational age, and preterm birth in African-American and white women. *Am J Epidemiol.* 2010, Jul 7;172(2):127–134.

92. Keene DE and Geronimus AT. "Weathering" HOPE VI: The importance of evaluating the population health impact of public housing demolition and displacement. *J Urban Health.* 2011, Jun;88(3):417–435.

93. Geronimus AT, Hicken M, Keene D, and Bound J. "Weathering" and age patterns of allostatic load scores among blacks and whites in the United States. *Am J Public Health.* 2006, May;96(5):826–833.

94. Geronimus AT. Understanding and eliminating racial inequalities in women's health in the United States: The role of the weathering conceptual framework. *J Am Med Womens Assoc.* 2001;56(4):133–136, 149–150.

95. Chakraborty J, Maantay JA, and Brender JD. Disproportionate proximity to environmental health hazards: Methods, models, and measurement. *Am J Public Health.* 2011, Dec;101 Suppl 1:S27–S36.

96. Williams PRD. Health risk communication using comparative risk analyses. *J Expo Anal Environ Epidemiol.* 2004, Nov;14(7):498–515.

97. Wetmore BA, Wambaugh JF, Ferguson SS, Sochaski MA, Rotroff DM, Freeman K et al. Integration of dosimetry, exposure and high-throughput screening data in chemical toxicity assessment. *Toxicol Sci.* 2011, Sep 9;125(1):157–174.

98. Wild CP. Complementing the genome with an "exposome": The outstanding challenge of environmental exposure measurement in molecular epidemiology. *Cancer Epidemiol Biomarkers Prev.* 2005, Aug;14(8):1847–1850.

99. Riess U, Tegtbur U, Fauck C, Fuhrmann F, Markewitz D, and Salthammer T. Experimental setup and analytical methods for the non-invasive determination of volatile organic compounds, formaldehyde and NO_x in exhaled human breath. *Anal Chim Acta.* 2010, Jun 6;669(1–2):53–62.

100. Rappaport SM. Implications of the exposome for exposure science. *J Expo Sci Environ Epidemiol.* 2011;21(1):5–9.

101. National Research Council NRC. *Exposure Science for the 21st Century: A Vision and a Strategy.* Washington, DC: National Academy Press, 2012.

102. Wang Z, Klipfell E, Bennett BJ, Koeth R, Levison BS, Dugar B et al. Gut flora metabolism of phosphatidylcholine promotes cardiovascular disease. *Nature.* 2011, Apr 4;472(7341):57–63.

103. Venet D, Dumont JE, and Detours V. Most random gene expression signatures are significantly associated with breast cancer outcome. *PLoS Comput Biol.* 2011, Oct;7(10):e1002240.

104. Lioy PJ. Exposure science: A view of the past and milestones for the future. *Environ Health Perspect.* 2010, Aug;118(8):1081–1090.

105. Meek MEB, Bolger M, Bus JS, Christopher J, Conolly RB, Lewis RJ et al. A framework for fit-for-purpose dose response assessment. *Regul Toxicol Pharmacol.* 2013, Apr 4;66(2):234–240.

106. Crump KS, Hoel DG, Langley CH, and Peto R. Fundamental carcinogenic processes and their implications for low dose risk assessment. *Cancer Res.* 1976, Sep;36(9 pt.1):2973–2979.

107. Hinderliter PM, Price PS, Bartels MJ, Timchalk C, and Poet TS. Development of a source-to-outcome model for dietary exposures to insecticide residues: An example using chlorpyrifos. *Regul Toxicol Pharmacol.* 2011, Oct;61(1):82–92.

108. Price PS, Schnelle KD, Cleveland CB, Bartels MJ, Hinderliter PM, Timchalk C, and Poet TS. Application of a source-to-outcome model for the assessment of health impacts from dietary exposures to insecticide residues. *Regul Toxicol Pharmacol.* 2011, Oct;61(1):23–31.

109. Smith JN, Timchalk C, Bartels MJ, and Poet TS. In vitro age-dependent enzymatic metabolism of chlorpyrifos and chlorpyrifos-oxon in human hepatic microsomes and chlorpyrifos-oxon in plasma. *Drug Metab Dispos.* 2011, Aug;39(8):1353–1362.

110. Deakin SP and James RW. Genetic and environmental factors modulating serum concentrations and activities of the antioxidant enzyme paraoxonase-1. *Clin Sci (Lond).* 2004, Nov;107(5):435–447.

111. Costa LG, Giordano G, and Furlong CE. Pharmacological and dietary modulators of paraoxonase 1 (PON1) activity and expression: The hunt goes on. *Biochem Pharmacol.* 2011, Feb 2;81(3):337–344.

112. Swenberg JA, Lu K, Moeller BC, Gao L, Upton PB, Nakamura J, and Starr TB. Endogenous versus exogenous DNA adducts: Their role in carcinogenesis, epidemiology, and risk assessment. *Toxicol Sci.* 2011, Mar;120 Suppl 1:S130–S145.

113. Fuchs P, Loeseken C, Schubert JK, and Miekisch W. Breath gas aldehydes as biomarkers of lung cancer. *Int J Cancer.* 2010, Jun 6;126(11):2663–2670.

114. Greenwald R, Fitzpatrick AM, Gaston B, Marozkina NV, Erzurum S, and Teague WG. Breath formate is a marker of airway S-nitrosothiol depletion in severe asthma. *PLoS One.* 2010;5(7):e11919 (accessed Jun 2, 2013).

115. Carmichael N, Bausen M, Boobis AR, Cohen SM, Embry M, Fruijtier-Pölloth C et al. Using mode of action information to improve regulatory decision-making: An ECETOC/ILSI RF/HESI workshop overview. *Crit Rev Toxicol.* 2011, Mar;41(3):175–186.

116. Scherer RW. 2.2 Evidence-based health care and the Cochrane Collaboration. *Hum Exp Toxicol.* 2009, Feb;28(2–3):109–111.

117. Sutton P, Woodruff TJ, Vogel S, and Bero LA. Conrad and Becker's "10 Criteria" fall short of addressing conflicts of interest in chemical safety studies. *Environ Health Perspect.* 2011, Dec;119(12):A506–A507; author reply A508–A509.

118. Schreider J, Barrow C, Birchfield N, Dearfield K, Devlin D, Henry S et al. Enhancing the credibility of decisions based on scientific conclusions: Transparency is imperative. *Toxicol Sci.* 2010, Jul;116(1):5–7.

119. Lutter R, Barrow C, Borgert CJ, Conrad JW, Edwards D, and Felsot A. Data disclosure for chemical evaluations. *Environ Health Perspect.* 2013, Feb;121(2):145–148.

120. Conrad JW and Becker RA. Enhancing credibility of chemical safety studies: Emerging consensus on key assessment criteria. *Environ Health Perspect.* 2011, Jun;119(6):757–764.

121. Barrow CS and Conrad JW. Assessing the reliability and credibility of industry science and scientists. *Environ Health Perspect.* 2006, Feb;114(2):153–155.

122. Maurissen JP, Gilbert SG, Sander M, Beauchamp TL, Johnson S, Schwetz BA et al. Workshop proceedings: Managing conflict of interest in science. A little consensus and a lot of controversy. *Toxicol Sci.* 2005, Sep;87(1):11–14.

123. Sass J. Credibility of scientists: Conflict of interest and bias. *Environ Health Perspect.* 2006, Mar;114(3):A147–A148; author reply A148.

124. Claxton LD. A review of conflict of interest, competing interest, and bias for toxicologists. *Toxicol Ind Health.* 2007, Nov;23(10):557 571.

125. Ong EK and Glantz SA. Constructing "sound science" and "good epidemiology": Tobacco, lawyers, and public relations firms. *Am J Public Health.* 2001, Nov;91(11):1749–1757.

126. Bitton A, Neuman MD, Barnoya J, and Glantz SA. The p53 tumour suppressor gene and the tobacco industry: Research, debate, and conflict of interest. *Lancet.* 2005;365(9458):531–540.

127. Baba A, Cook DM, McGarity TO, and Bero LA. Legislating "sound science": The role of the tobacco industry. *Am J Public Health.* 2005;95 Suppl 1:S20–S27.

128. *Society of Toxicology* (accessed May 30, 2013).

129. Fanelli D. Do pressures to publish increase scientists' bias? An empirical support from US States Data. *PLoS One.* 2010;5(4):e10271.

130. Song F, Parekh S, Hooper L, Loke YK, Ryder J, Sutton AJ et al. Dissemination and publication of research findings: An updated review of related biases. *Health Technol Assess.* 2010, Feb;14(8):iii, ix–xi, 1–193.

131. Song F, Eastwood AJ, Gilbody S, Duley L, and Sutton AJ. Publication and related biases. *Health Technol Assess.* 2000;4(10):1–115.

132. Rothman KJ. The ethics of research sponsorship. *J Clin Epidemiol.* 1991;44 Suppl 1:25S–28S.
133. Nickerson R. Confirmation bias: A ubiquitous phenomenon in many guises. *Rev Gen Psychol.* 1998;2:175–220.
134. Marsh DM and Hanlon TJ. Seeing what we want to see: Confirmation bias in animal behavior research. *Ethology.* 2007;113(11):1089–1098.
135. Bero L. Industry sponsorship and research outcome: A cochrane review. *JAMA Intern Med.* 2013, Apr 4;173(7):580–581.
136. Lundh A, Sismondo S, Lexchin J, Busuioc OA, and Bero L. Industry sponsorship and research outcome. *Cochrane Database Syst Rev.* 2012;12MR000033.
137. Wandall B, Hansson SO, and Rudén C. Bias in toxicology. *Arch Toxicol.* 2007;81(9):605–617.
138. Organisation for Economic Co-operation and Development (OECD). Website: *Series on Principles of Good Laboratory Practice (GLP) and Compliance Monitoring.* http://www.oecd-ilibrary.org/environment/oecd-series-on-principles-of-good-laboratory-practice-and-compliance-monitoring_2077785x (accessed May 30, 2013).
139. Thompson CM, Proctor DM, Haws LC, Hebert CD, Grimes SD, Shertzer HG et al. Investigation of the mode of action underlying the tumorigenic response induced in B6C3F1 mice exposed orally to hexavalent chromium. *Toxicol Sci.* 2011, Jun 6;123(1):58–70.
140. Horton R. The less acceptable face of bias. *Lancet.* 2000, Sep 9;356(9234):959–960.
141. Borgert CJ. Conflict of interest: Kill the messenger or follow the data? *Environ Sci Technol.* 2007, Feb 1;41(3):665.
142. Borgert CJ. Conflict of interest or contravention of science? *Regul Toxicol Pharmacol.* 2007, Jun;48(1):4–5.
143. Henry CJ and Conrad JW. Scientific and legal perspectives on science generated for regulatory activities. *Environ Health Perspect.* 2008, Jan;116(1):136–141.
144. Goozner M. Credibility of scientists: Industry versus public interest. *Environ Health Perspect.* 2006, Mar;114(3):A147; author reply A148.
145. Tweedale T. Enhancing credibility of chemical safety studies: No consensus. *Environ Health Perspect.* 2011, Dec;119(12):A507–A508; author reply A508–A509.
146. Guillette LJ, Gross TS, Masson GR, Matter JM, Percival HF, and Woodward AR. Developmental abnormalities of the gonad and abnormal sex hormone concentrations in juvenile alligators from contaminated and control lakes in Florida. *Environ Health Perspect.* 1994, Aug;102(8):680–688.
147. Guillette LJ, Gross TS, Gross DA, Rooney AA, and Percival HF. Gonadal steroidogenesis in vitro from juvenile alligators obtained from contaminated or control lakes. *Environ Health Perspect.* 1995, May;103 Suppl 4:31–36.
148. Guillette LJ, Pickford DB, Crain DA, Rooney AA, and Percival HF. Reduction in penis size and plasma testosterone concentrations in juvenile alligators living in a contaminated environment. *Gen Comp Endocrinol.* 1996, Jan;101(1):32–42.
149. Arnold SF, Klotz DM, Collins BM, Vonier PM, Guillette LJ, and McLachlan JA. Synergistic activation of estrogen receptor with combinations of environmental chemicals. *Science.* 1996, Jun 6;272(5267):1489–1492.
150. Ramamoorthy K. Potency of combined estrogenic pesticides. *Science.* 1997, Jan 1;275(5298):405–406.
151. McLachlan JA. Synergistic effect of environmental estrogens: Report withdrawn. *Science.* 1997, Jul 7;277(5325):462–463.
152. LaRosa JC. Tulane investigation completed. *Science.* 1999, Jun 18;284(5422):1932.

153. Kaiser J. Science policing: Tulane inquiry clears lead researcher. *Science*. 1999, Jun 6;284(5422):1905.
154. National Institutes of Health (NIH). Office of Extramural Research. Website: NIH Guide: Findings of Scientific Misconduct, October 15, 2001. http://grants.nih.gov/grants/guide/notice-files/NOT-OD-02-003.html (accessed May 30, 2013).
155. Colborn T, Dumanoski D, and Myers JP. *Our Stolen Future: Are We Threatening Our Fertility, Intelligence, and Survival?: A Scientific Detective Story*. New York: Dutton, 1996.
156. Byrd DM. Goodbye pesticides: The Food Quality Protection Act of 1996. *Regulation*. Fall, 1997; pp. 57–62. http://object.cato.org/sites/cato.org/files/serials/files/regulation/1997/10/reg20n4l.pdf
157. United States Environmental Protection Agency (USEPA). Website: *Pesticide Registration (PR) Notice 97-1, Agency Actions Under the Requirements of the Food Quality Protection Act/Regulating Pesticides/US EPA*. http://www.epa.gov/PR_Notices/pr97-1.html (accessed May 30, 2013).
158. National Research Council (NRC). *Pesticides in the Diets of Infants and Children*. Washington, DC: The National Academies Press, 1993. http://www.nap.edu/catalog.php?record_id=2126
159. Bader MDM, Purciel M, Yousefzadeh P, and Neckerman KM. Disparities in neighborhood food environments: Implications of measurement strategies. *Econ Geogr*. 2010;86(4):409–430.
160. Larson NI, Story MT, and Nelson MC. Neighborhood environments: Disparities in access to healthy foods in the U.S. *Am J Prev Med*. 2009, Jan;36(1):74–81.
161. Hosler AS, Rajulu DT, Fredrick BL, and Ronsani AE. Assessing retail fruit and vegetable availability in urban and rural underserved communities. *Prev Chronic Dis*. 2008, Oct;5(4):A123.
162. Popper KA. *The Logic of Scientific Discovery*. Cambridge, MA: International Society for Science and Religion, 2007.
163. Bailey LA, Prueitt RL, and Rhomberg LR. Hypothesis-based weight-of-evidence evaluation of methanol as a human carcinogen. *Regul Toxicol Pharmacol*. 2012, Mar;62(2):278–291.
164. Rhomberg LR, Bailey LA, and Goodman JE. Hypothesis-based weight of evidence: A tool for evaluating and communicating uncertainties and inconsistencies in the large body of evidence in proposing a carcinogenic mode of action—Naphthalene as an example. *Crit Rev Toxicol*. 2010, Sep;40(8):671–696.
165. Phillips CV. Commentary: Lack of scientific influences on epidemiology. *Int J Epidemiol*. 2008, Feb;37(1):59–64; discussion 65–68.
166. Kuntz M. The postmodern assault on science. If all truths are equal, who cares what science has to say? *EMBO Rep*. 2012, Oct;13(10):885–889.
167. Otto S. *Fool Me Twice: Fighting the Assault on Science in America*. New York: Rodale; distributed to the trade by Macmillan, 2011.
168. Sass JB, Castleman B, and Wallinga D. Vinyl chloride: A case study of data suppression and misrepresentation. *Environ Health Perspect*. 2005, Jul;113(7):809–812.
169. Guth JH, Denison RA, and Sass J. Require comprehensive safety data for all chemicals. *New Solut*. 2007;17(3):233–258.
170. Master Z and Resnik DB. Hype and public trust in science. *Sci Eng Ethics*. 2013, Jun;19(2):321–335.
171. Douglas H. Weighing complex evidence in a democratic society. *Kennedy Inst Ethics J*. 2012, Jun;22(2):139–162.

172. de Melo-Martín I and Intemann K. Interpreting evidence: Why values can matter as much as science. *Perspect Biol Med.* 2012;55(1):59–70.
173. Mikulak A. Mismatches between 'scientific' and 'non-scientific' ways of knowing and their contributions to public understanding of science. *Integr Psychol Behav Sci.* 2011, Jun;45(2):201–215.
174. Ropeik D. The perception gap: Recognizing and managing the risks that arise when we get risk wrong. *Food Chem Toxicol.* 2012, May;50(5):1222–1225.
175. Subcommittee on Investigations and Oversight Hearing. *EPA's IRIS Program: Evaluating the Science and Process Behind Chemical Risk Assessment.* 2011, Jul 14. http://science.house.gov/hearing/investigations-and-oversight-hearing-epas-iris-program (accessed October 22, 2013).
176. United States Environmental Protection Agency (USEPA). Website: Six Common Air Pollutants/Air & Radiation/US EPA. n.d. http://www.epa.gov/air/urbanair/ (accessed June 11, 2013).
177. Cox LA. Confronting deep uncertainties in risk analysis. *Risk Anal.* 2012, Oct;32(10):1607–1629.
178. Hardell L, Walker MJ, Walhjalt B, Friedman LS, and Richter ED. Secret ties to industry and conflicting interests in cancer research. *Am J Ind Med.* 2007, Mar;50(3):227–233.

Appendix A: Useful Methods and Algorithms for Probabilistic Risk Assessment

This appendix is meant as an introduction to probabilistic risk assessment (PRA). Many users of this text may already know this material. This appendix is not complete but provides a short introduction to some of the methods of PRA and is intended as help getting started for those relatively inexperienced in statistical methods. EPA's *Risk Assessment Guidance for Superfund Volume III: Guidance for Conducting Probabilistic Risk Assessment*[1] also provides much information.

DETERMINISTIC VERSUS PROBABILISTIC EQUATIONS

A model is a mathematical equation or system of equations that attempts to describe or predict the behavior of some system. A model can be simply written as a function of its inputs as follows:

$$Y = f(A, B, C, D)$$

For a set of model inputs (A, B, C, and D), the model calculates the predicted outcome (Y).

Each of the inputs into a model can be classified into one of two categories: Constants, which might include

- The speed of light in a vacuum ($1.86E + 05$ miles/s)
- The number of feet in a mile (5280)

Variables, which might include

- Wind speed over a source area
- Rate and/or direction of groundwater flow
- Amount of soil ingested by a child in a day
- Body weight of a ground squirrel

Variables can be described using frequency distributions. For example, the body weight of adult humans may be considered to occur in a normal distribution ("bell curve") with a mean of 70 kg and a standard deviation of 15 kg.

If one or more inputs to a model are described by frequency distributions, then the output of the model will also be a distribution.

METHODS FOR MONTE CARLO SIMULATION

A number of software programs perform Monte Carlo (MC) simulation. These include Crystal Ball from Oracle, ModelRisk from Vose Software, @RISK from Palisade, and RiskAMP from Structured Data. MC simulation can also be performed with general mathematical software such as MATLAB or Mathematica. These methods and software programs are conceptually simple.

Consider the model

Y = f (A, B, C), where A, B, and C are random variables.

Here's what these software packages do:

Step 1: Selects a value at random for each specified variable (A, B, and C)
Step 2: For the selected values of A, B, and C, calculates the result $Y_i = f$ (A_i, B_i, C_i)
Step 3: Records the result, along with the values of the inputs
Step 4: Repeats steps 1–3 as many times as you wish (each repeat is called an iteration)
Step 5: After completing many iterations, provides the results, which may include the following (you choose):
  • Raw data (each of the results along with each of the inputs)
  • Summary statistics (mean, stdev, percentiles, etc.) of inputs and outputs
  • Graph (PDF and/or CDF) of inputs and outputs

Monte Carlo versus Latin Hypercube

In MC sampling, the number of samples drawn from any particular part of the probability density function (PDF) is proportional to the density for that section; hence, for a normal distribution, most of the samples will occur near the mean value. This method has a disadvantage: you have to draw a lot of samples to get enough values in the tails to define the tails with any accuracy. In contrast, the middle becomes defined much faster, especially for skewed distributions.

The Latin hypercube sampling method provides a fix. The PDF is divided into sections of equal probability (e.g., 0%–5%, 5%–10%, 10%–15%), and equal numbers of samples are drawn from each probability section. Hence, both the middle and tails of the distribution tend to converge at about the same rate because of equal sampling densities.

Sensitivity Analysis

Sensitivity analysis (SA) can provide information about the degree to which the various variables affect the results of a calculation. Hence, an important aspect of any MC simulation is an investigation of how the results depend on the inputs. There are a number of methods for SA. Simple ones include rank-order correlation and contribution to variance. These are implemented in the software packages mentioned earlier.

DEPENDENCIES BETWEEN INPUTS

This aspect of PRA is the red-headed stepchild of risk assessment that no one wants to talk about. A risk assessor about to embark on an MC simulation may have carefully determined distributions for all variables in the calculation. What may be forgotten are the dependencies or correlations between those variables. Most of the time, these dependencies are not known.

For example, do consumers of self-caught fish eat large meals of self-caught fish very often? A positive correlation between meal size and meal frequency could obviously drive up the upper tail of this distribution of the overall fish consumption rate expressed in g/day.

With little information on these dependencies, the wisest course of action is to assume several values for these unknown correlations coefficients, run the model several times using alternate value, and determine to what extent the dependency could potentially affect the result.

ALGORITHMS FOR GENERATING VARIATES FROM COMMON PROBABILITY DENSITY FUNCTIONS

One or more random numbers are used in these algorithms. MS-Excel has a passable random number generator. Enter = RAND() in any cell in an Excel worksheet and a random number between 0 and 1 will be produced. This method was used in the bootstrap example in Chapter 3.

UNIFORM PDF (MIN, MAX)

As noted, Excel provides random variates from a uniform PDF with min = 0 and max = 1. For integers, the Excel function = RANDBETWEEN(min, max) provides an integer value in the specified range.

User inputs

1. Minimum (min)
2. Maximum (max)

Steps

1. Select random variate from Uniform (0, 1) and save as "u."
2. Calculate the uniform variate "X" as

$$X = min + u \times (max - min)$$

Normal PDF (μ, σ)

User inputs

1. Arithmetic mean (μ)
2. Arithmetic standard deviation (σ)

Steps

1. Select random variate from Uniform (0, 1) and save as "u_1."
2. Select a second random variate from Uniform (0, 1); save as "u_2."
3. Calculate "z" as

$$z = \mathrm{sqrt}(-2 \times \ln(u_1)) \times \sin(2 \times \mathrm{pi} \times u_2)$$

4. Calculate the normal variate "X" as

$$X = \mu + z \times \sigma$$

Lognormal PDF: Either (AM, ASD), (GM, GSD), or (μ, σ)

User inputs

1. Arithmetic mean (AM)
2. Arithmetic standard deviation (ASD)

Steps

1. Select random variate from Uniform (0, 1); save as "u_1."
2. Select a second random variate from Uniform (0, 1); save as "u_2."
3. Calculate the geometric mean (GM) and geometric standard deviation as follows:

$$GM = \frac{AM^2}{\sqrt{AM^2 + ASD^2}}$$

$$GSD = EXP\left(SQRT\left(LN\left(1 + \frac{ASD^2}{AM^2} \right) \right) \right)$$

4. Obtain the natural logarithms of the GM and GSD. These will be referred to as μ and σ.
5. Generate two random numbers and calculate "z" as for the normal variate.
6. Calculate the lognormal variate "Y" as

$$Y = \exp(\mu + z \times \sigma)$$

Triangular (min, mode, max)

User inputs

1. Minimum (a)
2. Mode (b)
3. Maximum (c)

Steps

1. Select random variate from Uniform (0, 1) and save as "u."
2. If u = 1.0 (the maximum possible), reset to u_1 = 0.999999.
3. Calculate the interim variable "delta" as

$$delta = \frac{b-a}{c-a}$$

4. If u $\leq$ delta, then calculate "X" from the triangular PDF as

$$X = a + 0.5 \times SQRT(4 \times a^2 - 4 \times (a^2 - u \times (c - a) \times (b - a))$$

5. If u_1 > delta, then calculate "X" from the triangular PDF as

$$X = c - 0.5 \times SQRT(4 \times c^2 + 4 \times ((b - a) \times (c - b) - 2 \times c \times b + b^2 - u \times (c - a) \times (c - b))$$

ESTIMATING CONFIDENCE INTERVALS USING THE BOOTSTRAP

The bootstrap was demonstrated in Chapter 3. This appendix provides more information and background. Here, an estimate of the standard error or numerical uncertainty of a statistic, such as the mean, is calculated using several methods. Bootstrapped confidence intervals (CIs) for any statistic can be obtained from a dataset that is resampled with replacement to form bootstrap samples, each containing the same number of data points as the original sample.

Think of a jar of red and black jellybeans. To estimate the proportion of red jellybeans, you could select 10 of them, replacing the jellybean back in the jar each time. The number of red ones out of 10 selections would provide an estimate of this proportion. By doing this 1000 times, you could obtain CIs for that proportion.

The statistic of interest calculated from each of the bootstrap samples is called a replicate. A CI for the statistic is determined using the standard error of the replicates or from the cumulative distribution function of the replicates. The number of bootstrap samples generated depends on the statistic of interest and the acceptable error in the bootstrap CI. Generally, 1000 bootstrap iterations or more should be used when estimating CIs.

Two advantages of the bootstrap method are as follows: (1) It does not require an assumption of the distribution of the data (or, depending upon which bootstrap method is used, the distribution of the means) and (2) it is relatively easy to implement on a computer.

There are many bootstrap methods available. EPA's software ProUCL implements several of these methods and they are described in detail in the ProUCL manual.

REFERENCE

1. United States Environmental Protection Agency (USEPA). *Risk Assessment Guidance for Superfund (RAGS) Volume III - Part A: Process for Conducting Probabilistic Risk Assessment*. EPA 540-R-02-002. OSWER 9285.7-45. PB2002 963302. Washington, DC, December, 2001. http://www.epa.gov/oswer/riskassessment/rags3adt/ (accessed September 1, 2012).

Index